DNA METHYLATION RESEARCH TRENDS

DNA Methylation Research Trends

Taku B. Kobayashi
Editor

Nova Biomedical Books
New York

For permission to use material from this book please contact us:
Telephone 631-231-7269; Fax 631-231-8175
Web Site: http://www.novapublishers.com

Library of Congress Cataloging-in-Publication Data

DNA methylation research trends / Taku B. Kobayashi, editor.
p. ; cm.
Includes bibliographical references and index.
ISBN-13: 978-1-60021-721-0 (hardcover)
ISBN-10: 1-60021-721-4 (hardcover)
1. DNA--Methylation--Research. I. Kobayashi, Taku B.
[DNLM: 1. DNA Methylation. 2. Neoplasms. QU 58.5 D62916 2007]
QP624.5.M46D553 2007
572.8'6--dc22

2007014432

Published by Nova Science Publishers, Inc. ✦ New York

Contents

Preface

DNA methylation is a type of chemical modification of DNA that can be inherited without changing the DNA sequence. As such, it is part of the epigenetic code and is the most characterized epigenetic mechanism. DNA methylation involves the addition of a methyl group to DNA — for example, to the number 5 carbon of the cytosine pyrimidine ring. DNA methylation is probably universal in eukaryotes. In humans, approximately 1% of DNA bases undergo DNA methylation. In adult somatic tissues, DNA methylation typically occurs in a CpG dinucleotide context; non-CpG methylation is prevalent in embryonic stem cells.In plants, cytosines are methylated both symmetrically (CpG or CpNpG) and asymmetrically (CpNpNp), where N can be any nucleotide. The methylation status of specific cytosines can be determined using methods based on bisulfite sequencing. This book presents new and important research from around the world in this field.

Chapter 1 - CpG dinucleotides are globally methylated at the 5 position on the cytosine ring (5^{m}CpGs) in mammalian genomes. However, they often remain unmethylated in CpG islands. The hypermutability of 5^{m}CpG to TpG, which is known as the *CpG effect*, results in a strong deficiency of CpG dinucleotides and a decrease of GC content in mammalian genomes over evolutionary time. Disruption of DNA methylation system leads to hypermethylation in the promoter-associated CpG islands of tumor suppressor genes and hypomethylation in other global regions.

While the CpG effect has been well documented, its quantitative measurement could not be performed until the recent releases of several mammalian genomes, their genome-wide polymorphism data, and large-scale methylation status at CpG sites. In the present study, the authors performed a systematic examination of mutation direction in human genome using over one million high-quality single nucleotide polymorphism (SNP) data. The frequencies of nucleotide changes G:C→A:T were the highest among all types, which were mainly attributed to the hypermutability of 5^{m}CpG→TpG. The authors comparison of mutational spectra in genomic categories revealed that exons and their four-fold degenerate sites had the strongest impact of methylation-related mutation (i.e., 5^{m}CpGs→TpGs/CpAs), whereas CpG islands had the weakest, suggesting strong CpG effect in exons and a lack of CpG effect in CpG islands.

Next, the authors quantitatively measured the CpG effect by estimating the methylation-dependent transition rate (5^{m}C deamination rate), the difference between CpG transition rate and GpC transition rate. The authors found that the $5^{m}C$ deamination rates were in a 90-fold range higher than other transitional changes and were strongly dependent on the local GC content.

The authors also studied the evolution of CpG islands in the promoter regions of genes. Our results suggest that both human and mouse genomes have been under loss of CpG islands, with a faster rate in mice, due to *de novo* methylation in some CpG islands. Gene Ontology (GO) term analysis found that genes maintaining CpG islands tend to be essential for cell life while genes losing CpG islands tend to be differentially regulated.

Finally, the authors investigated the association of aberrant methylation in the promoter-associated CpG islands with disruption (silence) of gene function. Experimental work has demonstrated that the methylation changes are associated with tumorigenesis, abnormal development, and mental disorders.

Chapter 2 - Although a number of studies on CpG islands in humans have been reported, knowledge on plant CpG islands was quite limited. To learn whether plants contain chromosomal regions that can be regarded as CpG islands, the authors screened plant genomic sequences for CpG island-like chromosomal regions. Analysis of plant DNA sequence data, primarily from rice, revealed the presence of CpG-rich DNA regions in the genome. These CpG-rich clusters were structurally similar to human CpG islands. In rice, about one-third of the CpG clusters were found around the 5'-terminus of genes. Rice genes having this type of a CpG cluster differed from other rice genes in their expression pattern; they tended to be expressed in a wider range of tissues. This expression pattern resembles reported human genes having a CpG island, which suggests the presence of a functional similarity between plant CpG clusters and human CpG islands. In contrast to these similarities, the DNA methylation pattern of plant genes having CpG clusters seemed to differ from that of mammalian genes; in plants, gene-rich chromosomal regions were revealed to be generally undermethylated, whether they were within or outside the CpG clusters, whereas in humans undermethylation is known to be limited within CpG island regions. This difference in the DNA methylation pattern implies the difference in the role of DNA methylation on the epigenetic gene regulation of plants and mammals. In rice, genes having a CpG cluster were clustered into discrete chromosomal segments, creating mosaics comprising chromosomal regions either rich in genes having a CpG cluster or rich in genes not having it. This segmental distribution of genes was expected to cause a non-uniform distribution of expressed genes. Actually, the EST-based analysis on the distribution of expressed genes showed a preferential distribution of expressed genes within the CpG cluster-rich chromosomal regions.

Chapter 3 - Cytosine methylation plays a major role in the regulation of sequential and tissue-specific expression of genes. *De novo* aberrant DNA methylation and demethylation are also crucial processes in tumorigenesis and tumor progression. The mechanisms of how and when such aberrant methylation and demethylation occur in tumor cells are still obscure, however, in this regard, the utility of archival formalin-fixed, paraffin-embedded pathological specimens for molecular biological analysis has become increasingly important in not only the experimental but also the diagnostic field. In this article, the authors first introduce technical aspects of morphology-oriented epigenetic research by a modified agarose-bead-mediated bisulfite mapping method with the use of microdissected fragments of paraffin sections. The authors then present our observation of the unusual upregulation of TrkA, a receptor of nerve growth factor, gene expression by methylation around the negatively regulating elements through the direct interference of c-Jun binding. Recently, we cloned and characterized the TrkA gene promoter region to clarify the molecular mechanism that controls TrkA gene expression during pancreatic cancer progression. By microdissecting a subset of cancer cells showing positive TrkA immunohistochemistry from pancreatic cancer specimens obtained at autopsy and surgery, we noticed that methylation of non-CpG islands around the negatively regulating AP-1 site in the 5'-untranslated region of the TrkA gene was positively related to the expression of the gene. During carcinogenesis and cancer progression, gene silencing by global hypermethylation of CpG islands in the promoter region is observed in many genes related to the cell cycle ($p16^{INK4a}$ and $p15^{INK4b}$), DNA repair (hMLH1, MGMT, and BRCA1), cell adherence and metastatic processes (CDH1, TIMP3, DAPK), the p53 network ($p14^{ARF}$ and p73), metabolic enzymes (GSTP1), and the APC/β-catenin pathway (APC). Global genomic hypomethylation, which is linked to DNA instability and oncogene activation, has also been found in many types of human cancers including prostate, liver, and uterine tumors. Unlike such global epigenetic alteration at CpG islands that leads to drastic alteration of gene expression, methylation at non-CpG islands has been regarded as a meaningless, random and incidental event and has, therefore, drawn little interest. By selecting a subset of cells based on characteristic morphological features revealed by microdissection, we discovered that methylation at certain CpG sites in non-CpG islands would play a crucial epigenetic role in the versatility and flexibility of gene expression during cancer progression. In the final part of this article, we introduce a novel MeCP2-mediated gene silencing mechanism by the addition of methylation of a single CpG-locus upstream of the TATA-box of the Receptor activator of nuclear factor-κB ligand (RANKL) gene promoter.

Chapter 4 - Aberrant DNA methylation is an epigenetic event, which has an important role in carcinogenesis by silencing tumor suppressor genes in many cancers. Therefore, DNA methyltransferase (DNMT) inhibitors have been investigated as anti-cancer agents. These inhibitors transiently induce DNA hypomethylation of tumor suppressor genes, such as p15INK4B, as observed in clinical studies, as well as, in vitro and in vivo studies. Furthermore, two DNMTs, azacytidine (5-azacytidine) and decitabine (5-aza-2'-deoxycytidine), have shown promising clinical results in hematological malignancies, including myelodysplastic syndrome (MDS), acute leukemia, and chronic myelogenous leukemia. Recently, these drugs have been approved by the FDA for the treatment of MDS. Because of their toxicity, they have been used at low doses; however, the optimization of dose and duration has improved the epigenetic modulation as well as clinical response. Moreover, some studies have demonstrated the improved clinical responses of these drugs in combination with other clinically useful drugs, especially histone deacetylation inhibitors. Therefore, DNA methyltransferase inhibitors, alone or in combination with other drugs, is a promising molecular targeted therapy in hematological malignancies.

Chapter 5 - A cumulative series of genetic and epigenetic alterations leading to unregulated cell growth is the foundation of tumorigenesis. Changes in DNA methylation patterns occur early in the progression process and often precede malignancy. Global hypomethylation of the genome is observed and has been suggested to initiate and propagate oncogenesis by inducing chromosome instabilities and transcriptional activation of oncogenes. The overall decrease in DNA methylation is accompanied by a region- and gene-specific hypermethylation of CpG islands. Genes involved in DNA repair, detoxification, cell cycle regulation, and apoptosis are often inappropriately inactivated. It has been suggested that epigenetic lesions in normal tissue set the stage for neoplasia. Recent reports of DNA methylation analyses carried out on tumor biopsies and matched DNA samples extracted from body fluids such as serum, plasma, urine and sputum have demonstrated a high level of concordance. Sensitive technologies detecting low levels of aberrant promoter methylation in easily accessible biological material are therefore of great interest for the early detection of cancer, which is the most efficient way to improve survival of patients, as well as for the detection of cancer recurrence. Methylation patterns can be shared by different types of tumors as well as being tumor type-specific and the extent of hypo- and hypermethylation often correlates with the malignancy grade and/or the disease stage. Classifications based on the methylation profile are able to predict the metastatic potential of a cancer as well as the response to a specific anticancer drug or therapy. As DNA methylation is a non-mutational and therefore reversible modification, it can be used as point of departure for anticancer treatment by chemically or antisense oligonucleotides induced demethylation. Inhibitors of DNA methyltransferases such as the FDA approved 5-azacytidine or 5-aza-2'deoxycytidine and their analogues are incorporated into *de novo* synthesized DNA strands during replication. In promising clinical trials, they have been shown to induce stable re-expression of some previously epigenetic silenced tumor suppressor genes in hematological malignancies and to sensitize solid tumors to conventional cytotoxic drugs. DNA methylation based biomarkers promise to have a large impact on cancer diagnostics and treatment in the near future.

Chapter 6 - For a long time, cancer has been regarded as a genetically driven disease involving mutations of tumor suppressor genes and oncogenes as well as chromosomal aberrations. New concepts have recognized the fundamental role of the epigenome in the pathogenesis of human malignancies. Epigenetic changes refer to a heritable change in the pattern of gene expression without alterations in the primary nucleotide sequence. DNA methylation of cytosines within CpG dinucleotides and post-translational acetylation and methylation of histones are the most common epigenetic modifications with an impact on chromatin structure and transcriptional activity. Hypermethylation of CpG islands in the promoter region of genes is a well characterized epigenetic modification associated with transcriptional silencing of cancer-related genes and plays a crucial role in carcinogenesis. In solid tumors and hematopoietic malignancies, a large number of genes involved in different pathways have been described to be affected by CpG island methylation in association with transcriptional inactivation. Altered DNA methylation patterns may be clinically useful as biomarkers for cancer detection, assessment of prognosis and prediction of response to therapy. In contrast to genetic alterations, epigenetic changes in cancer are potentially reversible by pharmacological inhibition of DNA methylation and histone deacetylation. Thus, the increasing knowledge about the altered epigenome in malignant cells offers new strategies in targeted anti-cancer therapy. Clinical trials using epigenetically targeted therapies have yielded promising results in acute and chronic myelogenous leukemia as well as in myelodysplastic syndromes. The exploration of our growing knowledge about epigenetic aberrations will contribute to develop novel strategies in diagnosis and treatment of hematopoietic malignancies for the future.

Chapter 7 - DNA methylation is currently the only known natural base modification of the DNA molecule occurring under the control of a group of nuclear enzymes called DNA methylases, the principal one of these being DNA methyltransferase 1 (DNMT1). Methylation most commonly involves conversion of the base cytosine to 5'-methylcytosine (5'-MeC) and is a fundamental 'epigenetic' pathway involved in 'gene silencing'. Such events occur preferentially at specific loci, designated CpG areas (cytosine-guanine pairs or doublets) located throughout DNA coding areas. Runs of CpG doublets called CpG islands occur in the promoter regions of a number of important genes, though these remain unmethylated under most normal circumstances.

Along with regulating such important events like daughter strand methylation, genomic imprinting of paternal and maternal strands/allele selection and X-chromosome inactivation in females, methylation is also a key regulator of tumor promoter or tumor suppressor genes by virtue of its ability to methylate (hence 'silence') or demethylate (hence 'activate') these genes. Under normal conditions, this delicate and exquisitely balanced system helps to maintain normal mitosis and interphase; however abnormal methylation has been documented as a key aberration in a wide variety of malignancies, by far the commoner pattern being hypermethylation of tumor suppressor genes (such as p15, p16, p53, etc.) accompanied by global genomic hypomethylation. Most of the hypermethylation involves the CpG islands of the promoter regions of tumor suppressor genes, so that these genes are rendered functionally inactive or 'hypoactive'. This abnormal methylation pattern confers upon the cell a certain degree of abnormal 'unstable' genomic configuration that leads initially to the phenotype of the 'transformed' cell and later on to the fully malignant cell after several abnormal mitotic cycles. Studies of normal and abnormal methylation patterns have thus become central to the approach towards informed studies of the progression of a normal cell to a malignant one and also the progressive acquisition of greater malignant properties by neoplastic cells.

This chapter will attempt to provide a comprehensive overview of the processes of normal and abnormal DNA methylation, the genes involved in methylation processes and the roles of the major tumor suppressor genes in malignant transformation and progression, as they become progressively deranged in both structure and function (by abnormal methylation-related changes) during oncogenesis. It will also provide a view of current therapeutic strategies based on insights into methylation-related epigenetic pathways that can be exploited to therapeutic advantage for effective tumor control in cancer patients. It remains as yet for future researchers to develop those strategies that offer the promise of adopting a preventive approach to the cancer problem, particularly for high-risk individuals and for the population in general.

Chapter 8 - The recent discoveries of the RNA-mediated interference system in cells could explain all of the known features of human carcinogenesis, including methylation. A key, novel idea, proposed here, is that the cell has the ability to recognise a mutated protein and/or mRNA. Secondly, the cell can generate its own short interfering RNA (siRNA) using an RNA polymerase to destroy mutated mRNA, even when only a single base pair in the gene has mutated. The anti-sense strand of the short RNA molecule (called sicRNA), targets the mutated mRNA of an oncogene or a tumour suppressor. The resulting double stranded RNA, using the RNA-induced silencing complex (RISC) in the cytoplasm dices the mutated mRNA. During cell mitosis, the sicRNA complex can move into the nucleus to target the mutated gene. The sicRNA triggers the assembly of protein complexes leading to epigenetic modification of the promoter site of the mutant gene. In some instances, instead of methylation, the homologous DNA is degraded, leading to loss of heterozygosity (LOH). The factors controlling these two actions are unknown but the result is gene silencing by methylation or physical destruction of the mutant gene. The cell survives dependent on the functioning of the single, wild-type allele. A pretumor begins when CpG Methylation and/or LOH disrupt a key microRNA (miRNA) specific for that tissue. This produces a change in the profile of correct proteins for that tissue, resulting in dedifferentiation. The variability of the cancer response may depend on which specific miRNA is being transcribed at the time the sicRNA enters the nucleus. The driving force of carcinogenesis is the sequential loss of active miRNAs.

Chapter 9 - DNA methylation is an epigenetic modification characterized by the incorporation of methyl group at the 5-position of cytosine within of CpG dinucleotides. Epigenetic regulation plays a relevant role in development and DNA methylation is the mechanism that ensures cell-type specific expression of a limited range of genes. Different diseases, among them cancer, present altered methylation patterns that change the perfect epigenetic balance of normal cells. In cancer cells, the aberrant pattern is characterized by a global genomic hypomethylation and promoter hypermethylation. Such processes lead to genomic instability and gene silencing, respectively. Transcriptional repression associated with aberrant methylation of CpG islands on promoter areas is a consequence of several chromatin structural changes produced by the interaction of methylcytosines with different protein complexes that recruit histone-modifying enzymes, such as histone acetylases and methylases. Genes involved in different metabolic pathways may undergo aberrant methylation, including DNA repair genes, cellular cycle regulators, genes associated with apoptosis, cell invasion and cellular detoxification among others. Analyses of DNA methylation status has enabled methylator phenotypes associated with specific tumor groups to be established, advances that were partly possible due to the surge of new methodologies in this field of research. The main areas of clinical application that can benefit from biomarkers based on aberrant DNA methylation patterns are those related to diagnosis, tumor behavior and cancer treatment. Since methylation changes often precede apparent malignant changes in carcinogenesis, they can be used in the early diagnosis of cancer detecting tumor cells from biological samples. In addition, methylated genes that alter different metabolic pathways may be used as prognostic indicators, for example those related to angiogenesis (TIMP family) or cellular invasion (cadherins). Furthermore, the nature of the affected genes may help to predict the chemotherapy and/or hormonotherapy response. Finally, the potential to reverse DNA methylation and reactivate those genes affected is an attractive option for its use as a new therapeutic target in cancer treatment. In oncology research, the study of factors that determine the pathogenesis has enabled different neoplasias to be characterized and thus advances in the management of cancer patients. In this respect, understanding epigenetic mechanisms and the processes that they involve can bring new and promising expectations for cancer treatment, not forgetting that cancer is also an epigenetic disease.

In: DNA Methylation Research Trends
Editor: Taku B. Kobayashi, pp. 1-24

ISBN: 978-1-60021-721-0

Chapter I

Investigation of the CpG Effect in Mammalian Genomes

***Zhongming Zhao**[*1,2] **and Cizhong Jiang**[1]*
[1] Virginia Institute for Psychiatric and Behavioral Genetics, Virginia Commonwealth University, P.O. Box 980126, Richmond, VA 23298, USA
[2] Center for the Study of Biological Complexity, Virginia Commonwealth University, Richmond, VA 23284, USA

Abstract

CpG dinucleotides are globally methylated at the 5 position on the cytosine ring (5^mCpGs) in mammalian genomes. However, they often remain unmethylated in CpG islands. The hypermutability of 5^mCpG to TpG, which is known as the *CpG effect*, results in a strong deficiency of CpG dinucleotides and a decrease of GC content in mammalian genomes over evolutionary time. Disruption of DNA methylation system leads to hypermethylation in the promoter-associated CpG islands of tumor suppressor genes and hypomethylation in other global regions.

While the CpG effect has been well documented, its quantitative measurement could not be performed until the recent releases of several mammalian genomes, their genome-wide polymorphism data, and large-scale methylation status at CpG sites. In the present study, we performed a systematic examination of mutation direction in human genome using over one million high-quality single nucleotide polymorphism (SNP) data. The frequencies of nucleotide changes G:C→A:T were the highest among all types, which were mainly attributed to the hypermutability of 5^mCpG→TpG. Our comparison of mutational spectra in genomic categories revealed that exons and their four-fold

* Address correspondence to: Zhongming Zhao, PhD; Virginia Institute for Psychiatric and Behavioral Genetics; Virginia Commonwealth University; PO Box 980126; Richmond, VA 23298-0126, USA ; Phone: (804) 828-8129; FAX: (804) 828-1471; Email: zzhao@vcu.edu

degenerate sites had the strongest impact of methylation-related mutation (i.e., 5^{m}CpGs→TpGs/CpAs), whereas CpG islands had the weakest, suggesting strong CpG effect in exons and a lack of CpG effect in CpG islands.

Next, we quantitatively measured the CpG effect by estimating the methylation-dependent transition rate (5^{m}C deamination rate), the difference between CpG transition rate and GpC transition rate. We found that the 5^{m}C deamination rates were in a 90-fold range higher than other transitional changes and were strongly dependent on the local GC content.

We also studied the evolution of CpG islands in the promoter regions of genes. Our results suggest that both human and mouse genomes have been under loss of CpG islands, with a faster rate in mice, due to *de novo* methylation in some CpG islands. Gene Ontology (GO) term analysis found that genes maintaining CpG islands tend to be essential for cell life while genes losing CpG islands tend to be differentially regulated.

Finally, we investigated the association of aberrant methylation in the promoter-associated CpG islands with disruption (silence) of gene function. Experimental work has demonstrated that the methylation changes are associated with tumorigenesis, abnormal development, and mental disorders.

Introduction

DNA methylation is globally dispersed over mammalian genomes [1]. It typically occurs at the CpG sites by transferring a methyl group to the 5 position on the cytosine ring. This methylation process is catalyzed by enzymes in cells. Experiments have found that three DNA methyltransferases (DNMT1, DNMT3A, DNMT3B) may cooperate to add and maintain the methylation state in mammalian genomes [1-3]. One remarkable feature in methylation is that the 5-methylcytosine (5^{m}C) is vulnerable to deamination, which leads to a thymine [4, 5]. Unlike the uracil that results from the unmethylated cytosine by deamination, the thymine from 5^{m}C deamination cannot be removed and thus repaired by uracil glycosylase. Consequently, the mutation rate of methylated CpG to TpG/CpA is high; for example, it was estimated to be about 10-50 times higher than other transitional mutations [4, 6]. The hypermutability of 5^{m}CpG→TpG/CpA is known as the *CpG effect.*

The CpG effect results in a deficiency of CpG dinucleotides and, correspondingly, an excess of TpG and CpA dinucleotides in mammalian genomes [7]. It is estimated that CpG dinucleotides occur only about a fourth as often as would be expected under a random mutation model in the human genome [5, 8]. This deficiency is even greater in the mouse genome: the CpG dinucleotides occur only ~19% as often as expected [9]. Figure 1 shows a comparison of the frequencies of the 16 possible dinucleotides in the human and mouse genomes. The CpG effect has changed the genome compositions and structure over the course of evolution in two ways. First, the CpG effect has driven the GC content of the genomes to be AT-rich. The GC content of the human genome has been reduced by ~10% due to the deamination of 5^{m}C [10]. Second, the CpG effect has diminished the isochore (long DNA segment homogeneous in base composition) structure in vertebrates, such as the spatial distribution of GC content. Recent investigations suggest that GC-rich isochors are vanishing in mammalian genomes due to a large excess of GC→AT substitutions [11-13]. While the subject of genome composition is still under debate, it is now clear that the hypermutability of methylated CpG dinucleotides is correlated with the disequilibrium of GC content in vertebrate genomes and plays an important role in the evolution of isochors [10, 14].

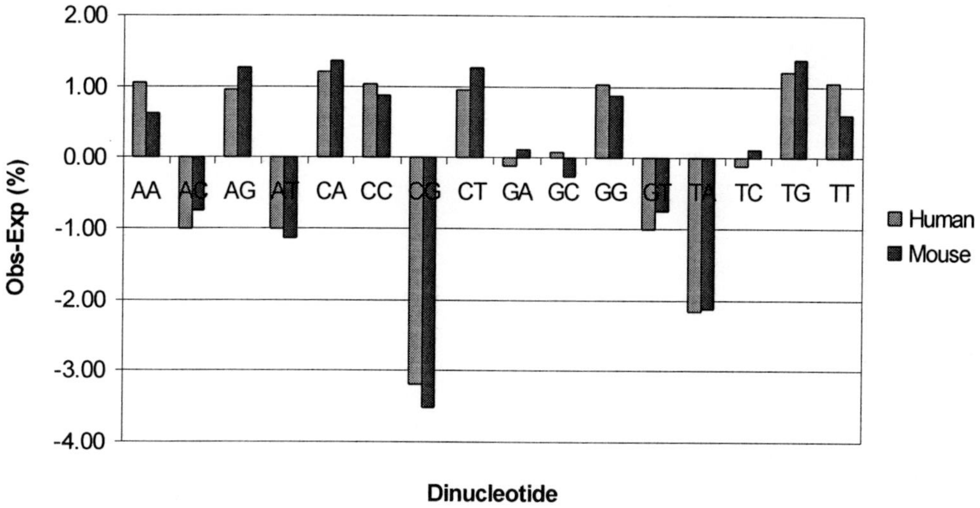

Figure 1. Difference of the observed and expected frequencies of the 16 dinucleotides in the human and mouse genomes. The frequencies were calculated based on the assembled human genome sequence (build 35) and the assembled mouse genome sequence (build 33), both were downloaded from the National Center for Biotechnology Information (ftp://ftp.ncbi.nih.gov/genomes/). It is interesting to note that besides dinucleotides "CA" and "TG", four dinucleotides "AA", "CC", "GG", and "TT" had higher frequencies than their expected frequencies in both the human and mouse genomes.

In contrast to the invertebrates, the vertebrate genome is often globally methylated. It is estimated that about 80% of the CpG dinucleotides in the genomes are methylated at the cytosine ring, though some regions stay free of methylation [15]. These unmethylated regions tend to have high CpG frequency and favor G+C in the base composition, thus forming CpG islands [15-17]. Most CpG islands are located in the promoters or 5' ends of genes, although they also may be found in introns, 3' untranslated regions (UTRs), or intergenic regions [18, 19]. Maintenance of an unmethylated state in promoter-associated CpG islands is essential to the expression of the associated genes [20]. While CpG islands are often unmethylated in normal cells, their methylation state is not always stable. CpG islands may be methylated progressively by some unknown mechanisms. Once some CpG dinucleotides in CpG islands become methylated, the high mutation rate of $5^{m}C$ to T gives rise to TpGs and CpAs. This would lead to a lower CpG frequency and GC content and would cause the erosion of CpG islands over the course of evolution. Two case studies demonstrated that α- and β-globin genes lost their CpG islands in chicken and mouse by acquiring methylation in germ-line cells [21, 22]. Further pairwise comparison of human-mouse orthologous genes provide evidence of the erosion of mouse CpG islands since the split of the human and rodent lineages [19, 23]. So far, the exact process of *de novo* methylation and loss of CpG islands remains largely unclear.

With recent advances in molecular technologies and the release of several mammalian genomes, DNA methylation studies—especially in the epigenetics field—have proliferated. Methylation changes in promoter CpG islands are associated with transcriptional silencing and disruption of gene function. Importantly, aberrant hypermethylation in promoter-associated CpG islands of tumor suppressor genes may cause tumorigenesis. One study found that promoter hypermethylation of *E-cadherin*, *p15,* and *p16* was associated with gastric carcinogenesis [24]. Moreover, aberrant promoter methylation of the genes *VHL*, *BRCA1*, and *STK11/LKB1* causes renal [25], breast [26], and colon cancers [27], respectively. More tumor-suppressor genes that are silenced by promoter hypermethylation and then cause cancers are expected to be discovered in the next few years. Moreover, aberrant promoter methylation has been found to be associated with mental disorders [28-30], developmental defects [31-33], and autoimmune diseases [34].

The recent completion of several mammalian genomes and the discovery of millions of single nucleotide polymorphisms (SNPs) in these genomes provide an unprecedented opportunity to systematically investigate the CpG effect in mammalian genomes, and, in particular, in the human genome. In this study, we performed a comprehensive analysis of mutation patterns in the human genome using the SNP data publicly available in the National Center for Biotechnology Information (NCBI) dbSNP database. On the basis of the frequencies of mutations normalized by the nucleotide content, we obtained the mutational spectrum in the genome and the different genomic regions, including noncoding regions, exons and their four-fold degenerate (FFD) sites, and CpG islands. To further investigate the influence of methylation on point mutations, we quantitatively measured the $5^{m}C$ deamination rate in methylated CpG dinucleotides and we estimated the range of the deamination rates. Moreover, we attempted to demonstrate that loss of CpG islands is a common feature in mammalian genomes by comparing the presence (or absence) of CpG islands in the human-mouse orthologous genes. We also examined the functional categories of the genes that might lose CpG islands. Finally, we investigated the epigenetic consequence of the aberrant methylation in the promoter-associated CpG islands of genes.

Materials and Methods

SNP Data and Genomic Sequences

The assembled human and chimpanzee genomic sequences were downloaded from ftp://ftp.ncbi.nih.gov/genomes/ (NCBI database, human build 35 version 1, chimpanzee build 1 version 1). A total of 8,353,499 human reference SNPs and their annotations were downloaded from ftp://ftp.ncbi.nih.gov/snp/ (dbSNP database, Build 124). We extracted 4,934,380 SNPs that were biallelic, uniquely mapped in the genome, and validated. We next removed those SNPs mapped in repetitive sequences. This retained 2,646,337 SNPs, among which were 2,632,415 SNPs whose flanking sequences were at least 100-nucleotide long at each side. These SNPs were formatted to have 100 nucleotides (nt) at each side for the following estimation of mutation direction. A flow chart of the SNP process and analysis is shown in figure 2.

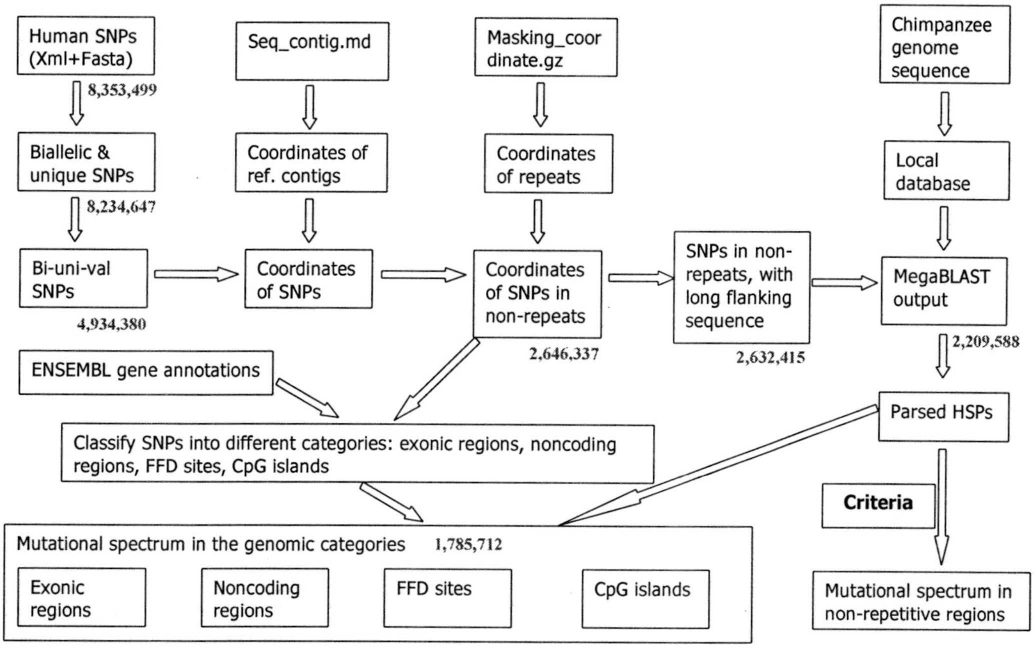

Figure 2. Flow chart of SNP process and analysis. HSP: high scoring pairs; FFD sites: four-fold degenerate sites. Bi-uni-val SNPs: SNPs that are annotated as "biallelic, uniquely mapped in the genome, and validated".

The human gene annotation information was retrieved and parsed from the ENSEMBL database (ftp://ftp.ensembl.org/pub/, version 32.35e). We classified SNPs in genomic regions by comparing the locations of SNPs and the coordinates of well-categorized (e.g., known) exonic, intronic, and intergenic regions on the assembled chromosomes. The CpG islands in the human genome were identified by using the CpG island search program CpGi130 [35]. We used the search criteria for GC content $\geq$ 55%, ratio of observed to expected CpG dinucleotides (termed Obs_{CpG}/Exp_{CpG}) $\geq$ 0.65, and length $\geq$ 500 bp. The criteria above effectively identify CpG islands associated with genes and exclude the universal *Alu* repeats, which typically have a sequence length of 300 bp, GC content of 53%, and Obs_{CpG}/Exp_{CpG} ratio of 0.62 [36, 37]. SNPs in CpG islands were then identified by comparing the locations of SNPs and CpG islands on the assembled chromosomes (figure 2).

Four-fold degenerate sites in codons are generally considered selectively neutral because nucleotide changes at these sites are synonymous (i.e., no change of amino acids). They are always located at the third position of a codon; half of the 64 codons is four-fold degenerate. We wrote a Perl script to identify the FFD and non-FFD sites in exons and to identify the SNPs in these two categories by comparing their locations in the genome, respectively.

Inference of Mutation Direction

We inferred the mutation direction of human SNPs by comparing their outgroup (i.e., ancestral) sequences in the chimpanzee genome. The MegaBLAST program was used to map human SNPs and their flanking sequences to the chimpanzee genome. A SNP was mapped in the chimpanzee genome when its high-scoring segment pairs (HSPs) in the MegaBLAST output satisfied the following stringent criteria: (1) the identity score was ≥ 95%; (2) the length of the alignment was between 196~206 bp; (3) the polymorphic site was located at positions 96~106 in the alignment (i.e., around the center of the alignment); (4) the matching chimpanzee base was one of the SNP alleles; (5) the immediate adjacent five nucleotides at the both sides of the SNP were identical to the aligned chimpanzee sequences; and (6) only one HSP satisfied the criteria above. A total of 1,785,712 SNPs satisfied these criteria, and their ancestral alleles (i.e., the matching chimpanzee base) were inferred (figure 2). The mutation direction was inferred on the basis of the parsimony principle of evolution. For example, if a human SNP is C/T at its polymorphic site and is aligned with C in the chimpanzee genome, the allele in the common ancestor of humans and chimpanzees was probably C, which was mutated to T in humans.

Normalized Frequencies of Point Mutation

After the inference of mutation direction, we estimated the mutational spectrum as follows. Let $n_{i->j}$ be the counts of point mutations from nucleotide *i* to nucleotide *j* (*i, j* = A, C, G or T) and N_i be the total counts of nucleotide *i* in the sequences. The normalized frequency of each nucleotide mutation was calculated by

$$f_{i->j} = \frac{n_{i->j} / N_i}{\sum_i \sum_{j \neq i} (n_{i->j} / N_i)} \times 100\%$$

where $f_{i \to j}$ is the expected number of nucleotide changes from nucleotide *i* to *j* in a random sequence. That is, the frequency of A, C, G or T is assumed to be the same in a random sequence.

Transition Rate in CpG and GpC Dinucleotides

We identified 292,216 SNPs involved in transition mutations that occurred in ancestral (chimpanzee) CpG or GpC dinucleotides. These SNPs were formatted to have 50, 100, 200, 300, and 500 nucleotides at each side, respectively. Thus, we have a set of length categories for each SNP, i.e., 101, 201, 401, 601, and 1,001 nt. For each length category, we counted the CpG and GpC dinucleotides and calculated the GC content in the SNP flanking sequences. Finally, we calculated the CpG transition rate by the number of CpG→TpG/CpA per CpG dinucleotide and GpC transition rate by the number of GpC→GpT/ApC per GpC dinucleotide.

Identification of CpG Islands in Human-Mouse Orthologues

We used the 3,750 human-mouse orthologue gene pairs prepared by Iwama and Gojobori [38]. We extracted 12-kb sequence in the 5' end of each gene. This 12-kb sequence includes a 2-kb sequence immediately upstream of the start codon and a 5-kb sequence both upstream and downstream of this 2-kb sequence. For those genes having alternative transcripts, we selected the transcript that uses the first start codon. Our analysis confirmed that no genic sequences were within the 7-kb upstream stretch of genes [38]. In our data preparation, we discarded those genes that did not have the "ATG" start codon or whose sequences were not consistent with those in Iwama and Gojobori [38] because of the updates in NCBI. As a result, we obtained a total of 3,197 human-mouse orthologue gene pairs.

To date, the definition of promoter regions is an unresolved issue. However, they often reside in a short region (e.g., 500 bp) immediately upstream of the transcription start site (TSS) [39-41]. Further, the average length of 5' untranslated regions (UTRs) is ~300 bp [5, 42]. Using the stringent criteria described as above, we identified CpG islands within or that overlapped the 2-kb sequences upstream of the first start codon. These CpG islands were considered promoter-associated.

Gene Ontology (GO) Analysis

For the 3,197 human and mouse orthologue gene pairs, we obtained their GO annotations from the EBI database (ftp://ftp.ebi.ac.uk/pub/databases/GO/goa/). A total of 2,976 human genes and 2,977 mouse genes had GO terms in EBI. The GO terms of these genes were mapped to the goslim-generic subset (ftp://ftp.geneontology.org/pub/go/GO_slims/goslim_generic.obo), in which the GO terms have been presented in a hierarchical style.

Results and Discussion

CpG Effect on Mutation Direction

There were a total of 1,785,712 human SNPs whose ancestral alleles could be inferred by mapping them to the chimpanzee genome. Table 1 summarizes the mutation direction normalized with the nucleotide content in the human genome and categorized genomic regions. The first row in table 1 shows the frequencies of 12 possible types of nucleotide change in the human genome. Overall, these frequencies varied greatly, indicating the nonrandomness of nucleotide changes in the human genome. Notably, transitional mutations, where nucleotides change between purines (A and G) and purines or between pyrimidines (C and T) and pyrimidines, occurred 2.6-8.0 times more often than transversional mutations, which are nucleotide changes between purines and pyrimidines. Among all nucleotide changes, C→T and G→A changes had the highest frequencies: 20.7% and 20.6%, respectively. This indicates a strong CpG effect on mutation direction.

The frequencies of each complementary pair of nucleotide changes, such as A→G and T→C, were nearly the same, reflecting DNA strand symmetry. This feature was similarly observed in most of the categorized genomic regions, indicating that the data size is sufficient to examine and compare the mutation direction. Hereafter, we will use A:T→G:C to denote the pair of mutations A→G and T→C. The other pairs follow the same rule. We will use A/T→C/G to denote the changes from A or T to C or G.

The mutation spectrum in intergenic regions and introns was similar; therefore, these two regions were combined as noncoding regions (table 1). The mutational spectrum in noncoding regions was similar to that in the whole genome, mainly because of the large sample size (87% of the 1,785,712 SNPs were in noncoding regions). Strikingly, the frequencies of nucleotide changes G→A and C→T in the exons (26.1% and 24.8%) were notably higher than those in the noncoding regions (20.5% and 20.6%). We further examined the frequencies of nucleotide changes in the FFD and non-FFD sites because the FFD sites are often considered selectively neutral while the non-FFD sites are under natural selection. Interestingly, the frequency of G→A changes in the FFD sites (30.7%) was higher than that in the non-FFD sites (23.9%); conversely, the frequency of C→T changes in the non-FFD sites (26.7%) was higher that that in the FFD sites (22.6%). When we combined these two changes (G:C→A:T), the FFD sites had a higher frequency (53.3%) than the non-FFD sites (50.6%), whereas both the FFD and non-FFD sites had higher frequencies of such changes than did the whole genome (41.3%) or noncoding regions (41.1%) (table 1). The higher frequencies of G:C→A:T changes in exons suggest a stronger CpG effect in exons than in noncoding regions. Recently, Subramanian and Kumar [43] compared the synonymous mutations in exons and intergenic regions and found a 30-60% higher mutation rate in exons than in noncoding regions, largely due to the overabundance of synonymous sites involved in CpG dinucleotides. Given that those CpG dinucleotides in exons are mostly methylated, the higher mutation rate reflects a stronger CpG effect. The asymmetric frequencies between G→A and C→T changes in the FFD and non-FFD sites also reflect biases in the transcription-coupled repair as well as different selection pressure in these two sites [44, 45].

The mutational spectra in the CpG islands and promoter-associated CpG islands were opposite to those in the exons. The frequencies of nucleotide changes G→A and C→T in the CpG islands (18.7% and 19.0%, respectively) or promoter-associated CpG islands (18.2% and 18.1%, respectively) were lower than those in the noncoding regions or whole genome (table 1). Note that the frequency of G:C→A:T in the promoter-associated CpG islands were lower than in the CpG islands. These results indicate a weak or lack of CpG effect in the CpG islands, especially in the promoter-associated CpG islands.

We observed that the frequency of G/C→A/T changes was 51.1%, much higher than the frequency (33.2%) of A/T→G/C in the human genome (table 1). This indicates a trend to have more G/C to be changed to A/T in the human genome, which decreases the GC content. A similar pattern was found in other genomic regions. It should be noted that this trend is weaker in CpG islands. Our analysis suggests that the GC content is not at equilibrium in the human genome.

Table 1. Normalized frequencies (%) of nucleotide changes in the human genome and categorized genomic regions

Category	GC (%)	No. of SNPs	G→A	C→T	A→G	T→C	G→T	C→A	A→C	T→G	G→C	C→G	A→T	T→A
Genome	40.31	1,785,712	20.6	20.7	13.4	13.4	4.9	4.9	3.2	3.2	5.2	5.2	2.6	2.6
Noncoding regions	39.78	1,560,207	20.5	20.6	13.3	13.3	5.0	5.0	3.2	3.2	5.3	5.2	2.7	2.7
Exons	52.10	12,430	26.1	24.8	12.9	12.9	3.2	3.4	2.4	2.2	4.0	4.6	1.9	1.8
FFD sites	56.75	3,722	30.7	22.6	11.2	11.3	3.4	3.0	3.1	2.6	4.2	4.2	2.0	1.7
Non-FFD sites	49.32	8,708	23.9	26.7	13.0	13.2	3.0	3.8	2.1	2.0	3.8	5.0	1.8	1.7
CpG islands	62.07	25,913	18.7	19.0	12.6	12.7	5.5	5.2	3.7	3.9	6.6	6.7	2.6	2.8
Promoter CpG islands	63.38	2,731	18.2	18.1	11.4	13.3	5.5	5.4	4.2	3.9	7.2	6.9	2.9	2.8

Table 2. Normalized frequencies (%) of nucleotide changes in the human genome and the categorized genomic regions after excluding CpG→TpG/CpA SNPs

Category	No. of SNPs	G→A	C→T	A→G	T→C	G→T	C→A	A→C	T→G	G→C	C→G	A→T	T→A
Genome	1,580,178	16.0	16.0	15.5	15.5	5.7	5.7	3.7	3.7	6.1	6.0	3.1	3.1
Noncoding regions	1,387,825	16.0	16.1	15.3	15.3	5.8	5.8	3.7	3.7	6.1	6.0	3.1	3.1
Exons	8,939	16.6	15.8	17.7	17.7	4.3	4.7	3.4	3.0	5.4	6.4	2.6	2.4
FFD sites	2,156	16.3	12.3	17.4	17.0	5.4	4.8	4.8	3.9	6.6	6.3	2.6	2.5
Non-FFD sites	6,783	16.4	18.1	17.1	17.4	4.0	4.9	2.9	2.7	5.0	6.7	2.5	2.3

To test whether the notably high frequencies of mutations G:C→A:T were due to the hypermutability of 5-methylcytosines (5^{m}CpGs) to TpGs/CpAs (the CpG effect), we estimated the mutational spectrum by excluding CpG→TpG/CpA SNPs. The results indicate that the frequencies of mutations G:C→A:T decrease dramatically in all categories compared to the corresponding frequencies before the exclusion (table 2). Remarkably, these frequencies in exons became close to those in noncoding regions. This strongly suggests that the hypermutability of 5^{m}C to T in CpG dinucleotides is the major contributor to the high frequencies of mutations G:C→A:T in exons. It also confirms that the excess of G:C→A:T changes in exons is mainly due to the overabundance of 5^{m}CpG at the synonymous sites. These 5-methylatd CpG dinucleotides have a greater chance to mutate to TpGs/CpAs. Interestingly, while the frequencies of G→A in the FFD (16.3%) and non-FFD (16.4%) sites were moderately higher than those in the overall genome (16.0%) or noncoding regions (16.0%), the frequencies of C→T in the FFD (12.3%) and non-FFD (18.1%) sites were much different from those in the overall genome (16.0%) or noncoding regions (16.1%). The frequency of C→T in the FFD sites was the lowest among the genomic categories and decreased by 10.3% (22.6 - 12.3%) after excluding the CpG→TpG/CpA SNPs, indicating that the CpG effect is the strongest in the FFD sites (tables 1-2). Conversely, there was an excess of changes C→T over G→A in the non-FFD sites (table 2). This is consistent with a previous report [44]. The strand bias may reflect the selective pressure due to the functional effects of SNPs at the synonymous sites and the biases in codon usage. The different mutational spectra in the FFD sites and noncoding regions imply that genetic factors may act differently during evolution, suggesting that the FFD sites might not be selectively neutral as is generally thought [46].

Mutations in the Methylated CpG and Unmethylated GpC Dinucleotides

Cytosines in CpG dinucleotides are often methylated and are vulnerable to deamination [4, 5]. The hypermutability of 5^{m}C to T results in a genome-wide excess of G:C→A:T changes as described above. In contrast, cytosines in GpC dinucleotides are not methylated in mammalian genomes [47]. To explore the different methylation and mutation pattern between CpG and GpC dinucleotides, we counted the number of all possible mutations occurred in the CpG or GpC sites. Next, we calculated the proportion of CpG→TpG/CpA changes among all mutations in the CpG sites and, similarly, we calculated the proportion of GpC→GpT/ApC changes among all mutations in the GpC sites. The difference was >20% between the proportion of changes CpG→TpG/CpA in CpG dinucleotides and the proportion of changes GpC→GpT/ApC in GpC dinucleotides in the human genome and noncoding regions (intergenic regions and introns) (figure 3a). On the other hand, this difference of proportion between CpG and GpC dinucleotides was ~10-15% in the exons, FFD sites, and non-FFD sites. The smaller difference in the exonic regions might be partially caused by the relatively abundance of CpGpC trinucleotides in GC-rich regions such as exons (GC content = 52.10%). However, when we examined the pattern in CpG islands, the difference of proportion between CpG and GpC dinucleotides was nearly the same in both the CpG islands and the promoter-associated CpG islands (figure 3a). Although the CpG islands had the highest GC content among the categorized genomic regions we compared, the proportion of CpG→TpG/CpA changes in the CpG islands was ~60%, much lower than that (~90%) in other regions. In summary, cytosines in CpG dinucleotides are highly methylated in all genomic regions except CpG islands.

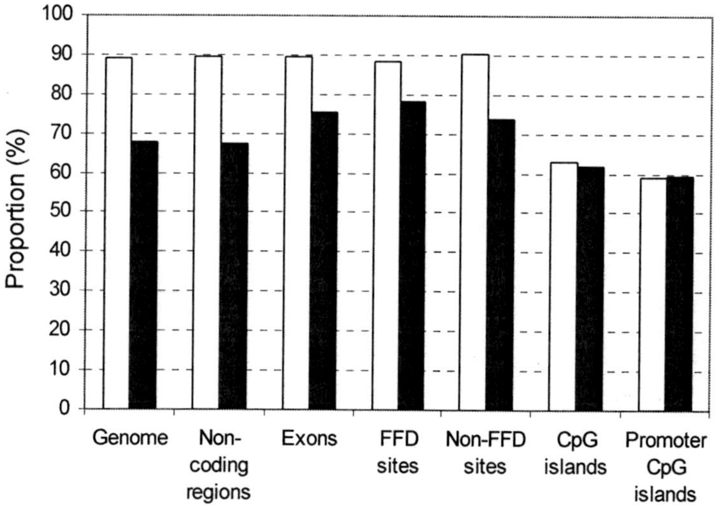

a.

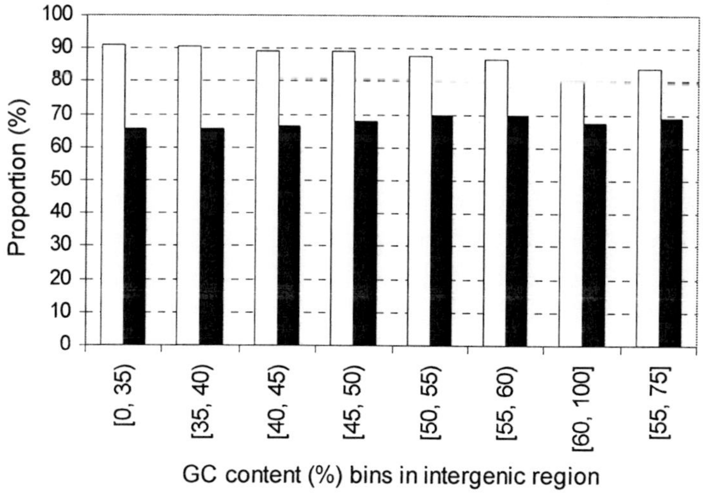

b.

Figure 3. Proportion of G:C→A:T changes in CpG and GpC dinucleotides. (A) Proportion of CpG→TpG/CpA changes in CpG dinucleotides (white bar) and proportion of GpC→GpT/ApC changes in GpC dinucleotides (black bar) in different genomic categories. (B) Proportion of CpG→TpG/CpA changes in CpG dinucleotides (white bar) and proportion of GpC→GpT/ApC changes in GpC dinucleotides (black bar) in the non-repetitive intergenic regions separated into different GC content bins. GC content in the intergenic region was calculated in a window with size of 500 nucleotides. Square brackets denote inclusive boundary and parentheses denote exclusive boundary.

To examine whether GC content influences the nucleotide changes in the CpG and GpC sites, we performed a similar analysis in intergenic regions which were separated into different GC content. The results indicate that when GC content was lower than 60%, about 90% of CpG dinucleotides were mutated to TpGs/CpAs in intergenic regions and the proportion of changes G:C→A:T in CpG dinucleotides was still ~20% more than that in GpC dinucleotides (figure 3b). When we compared the intergenic regions that had similar GC content (50-55%) with the exons, the proportion of changes G:C→A:T in CpG dinucleotides in the exons was slightly higher than that in the intergenic regions. This again supports our finding (above) of a higher proportion of methylated CpG dinucleotides in exons. Further, when we compared the intergenic regions that had similar GC content (55-75%) with the CpG islands, we found a much higher proportion of methylated CpG dinucleotides in the intergenic regions than that in the CpG islands (figure 3).

Quantitative Measurement of 5-Methylcytosine Deamination Rates

Deamination of 5^{m}CpG leads to TpG or CpA, or CpG turnover. We defined CpG transition rate (CpG_{trans}) as the number of CpG→TpG/CpA changes per CpG dinucleotide in a sequence. Similarly, we defined GpC transition rate (GpC_{trans}) as the number of GpC→GpT/ApC changes per GpC dinucleotide. Because GpC dinucleotides are not methylated in mammalian genomes [47], the mathematical difference of transition rate CpG_{trans} and GpC_{trans} represents the CpG turnover rate or 5' methylcytosine deamination rate [48]. We calculated the GC content in the stretch of nucleotides surrounding the SNPs and then compared the transition rates in CpG and GpC dinucleotides in different GC content bins. Our results indicate that both the CpG and GpC transition rates were dependent on the local GC content; however, the dependence of CpG rates on GC content was much stronger than that of GpC rates (figure 4a). Therefore, our results revealed that 5^{m}C deamination rates are strongly dependent on the GC content. Furthermore, deamination rates varied over a 90-fold range (figure 4b). This finding indicates that the previous estimate of 10- to 50-fold higher than other transitional changes, which is often cited in the literature [4, 6, 48, 49], may be conservative.

We also plotted $\log_{10}(5^{m}$C deamination rate) versus the GC content and found $\log_{10}(5^{m}$C deamination rate) was linearly correlated with the GC content (figure 4c). The linear correlation revealed that 5^{m}C deamination rate is exponentially dependent on the local GC content of the sequences. The linear regression of $\log_{10}(5^{m}$C deamination rate) on the GC content gave a best-fit slope of -3.3 (figure 4c). We used the length of SNP sequence 201 nt, including 100 nt at each flanking side and the polymorphic site, to obtain the GC content bins in the above analysis (see Materials and methods). Further analysis using different length of SNP sequences (e.g., 101, 401, 601, and 1,001 nt) revealed that the slopes were dependent on the lengths of the SNP sequences examined: the shorter SNP sequences, the larger slope values (or smaller in absolute values). Moreover, when we performed similar analysis using the SNPs in genomic categories, we found that methylation-dependent transition rates varied with the genomic regions. Of note, CpG islands always had much smaller slope values than other regions for each length category. This is consistent with the lack of methylation effect in the CpG islands.

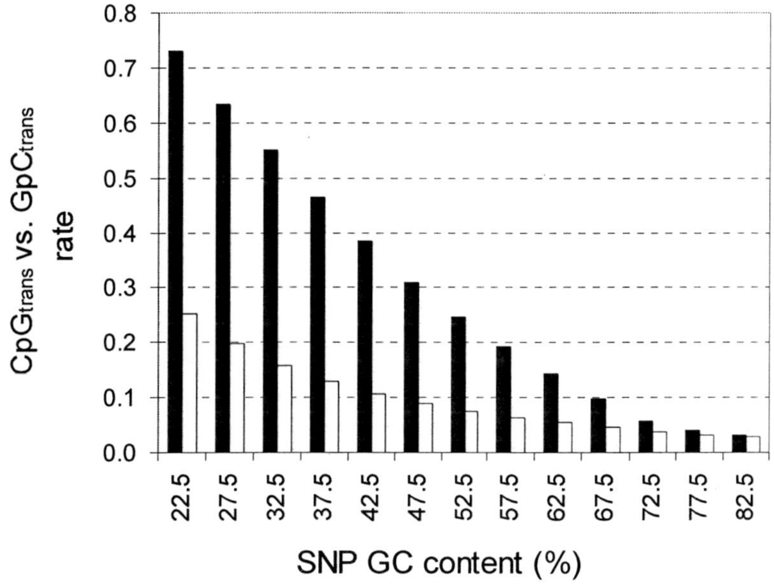

a.

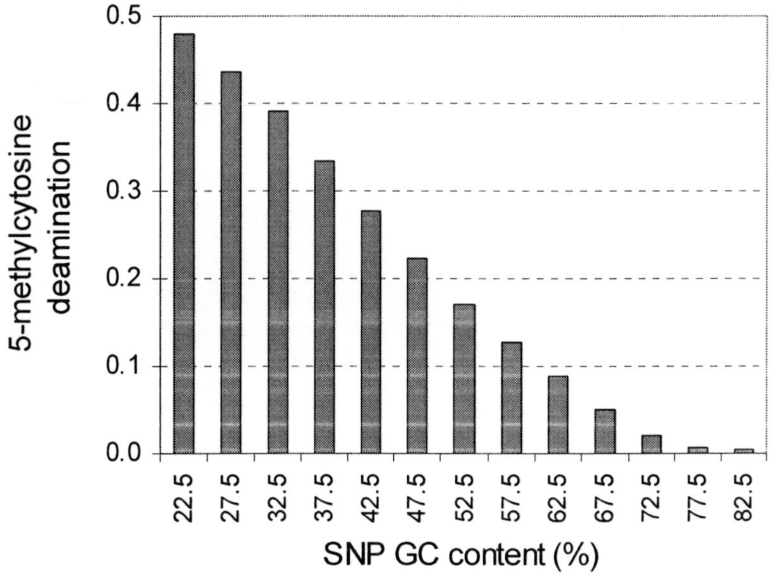

b.

Figure 4 (Continued)

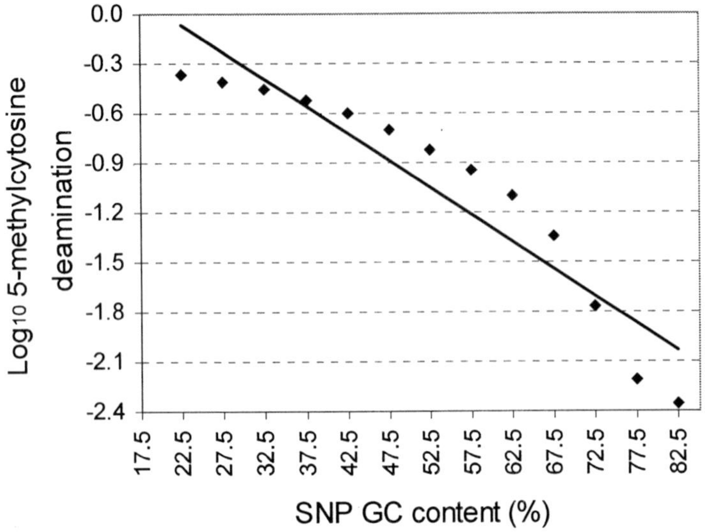

c.

Figure 4. 5^mC deamination rates are dependent on local GC content. (A) Black bars denote CpG transition rates (CpG_{trans}) calculated by the number of CpGs→TpGs/CpAs per CpG dinucleotide in the SNP sequences. White bars denote GpC transition rates (GpC_{trans}) calculated by the number of GpCs→GpTs/ApCs per GpC dinucleotide. The length of SNP sequences was set to 201 nt, including 100 nt at each flanking side of SNP and the polymorphic site. (B) 5^mC deamination rate, which is the difference between CpG_{trans} and GpC_{trans} in A. (C) Plot of $\log_{10}$(5^mC deamination rate) over local GC content in SNP sequences.

Loss of CpG Islands by *De Novo* Methylation

We identified 1,795 and 1,553 CpG islands in the promoter regions of the 3,197 human and mouse orthologous genes, respectively. Of the 3,197 gene pairs, 1,393 had at least one CpG island in both human and mouse genes (we denote it H+M+ group), 293 had at least one CpG islands in human genes but none in mouse genes (H+M- group), 87 had at least one CpG island in mouse genes but none in human genes (H-M+ group), and 1,424 had no CpG island in human or mouse genes (H-M- group). Among the 1,686 human genes that have CpG islands in the promoter regions, 17.4% of their mouse orthologues did not contain CpG islands. This proportion is close to a previous estimate (18.8%) based on 23 genes [19]. Notably, among the 1,480 mouse genes having CpG islands in the promoter regions, only 5.9% of their human orthologues had no CpG island presented. This comparison, which should be representative of the human and mouse genomes, clearly indicates that the mouse is more likely to lose its CpG islands. Recent studies found that some promoter-associated CpG islands may be methylated even in normal cells [50]. The hypermethylation of promoter-associated CpG islands may be introduced by a gradual and progressive process in cells [51]. Given this scenario, we would expect to detect the signature of CpG→TpG/CpA changes in the alignments of CpG island sequences, because the newly methylated CpGs in the CpG islands are likely to be deaminated. Indeed, we found a significant excess of TpGs and CpAs, especially in the categories of H+M- and H-M+. For example, in the category of

H+M-, there were 2,261 human CpGs that were aligned to mouse TpGs/CpAs, significantly more than the 771 mouse CpGs that were aligned to human TpGs/CpAs.

According to the data in Larsen et al. [52], all housekeeping genes have a CpG island covering the transcription start site (TSS), while only 25% of tissue-specific genes have a CpG island overlapping with the TSS. Other studies have shown that CpG islands are more frequently associated with broadly-expressed genes than with narrowly-expressed genes [37, 53]. We extended this analysis to investigate whether genes in different functional categories tend to lose CpG islands over the course of evolution. We examined the distribution of characteristic GO terms associated with genes in the H+M+, H-M+ and H+M- groups. Tables 3 and 4 shows the GO terms that were over-represented in the H-M+ or H+M- group compared with the H+M+ group. The GO terms that were significantly over-represented in H+M+ group include 'ion transport', 'protein modification', 'protein kinase activity', 'nucleus', 'transcription regulator activity'. These functions are generally essential for cell life. In contrast, the GO terms such as 'cytoplasmic membrane-bound vesicle', 'lipid binding', 'catalytic activity', 'physiological process', 'actin binding', and 'cytoskeletal protein binding' were significantly over-represented in the H+M- or H-M+ group. These functions tend to be differentially expressed in cells or characteristic of the highly regulated cells (tables 3-4).

Table 3. Comparison of GO terms between H+M+ and H-M+ genes using human data

GO code	GO term description	H+M+ (%)	H-M+ (%)	P value
Overrepresented in class H+M+				
B:0006811	Ion transport	177 (13.6)	3 (3.8)	0.013
B:0006464	Protein modification	293 (22.6)	10 (12.8)	0.043
C:0005886	Plasma membrane	278 (21.4)	9 (11.5)	0.037
M:0005216	Ion channel activity	111 (8.6)	1 (1.3)	0.023
M:0004672	Protein kinase activity	159 (12.3)	3 (3.8)	0.025
Overrepresented in class H-M+				
B:0006810	Transport	155 (12.0)	19 (24.4)	0.001
B:0006996	Organelle organization and biogenesis	44 (3.4)	6 (7.7)	0.049
C:0005694	Chromosome	32 (2.5)	12 (15.4)	3.06E-10
C:0016023	Cytoplasmic membrane-bound vesicle	25 (1.9)	5 (6.4)	0.008
M:0030234	Enzyme regulator activity	63 (4.9)	21 (26.9)	2.73E-15
M:0003682	Chromatin binding	5 (0.4)	5 (6.4)	1.19E-09
M:0003677	DNA binding	242 (18.7)	26 (33.3)	0.001
M:0005215	Transporter activity	125 (9.6)	15 (19.2)	0.007
M:0008289	Lipid binding	46 (3.5)	7 (9.0)	0.016
M:0003824	Catalytic activity	240 (18.5)	22 (28.2)	0.034

GO code includes three organizing principles of GO terms: cellular component (C), biological process (B), and molecular function (M). In the H+M+ and H-M+ columns, it lists the counts (frequency) of a GO term in the category. Human genes, including 1,297 H+M+ and 78 H-M+, were analyzed. P values were calculated by χ^2 test using 2 × 2 contingency tables. Only small P-values (≤0.05) are included.

Table 4. Comparison of GO terms between H+M+ and H+M- genes using mouse data

GO code	GO term description	H+M+ (%)	H+M- (%)	P value
Overrepresented in class H+M+				
B:0006950	Response to stress	110 (8.4)	10 (3.7)	0.007
B:0006464	Protein modification	318 (24.4)	47 (17.2)	0.010
B:0009719	Response to endogenous stimulus	48 (3.7)	2 (0.7)	0.011
B:0006259	DNA metabolism	85 (6.5)	9 (3.3)	0.041
C:0005634	Nucleus	555 (42.6)	86 (31.5)	0.001
C:0005737	Cytoplasm	265 (20.3)	40 (14.7)	0.031
C:0005739	Mitochondrion	123 (9.4)	15 (5.5)	0.036
M:0004672	Protein kinase activity	187 (14.4)	20 (7.3)	0.002
M:0030528	Transcription regulator activity	104 (8.0)	12 (4.4)	0.039
Overrepresented in class H+M-				
B:0007582	Physiological process	325 (24.9)	91 (33.3)	0.004
B:0006139	Nucleobase, nucleoside, nucleotide and nucleic acid metabolism	107 (8.2)	34 (12.5)	0.026
B:0009058	Biosynthesis	86 (6.6)	28 (10.3)	0.034
C:0005623	Cell	708 (54.3)	192 (70.3)	1.20E-06
C:0005941	Unlocalized protein complex	1 (0.1)	3 (1.1)	0.002
M:0004872	Receptor activity	241 (18.5)	72 (26.4)	0.003
M:0003779	Actin binding	34 (2.6)	16 (5.9)	0.005
M:0008092	Cytoskeletal protein binding	17 (1.3)	10 (3.7)	0.006
M:0030234	Enzyme regulator activity	72 (5.5)	26 (9.5)	0.013

GO code includes three organizing principles of GO terms: cellular component (C), biological process (B), and molecular function (M). In the H+M+ and H+M- columns, it lists the counts (frequency) of a GO term in the category. Humen genes, including 1,297 H+M+ and 78 H-M+, were analyzed. *P* values were calculated by χ^2 test using 2 × 2 contingency tables. Only small *P*-values (≤ 0.05) are included.

Epigenetic Effect of Aberrant Methylation

5-methylcytosines are biologically important, especially in epigenetics. Epigenetic changes are often caused by methylation and associated with inappropriate gene silencing. An important feature in epigenetics is that the aberrant hypermethylation in the promoter-associated CpG islands leads to a loss of gene function. Aberrant methylation of the promoter-associated CpG islands has been linked to many abnormal functions and phenotypes including gene silencing, genomic imprinting, X-chromosome inactivation, carcinogenesis, and the silencing of intragenomic parasites [36, 54]. Numerous studies suggest that dysregulation in CpG-island methylation control is likely involved in the early carcinogenesis process. Table 5 shows a list of genes with aberrant methylation in their promoter regions which link to diseases by our literature search. Most of them are tumor suppressor genes. In tumor suppressor genes, the transcription silencing by hypermethylation disrupts gene function or reduces gene expression, which subsequently leads to over-expression of oncogenes. Interestingly, DNA hypomethylation outside of CpG islands has also been reported to be correlated with diseases, such as Alzheimer [55], schizophrenia [29, 56], lupus erythematosis, and other autoimmune diseases [34].

Table 5. Category of genes with aberrant promoter methylation

Gene	Consequence of aberrant promoter methylation	Full name	Ref.
Tumorigenesis			
BCL2	Lung cancer	BCL2-antagonist of cell death	[69]
BRCA1	Breast cancer	Breast cancer 1, early onset	[26]
CDH1	Leukemia	E-cadherin	[70]
ESR1	Breast and colon cancer	Estrogen receptor 1	[57, 71]
GSTP1	Prostate cancer	Glutathione S-transferase pi	[72]
HIC1	Tumorigenesis	Hypermethylated in cancer 1	[73]
HoxA5	Tumorigenesis	Homeobox A5	[74]
MLH1	Colon cancer	mutL homologue 1	[75-77]
RASSF1	Lung and breast cancers	Ras association (RalGDS/AF-6) domain family 1	[78, 79]
STK11/LKB1	Colon cancer	Serine/threonine kinase 11	[27]
TIMP3	Tumorigenesis	Tissue inhibitor of metalloproteinase 3	[80]
VHL	Renal cancer	von Hippel-Lindau	[25]
CDKN2B	Tumorigenesis	Cyclin-dependent kinase inhibitor 2B	[81]
CDKN2A	Tumorigenesis	Cyclin-dependent kinase inhibitor 2A	[81-83]
TP53	Colorectal tumorigenesis	Tumor protein p53	[84]
Development			
IGF2	Beckwith-Wiedemann syndrome	Insulin-like growth factor 2	[31, 32, 85]
Mental disorder			
APP	Alzheimer	Amyloid precursor protein	[55]
COMT	Schizophrenia and bipolar disorder	Catechol-O-methyltransferase	[30]
DRD2	Schizophrenia	Dopamine receptor D2	[86]
FMR1	Psychiatric morbidity	Fragile X mental retardation 1	[87]
HTR2A	Schizophrenia	5-hydroxytryptamine (serotonin) receptor 2A	[56]
RELN	Schizophrenia and bipolar disorder	Reelin	[28]
Other			
CD4		CD4 antigen	[29]
CD8A		CD8 antigen	[29]
HBB		Hemoglobin beta	[29]
IFNG		Interferon, gamma	[29]
IL4		Interleukin 4	[29]
OXTR		Oxytocin receptor	[29]

So far, it is mostly unclear what causes aberrant methylation. Methylation of promoter-associated CpG islands was reported to be associated with ageing [57]. Such age-dependent promoter methylation could explain the ageing-cancer link. Many environmental factors, such as diet, proteins, drugs, and hormones, can disrupt or modify DNA methylation and cause disease syndromes [29, 58-60]. Since abnormal methylation has been reported to be associated with breast cancer, colon cancer, colorectal cancer, Werner syndrome, and other diseases (table 5) [61-63], it is necessary to investigate the methylation states and mutation patterns in cancer and other disease associated genes. The recently available data provide us this opportunity. Several databases have been recently established to annotate cancer genes;

some are specifically designed for tumor suppressor genes. These databases include National Cancer Institute Cancer (NCI) Gene Data Curation Project, Tumor Suppressor Gene Database, Breast Cancer Database, Tumor Gene Family Database, and Cancer Gene Expression Database (CGED). The existing CpG island search programs can be employed to scan the promoter-associated CpG islands in the cancer-related genes. Meanwhile, large-scale methylation status data have been made available recently. The most recent data are the methylation status information in ~30 Mb brain DNA sequences from 22 human autosomes [64] and a DNA methylation profile covering 1.9 million CpG methylation values on human chromosomes 6, 20, and 22 [65]. Further, technologies have been developed to detect global methylation status in the human genome. For example, Affymetrix, Inc. recently released CpG island microarrays for high-throughput methylation detection [66]. The methylation status in promoter-associated CpG islands can also be predicted by recently developed software [67, 68]. A systematic investigation of the methylation status and mutation pattern in the promoter regions of disease-related genes and comparison with the control genes will provide many insights on the molecular mechanisms of methylation initiation, modification, fine-tuning gene expression, and causes of diseases.

Conclusion

We performed a comprehensive analysis of the mutation direction using 1,785,712 high-quality human SNPs with ancestral information and examined methylation-related CpG effects on point mutation. The frequencies of nucleotide changes G→A and C→T were the highest among all types of changes, which were mainly influenced by the hypermutability of methylated cytosines that are globally dispersed over the genome. Our comparison of mutational spectra in genomic categories revealed that exons and their four-fold degenerate sites had the strongest impact of methylation-related mutation (i.e., 5^{m}CpGs→TpGs/CpAs), suggesting the strongest CpG effect. The opposite mutation pattern existed in the CpG islands, which are largely methylation free. We found that the 5^{m}C deamination rates at the CpG dinucleotides varied over a 90-fold range and are strongly dependent on the local GC content. Moreover, comparative genomics analysis of CpG islands in the promoter regions of the human-mouse orthologues indicates a loss of CpG islands in mammalian genomes by *de novo* methylation. GO term analysis found that genes having CpG islands tend to be essential for cell life while genes without CpG islands tend to be specific or differentiated. Finally, we discussed the association of aberrant methylation in the promoter-associated CpG islands with disruption of gene function and diseases such as cancer and mental disorders.

Acknowledgements

We are indebted to the people who collected, annotated, and managed those data for public access. We thank Leng Han and Matthew Keller for assistance. This project was supported by the Thomas F. and Kate Miller Jeffress Memorial Trust Fund.

References

[1] Bird, A. (2002). DNA methylation patterns and epigenetic memory. *Genes Dev.*, *16*: 6-21.

[2] Okano, M., Bell, D. W., Haber, D. A. and Li, E. (1999). DNA methyltransferases Dnmt3a and Dnmt3b are essential for de novo methylation and mammalian development. *Cell*, *99*: 247-257.

[3] Robertson, K. D., Uzvolgyi, E., Liang, G., Talmadge, C., Sumegi, J., Gonzales, F. A. and Jones, P. A. (1999). The human DNA methyltransferases (DNMTs) 1, 3a and 3b: coordinate mRNA expression in normal tissues and overexpression in tumors. *Nucleic Acids Res.*, *27*: 2291-2298.

[4] Sved, J. and Bird, A. (1990). The expected equilibrium of the CpG dinucleotide in vertebrate genomes under a mutation model. *Proc. Natl. Acad. Sci. USA*, *87*: 4692-4696.

[5] Lander, E. S., Linton, L. M., Birren, B., Nusbaum, C., Zody, M. C., Baldwin, J., Devon, K., Dewar, K., Doyle, M., FitzHugh, W., et al. (2001). Initial sequencing and analysis of the human genome. *Nature*, *409*: 860-921.

[6] Duncan, B. K. and Miller, J. H. (1980). Mutagenic deamination of cytosine residues in DNA. *Nature*, *287*: 560-561.

[7] Bird, A. P. (1980). DNA methylation and the frequency of CpG in animal DNA. *Nucleic Acids Res.*, *8*: 1499-1504.

[8] Zhao, Z. and Zhang, F. (2006). Sequence context analysis of 8.2 million single nucleotide polymorphisms in the human genome. *Gene*, *366*: 316-324.

[9] Zhao, Z. and Zhang, F. (2006). Sequence context analysis in the mouse genome: Single nucleotide polymorphisms and CpG island sequences. *Genomics*, *87*: 68-74.

[10] Fryxell, K. J. and Zuckerkandl, E. (2000). Cytosine deamination plays a primary role in the evolution of mammalian isochores. *Mol. Biol. Evol.*, *17*: 1371-1383.

[11] Duret, L., Semon, M., Piganeau, G., Mouchiroud, D. and Galtier, N. (2002). Vanishing GC-rich isochores in mammalian genomes. *Genetics*, *162*: 1837-1847.

[12] Belle, E. M., Duret, L., Galtier, N. and Eyre-Walker, A. (2004). The decline of isochores in mammals: an assessment of the GC content variation along the mammalian phylogeny. *J. Mol. Evol.*, *58*: 653-660.

[13] Duret, L., Eyre-Walker, A. and Galtier, N. (2006). A new perspective on isochore evolution. *Gene*, *385*: 71-74.

[14] Duret, L. (2006). The GC content of primates and rodents genomes is not at equilibrium: a reply to Antezana. *J. Mol. Evol.*, *62*: 803-806.

[15] Antequera, F. (2003). Structure, function and evolution of CpG island promoters. *Cell. Mol. Life. Sci.*, *60*: 1647-1658.

[16] Bird, A. P. (1986). CpG-rich islands and the function of DNA methylation. *Nature*, *321*: 209-213.

[17] Bird, A. P. (1987). CpG islands as gene markers in the vertebrate nucleos. *Trends Genet.*, *3*: 342-347.

[18] Ioshikhes, I. P. and Zhang, M. Q. (2000). Large-scale human promoter mapping using CpG islands. *Nat. Genet.*, *26*: 61-63.

[19] Antequera, F. and Bird, A. (1993). Number of CpG islands and genes in human and mouse. *Proc. Natl. Acad. Sci. USA*, *90*: 11995-11999.

[20] Tazi, J. and Bird, A. (1990). Alternative chromatin structure at CpG islands. *Cell*, *60*: 909-920.

[21] Bird, A. P., Taggart, M. H., Nicholls, R. D. and Higgs, D. R. (1987). Non-methylated CpG-rich islands at the human alpha-globin locus: implications for evolution of the alpha-globin pseudogene. *Embo. J.*, *6*: 999-1004.

[22] Gardiner-Garden, M. and Frommer, M. (1987). CpG islands in vertebrate genomes. *J. Mol. Biol.*, *196*: 261-282.

[23] Matsuo, K., Clay, O., Takahashi, T., Silke, J. and Schaffner, W. (1993). Evidence for erosion of mouse CpG islands during mammalian evolution. *Somat. Cell. Mol. Genet.*, *19*: 543-555.

[24] Leung, W. K., Yu, J., Ng, E. K., To, K. F., Ma, P. K., Lee, T. L., Go, M. Y., Chung, S. C. and Sung, J. J. (2001). Concurrent hypermethylation of multiple tumor-related genes in gastric carcinoma and adjacent normal tissues. *Cancer*, *91*: 2294-2301.

[25] Herman, J. G., Latif, F., Weng, Y., Lerman, M. I., Zbar, B., Liu, S., Samid, D., Duan, D. S., Gnarra, J. R., Linehan, W. M., et al. (1994). Silencing of the VHL tumor-suppressor gene by DNA methylation in renal carcinoma. *Proc. Natl. Acad. Sci. USA*, *91*: 9700-9704.

[26] Esteller, M., Silva, J. M., Dominguez, G., Bonilla, F., Matias-Guiu, X., Lerma, E., Bussaglia, E., Prat, J., Harkes, I. C., Repasky, E. A., et al. (2000). Promoter hypermethylation and BRCA1 inactivation in sporadic breast and ovarian tumors. *J. Natl. Cancer. Inst.*, *92*: 564-569.

[27] Esteller, M., Avizienyte, E., Corn, P. G., Lothe, R. A., Baylin, S. B., Aaltonen, L. A. and Herman, J. G. (2000). Epigenetic inactivation of LKB1 in primary tumors associated with the Peutz-Jeghers syndrome. *Oncogene*, *19*: 164-168.

[28] Abdolmaleky, H. M., Cheng, K. H., Russo, A., Smith, C. L., Faraone, S. V., Wilcox, M., Shafa, R., Glatt, S. J., Nguyen, G., Ponte, J. F., et al. (2005). Hypermethylation of the reelin (RELN) promoter in the brain of schizophrenic patients: a preliminary report. *Am. J. Med. Genet. B Neuropsychiatr. Genet.*, *134*: 60-66.

[29] Abdolmaleky, H. M., Smith, C. L., Faraone, S. V., Shafa, R., Stone, W., Glatt, S. J. and Tsuang, M. T. (2004). Methylomics in psychiatry: Modulation of gene-environment interactions may be through DNA methylation. *Am. J. Med. Genet. B Neuropsychiatr. Genet.*, *127*: 51-59.

[30] Abdolmaleky, H. M., Cheng, K. H., Faraone, S. V., Wilcox, M., Glatt, S. J., Gao, F., Smith, C. L., Shafa, R., Aeali, B., Carnevale, J., et al. (2006). Hypomethylation of MB-COMT promoter is a major risk factor for schizophrenia and bipolar disorder. *Hum. Mol. Genet.*, *15*: 3132-3145.

[31] Bell, A. C. and Felsenfeld, G. (2000). Methylation of a CTCF-dependent boundary controls imprinted expression of the Igf2 gene. *Nature*, *405*: 482-485.

[32] Maher, E. R. and Reik, W. (2000). Beckwith-Wiedemann syndrome: imprinting in clusters revisited. *J. Clin. Invest.*, *105*: 247-252.

[33] Watson, R. E. and Goodman, J. I. (2002). Epigenetics and DNA methylation come of age in toxicology. *Toxicol. Sci.*, *67*: 11-16.

[34] Attwood, J. T., Yung, R. L. and Richardson, B. C. (2002). DNA methylation and the regulation of gene transcription. *Cell. Mol. Life. Sci.*, *59*: 241-257.

[35] Takai, D. and Jones, P. A. (2003). The CpG island searcher: a new WWW resource. *In Silico Biol.*, *3*: 235-240.

[36] Takai, D. and Jones, P. A. (2002). Comprehensive analysis of CpG islands in human chromosomes 21 and 22. *Proc. Natl. Acad. Sci. USA*, *99*: 3740-3745.

[37] Ponger, L., Duret, L. and Mouchiroud, D. (2001). Determinants of CpG islands: expression in early embryo and isochore structure. *Genome Res.*, *11*: 1854-1860.

[38] Iwama, H. and Gojobori, T. (2004). Highly conserved upstream sequences for transcription factor genes and implications for the regulatory network. *Proc. Natl. Acad. Sci. USA*, *101*: 17156-17161.

[39] Xuan, Z., Zhao, F., Wang, J., Chen, G. and Zhang, M. Q. (2005). Genome-wide promoter extraction and analysis in human, mouse, and rat. *Genome Biol.*, *6*: R72.

[40] Yamashita, R., Suzuki, Y., Wakaguri, H., Tsuritani, K., Nakai, K. and Sugano, S. (2006). DBTSS: DataBase of Human Transcription Start Sites, progress report 2006. *Nucleic Acids Res.*, *34*: D86-89.

[41] Waterston, R. H., Lindblad-Toh, K., Birney, E., Rogers, J., Abril, J. F., Agarwal, P., Agarwala, R., Ainscough, R., Alexandersson, M., An, P., et al. (2002). Initial sequencing and comparative analysis of the mouse genome. *Nature*, *420*: 520-562.

[42] Pesole, G., Liuni, S., Grillo, G., Ippedico, M., Larizza, A., Makalowski, W. and Saccone, C. (1999). UTRdb: a specialized database of 5' and 3' untranslated regions of eukaryotic mRNAs. *Nucleic Acids Res.*, *27*: 188-191.

[43] Subramanian, S. and Kumar, S. (2003). Neutral substitutions occur at a faster rate in exons than in noncoding DNA in primate genomes. *Genome Res.*, *13*: 838-844.

[44] Qu, H. Q., Lawrence, S. G., Guo, F., Majewski, J. and Polychronakos, C. (2006). Strand bias in complementary single-nucleotide polymorphisms of transcribed human sequences: evidence for functional effects of synonymous polymorphisms. *BMC Genomics*, *7*: 213.

[45] Green, P., Ewing, B., Miller, W., Thomas, P. J. and Green, E. D. (2003). Transcription-associated mutational asymmetry in mammalian evolution. *Nat. Genet.*, *33*: 514-517.

[46] Kimura, M. (1983).*The neutral theory of molecular evolution*. New York: Cambridge University Press. 367.

[47] Razin, A. and Riggs, A. D. (1980). DNA methylation and gene function. *Science*, *210*: 604-610.

[48] Fryxell, K. J. and Moon, W. J. (2005). CpG mutation rates in the human genome are highly dependent on local GC content. *Mol. Biol. Evol.*, *22*: 650-658.

[49] Jiang, C. and Zhao, Z. (2006). Directionality of point mutation and 5-methylcytosine deamination rates in the chimpanzee genome. *BMC Genomics*, *7*: 316.

[50] Yamada, Y., Watanabe, H., Miura, F., Soejima, H., Uchiyama, M., Iwasaka, T., Mukai, T., Sakaki, Y. and Ito, T. (2004). A comprehensive analysis of allelic methylation status of CpG islands on human chromosome 21q. *Genome Res.*, *14*: 247-266.

[51] Graff, J. R., Gabrielson, E., Fujii, H., Baylin, S. B. and Herman, J. G. (2000). Methylation patterns of the E-cadherin 5' CpG island are unstable and reflect the

dynamic, heterogeneous loss of E-cadherin expression during metastatic progression. *J. Biol. Chem.*, *275*: 2727-2732.

[52] Larsen, F., Gundersen, G., Lopez, R. and Prydz, H. (1992). CpG islands as gene markers in the human genome. *Genomics*, *13*: 1095-1107.

[53] Robinson, P. N., Bohme, U., Lopez, R., Mundlos, S. and Nurnberg, P. (2004). Gene-Ontology analysis reveals association of tissue-specific 5' CpG-island genes with development and embryogenesis. *Hum. Mol. Genet.*, *13*: 1969-1978.

[54] Jones, P. A. and Baylin, S. B. (2002). The fundamental role of epigenetic events in cancer. *Nat. Rev. Genet.*, *3*: 415-428.

[55] West, R. L., Lee, J. M. and Maroun, L. E. (1995). Hypomethylation of the amyloid precursor protein gene in the brain of an Alzheimer's disease patient. *J. Mol. Neurosci.*, *6*: 141-146.

[56] Polesskaya, O. O. and Sokolov, B. P. (2002). Differential expression of the "C" and "T" alleles of the 5-HT2A receptor gene in the temporal cortex of normal individuals and schizophrenics. *J. Neurosci. Res.*, *67*: 812-822.

[57] Issa, J. P., Ottaviano, Y. L., Celano, P., Hamilton, S. R., Davidson, N. E. and Baylin, S. B. (1994). Methylation of the oestrogen receptor CpG island links ageing and neoplasia in human colon. *Nat. Genet.*, *7*: 536-540.

[58] Shen, L., Ahuja, N., Shen, Y., Habib, N. A., Toyota, M., Rashid, A. and Issa, J. P. (2002). DNA methylation and environmental exposures in human hepatocellular carcinoma. *J. Natl. Cancer Inst.*, *94*: 755-761.

[59] Bertino, L., Ruffini, M. C., Copani, A., Bruno, V., Raciti, G., Cambria, A. and Nicoletti, F. (1996). Growth conditions influence DNA methylation in cultured cerebellar granule cells. *Brain Res. Dev. Brain Res.*, *95*: 38-43.

[60] Wolff, G. L., Kodell, R. L., Moore, S. R. and Cooney, C. A. (1998). Maternal epigenetics and methyl supplements affect agouti gene expression in Avy/a mice. *Faseb. J.*, *12*: 949-957.

[61] Lind, G. E., Thorstensen, L., Lovig, T., Meling, G. I., Hamelin, R., Rognum, T. O., Esteller, M. and Lothe, R. A. (2004). A CpG island hypermethylation profile of primary colorectal carcinomas and colon cancer cell lines. *Mol. Cancer*, *3*: 28.

[62] Agrelo, R., Cheng, W. H., Setien, F., Ropero, S., Espada, J., Fraga, M. F., Herranz, M., Paz, M. F., Sanchez-Cespedes, M., Artiga, M. J., et al. (2006). Epigenetic inactivation of the premature aging Werner syndrome gene in human cancer. *Proc. Natl. Acad. Sci. USA*, *103*: 8822-8827.

[63] Esteller, M. (2006). Epigenetics provides a new generation of oncogenes and tumour-suppressor genes. *Br. J. Cancer*, *94*: 179-183.

[64] Rollins, R. A., Haghighi, F., Edwards, J. R., Das, R., Zhang, M. Q., Ju, J. and Bestor, T. H. (2006). Large-scale structure of genomic methylation patterns. *Genome Res.*, *16*: 157-163.

[65] Eckhardt, F., Lewin, J., Cortese, R., Rakyan, V. K., Attwood, J., Burger, M., Burton, J., Cox, T. V., Davies, R., Down, T. A., et al. (2006). DNA methylation profiling of human chromosomes 6, 20 and 22. *Nat. Genet.*, *38*: 1378-1385.

[66] Bibikova, M., Lin, Z., Zhou, L., Chudin, E., Garcia, E. W., Wu, B., Doucet, D., Thomas, N. J., Wang, Y., Vollmer, E., et al. (2006). High-throughput DNA methylation profiling using universal bead arrays. *Genome Res.*, *16*: 383-393.

[67] Das, R., Dimitrova, N., Xuan, Z., Rollins, R. A., Haghighi, F., Edwards, J. R., Ju, J., Bestor, T. H. and Zhang, M. Q. (2006). Computational prediction of methylation status in human genomic sequences. *Proc. Natl. Acad. Sci. USA*, *103*: 10713-10716.

[68] Fang, F., Fan, S., Zhang, X. and Zhang, M. Q. (2006). Predicting methylation status of CpG islands in the human brain. *Bioinformatics*, *22*: 2204-2209.

[69] Nagatake, M., Osada, H., Kondo, M., Uchida, K., Nishio, M., Shimokata, K. and Takahashi, T. (1996). Aberrant hypermethylation at the bcl-2 locus at 18q21 in human lung cancers. *Cancer Res.*, *56*: 1886-1891.

[70] Cameron, E. E., Baylin, S. B. and Herman, J. G. (1999). p15(INK4B) CpG island methylation in primary acute leukemia is heterogeneous and suggests density as a critical factor for transcriptional silencing. *Blood*, *94*: 2445-2451.

[71] Nass, S. J., Herman, J. G., Gabrielson, E., Iversen, P. W., Parl, F. F., Davidson, N. E. and Graff, J. R. (2000). Aberrant methylation of the estrogen receptor and E-cadherin 5' CpG islands increases with malignant progression in human breast cancer. *Cancer Res.*, *60*: 4346-4348.

[72] Lin, X., Asgari, K., Putzi, M. J., Gage, W. R., Yu, X., Cornblatt, B. S., Kumar, A., Piantadosi, S., DeWeese, T. L., De Marzo, A. M., et al. (2001). Reversal of GSTP1 CpG island hypermethylation and reactivation of pi-class glutathione S-transferase (GSTP1) expression in human prostate cancer cells by treatment with procainamide. *Cancer Res.*, *61*: 8611-8616.

[73] Wales, M. M., Biel, M. A., el Deiry, W., Nelkin, B. D., Issa, J. P., Cavenee, W. K., Kuerbitz, S. J. and Baylin, S. B. (1995). p53 activates expression of HIC-1, a new candidate tumour suppressor gene on 17p13.3. *Nat. Med.*, *1*: 570-577.

[74] Watson, R. E., Curtin, G. M., Hellmann, G. M., Doolittle, D. J. and Goodman, J. I. (2004). Increased DNA methylation in the HoxA5 promoter region correlates with decreased expression of the gene during tumor promotion. *Mol. Carcinog.*, *41*: 54-66.

[75] Kane, M. F., Loda, M., Gaida, G. M., Lipman, J., Mishra, R., Goldman, H., Jessup, J. M. and Kolodner, R. (1997). Methylation of the hMLH1 promoter correlates with lack of expression of hMLH1 in sporadic colon tumors and mismatch repair-defective human tumor cell lines. *Cancer Res.*, *57*: 808-811.

[76] Herman, J. G., Umar, A., Polyak, K., Graff, J. R., Ahuja, N., Issa, J. P., Markowitz, S., Willson, J. K., Hamilton, S. R., Kinzler, K. W., et al. (1998). Incidence and functional consequences of hMLH1 promoter hypermethylation in colorectal carcinoma. *Proc. Natl. Acad. Sci. USA*, *95*: 6870-6875.

[77] Nakagawa, H., Nuovo, G. J., Zervos, E. E., Martin, E. W., Salovaara, R., Aaltonen, L. A. and de la Chapelle, A. (2001). Age-related hypermethylation of the 5' region of MLH1 in normal colonic mucosa is associated with microsatellite-unstable colorectal cancer development. *Cancer Res.*, *61*: 6991-6995.

[78] Burbee, D. G., Forgacs, E., Zochbauer-Muller, S., Shivakumar, L., Fong, K., Gao, B., Randle, D., Kondo, M., Virmani, A., Bader, S., et al. (2001). Epigenetic inactivation of

RASSF1A in lung and breast cancers and malignant phenotype suppression. *J. Natl. Cancer. Inst.*, *93*: 691-699.

[79] Dammann, R., Li, C., Yoon, J. H., Chin, P. L., Bates, S. and Pfeifer, G. P. (2000). Epigenetic inactivation of a RAS association domain family protein from the lung tumour suppressor locus 3p21.3. *Nat. Genet.*, *25*: 315-319.

[80] Rhee, I., Bachman, K. E., Park, B. H., Jair, K. W., Yen, R. W., Schuebel, K. E., Cui, H., Feinberg, A. P., Lengauer, C., Kinzler, K. W., et al. (2002). DNMT1 and DNMT3b cooperate to silence genes in human cancer cells. *Nature*, *416*: 552-556.

[81] Magdinier, F. and Wolffe, A. P. (2001). Selective association of the methyl-CpG binding protein MBD2 with the silent p14/p16 locus in human neoplasia. *Proc. Natl. Acad. Sci. USA*, *98*: 4990-4995.

[82] Myohanen, S. K., Baylin, S. B. and Herman, J. G. (1998). Hypermethylation can selectively silence individual p16ink4A alleles in neoplasia. *Cancer Res.*, *58*: 591-593.

[83] Nuovo, G. J., Plaia, T. W., Belinsky, S. A., Baylin, S. B. and Herman, J. G. (1999). In situ detection of the hypermethylation-induced inactivation of the p16 gene as an early event in oncogenesis. *Proc. Natl. Acad. Sci. USA*, *96*: 12754-12759.

[84] Esteller, M., Risques, R. A., Toyota, M., Capella, G., Moreno, V., Peinado, M. A., Baylin, S. B. and Herman, J. G. (2001). Promoter hypermethylation of the DNA repair gene O(6)-methylguanine-DNA methyltransferase is associated with the presence of G:C to A:T transition mutations in p53 in human colorectal tumorigenesis. *Cancer Res.*, *61*: 4689-4692.

[85] Issa, J. P. and Baylin, S. B. (1996). Epigenetics and human disease. *Nat. Med.*, *2*: 281-282.

[86] Popendikyte, V., Laurinavicius, A., Paterson, A. D., Macciardi, F., Kennedy, J. L. and Petronis, A. (1999). DNA methylation at the putative promoter region of the human dopamine D2 receptor gene. *Neuroreport.*, *10*: 1249-1255.

[87] Tassone, F., Hagerman, R. J., Loesch, D. Z., Lachiewicz, A., Taylor, A. K. and Hagerman, P. J. (2000). Fragile X males with unmethylated, full mutation trinucleotide repeat expansions have elevated levels of FMR1 messenger RNA. *Am. J. Med. Genet.*, *94*: 232-236.

In: DNA Methylation Research Trends
Editor: Taku B. Kobayashi, pp. 25-41

ISBN: 978-1-60021-721-0

Chapter II

CpG Islands in Plants

Ikuo Ashikawa*
National Institute of Crop Science
Kannondai, Tsukuba, Ibaraki, 305-8518 Japan.

Abstract

Although a number of studies on CpG islands in humans have been reported, knowledge on plant CpG islands was quite limited. To learn whether plants contain chromosomal regions that can be regarded as CpG islands, we screened plant genomic sequences for CpG island-like chromosomal regions. Analysis of plant DNA sequence data, primarily from rice, revealed the presence of CpG-rich DNA regions in the genome. These CpG-rich clusters were structurally similar to human CpG islands. In rice, about one-third of the CpG clusters were found around the 5'-terminus of genes. Rice genes having this type of a CpG cluster differed from other rice genes in their expression pattern; they tended to be expressed in a wider range of tissues. This expression pattern resembles reported human genes having a CpG island, which suggests the presence of a functional similarity between plant CpG clusters and human CpG islands. In contrast to these similarities, the DNA methylation pattern of plant genes having CpG clusters seemed to differ from that of mammalian genes; in plants, gene-rich chromosomal regions were revealed to be generally undermethylated, whether they were within or outside the CpG clusters, whereas in humans undermethylation is known to be limited within CpG island regions. This difference in the DNA methylation pattern implies the difference in the role of DNA methylation on the epigenetic gene regulation of plants and mammals. In rice, genes having a CpG cluster were clustered into discrete chromosomal segments, creating mosaics comprising chromosomal regions either rich in genes having a CpG cluster or rich in genes not having it. This segmental distribution of genes was expected to cause a non-uniform distribution of expressed genes. Actually, the EST-based analysis on the distribution of expressed genes showed a preferential distribution of expressed genes within the CpG cluster-rich chromosomal regions.

* Ikuo Ashikawa: e-mail: ***ashikawa@affrc.go.jp***; TEL: +81-29-838-8861; FAX: +81-29-838-8861

Introduction

CpG islands are DNA regions found in mammalian chromosomes, where, in contrast to other chromosomal regions, the occurrence of CpG dinucleotides is not suppressed and cytosines are undermethylated [1-4]. The presence of a CpG island within gene regions and their *de novo* methylation were reported to influence expression of the associated genes [5-6]. Studies on CpG islands have been done mostly for humans, whereas in plants we had little knowledge on a chromosomal component corresponding to the mammalian CpG island. It is believed that development and differentiation in plants occur distinctly from mammals. Thus, there would be differences between plants and mammals in the gene regulation underlying these processes. Knowledge about how plant CpG islands and their methylation are involved in the gene regulation would be indispensable to understand the molecular mechanisms specifically regulating plant development and differentiation. For the last several years, the advent of the plant genome research has produced DNA sequence data on the whole genome of Arabidopsis and rice [7-8], and this information has made it possible to search plant genome sequences for a DNA component analogous to the human CpG island. In this chapter, I summarize the studies that have been done to characterize CpG island-like regions in the plant genomes and describe their similarities to and differences from the mammalian CpG islands

CpG Island in Humans

In the human genome, CG dinucleotides are not uniformly distributed and many of them occur in discrete regions, creating CpG-rich clusters. In contrast, in most of the regions surrounding the clusters, CG dinucleotids are significantly underrepresented compared to other dinucleotides. These CpG-rich clusters are called CpG islands [1-4]. In the islands, CG dinucleotides are approximately as frequent as one would expect from the G+C content. Typically, the size of these CG-rich regions is ca. 1 kb. Analysis of DNA sequence data from humans revealed that about 29,000 CG-rich regions, which satisfied the criteria of the CpG island, existed in the genome [9]. On average, 5 to 15 islands were found in 1 Mb, although the density of the island in the genome varied among chromosomes; in the Y chromosome, the frequency of the island was only 2.9/Mb, whereas in the chromosomes 16, 17, and 22, 19 to 22 islands were found per Mb. Although the relative position of these islands to genes in the genome varied, the majority (ca. 60%) of the islands were found within or close to genes, mostly at the 5'-terminus [10-13].

CpG-Rich Regions in Plants

The existence of CpG islands in plants had been predicted from analyses of DNA sequences of plant gene regions [14-17] and from the investigations about the digestion patterns of plant genomic DNA after treatment with methylation-sensitive restriction enzymes [18-20]. Unfortunately, because DNA sequence data on large contiguous genomic fragments had not been available until the 2000's, it was not possible to conclude whether CpG island-like regions were really present in plant DNA. In the last 7 years, thanks to the efforts of plant genome research, DNA sequence data necessary for analyzing the CG dinucleotide distribution in the genome have become available [7-8]. Using DNA sequence data mainly from rice and Arabidopsis, we screened for chromosomal regions analogous to mammalian CpG islands.

The CG frequency plots calculated for some DNA segments of rice, Arabidopsis, and sorghum showed that CG dinucleotides were non-uniformly distributed in these genomes (figure 1) and many of the CG dinucleotides occurred in discrete regions, creating CpG-rich clusters [21]. Figure 1 shows parts of these profiles, comprising 40 or 50 kb DNA segments from the analyzed regions. These plant CpG-rich clusters structurally resemble human CpG islands. Human CpG islands were defined as DNA regions that fulfill the following three conditions [3]: 1) the average of GC contents is greater than 50%, 2) the average of the ratio of observed to expected CpG (Obs/Exp CpG) is greater than 0.6, 3) more than several hundred bp, generally 200 to 500 bp, in length. The plant CpG-rich clusters satisfied this definition of the CpG island, although peaks in the (O/E)CpG graphs of plant CpG clusters were not so apparent as those of human CpG islands. Especially in a dicot plant, Arabidopsis, which has a fairly low C+G content, it was rather difficult to clearly define CpG islands in the genome from the (O/E)CpG plot (figure 1B) [22].

The frequency of occurrence of these plant CpG clusters in the genome differed from that of humans; in humans, CpG islands occur in the genome almost every 100 kb [11-12], whereas plant CpG clusters occurred much more densely in the genome and were detected every 4 to 5 Kb (figure 1). Because, as described below, a large percent of CpG-rich regions in the genome are associated with genes, this difference in the frequency of CpG clusters/islands between plants and humans may be relevant to the difference in the frequency of occurrence of genes in these two distinct organisims.

Figure 2A shows the size distribution of the examined plant CpG-rich clusters, as compared to the human CpG islands [21]. Plant CpG clusters were generally smaller than human CpG islands; the sizes of the majority of the plant CpG clusters ranged from 0.2 to 1.0 kb, whereas human CpG islands were larger in size, and the sizes of a large part of the human islands were distributed from 1.0 to 1.2 kb. Figure 2B shows the frequency of CpG dinucleotides in the clusters/islands of rice, Arabidopsis, and sorghum, together with humans. In rice, sorghum, and humans, one CpG cluster/island contained 15 to 20 CpGs, whereas in Arabidopsis most CpG-rich clusters contained up to 15 CpGs. This may be relevant to the fact that the genome of Arabidopsis is fairly AT-rich, as compared to the other plants or humans.

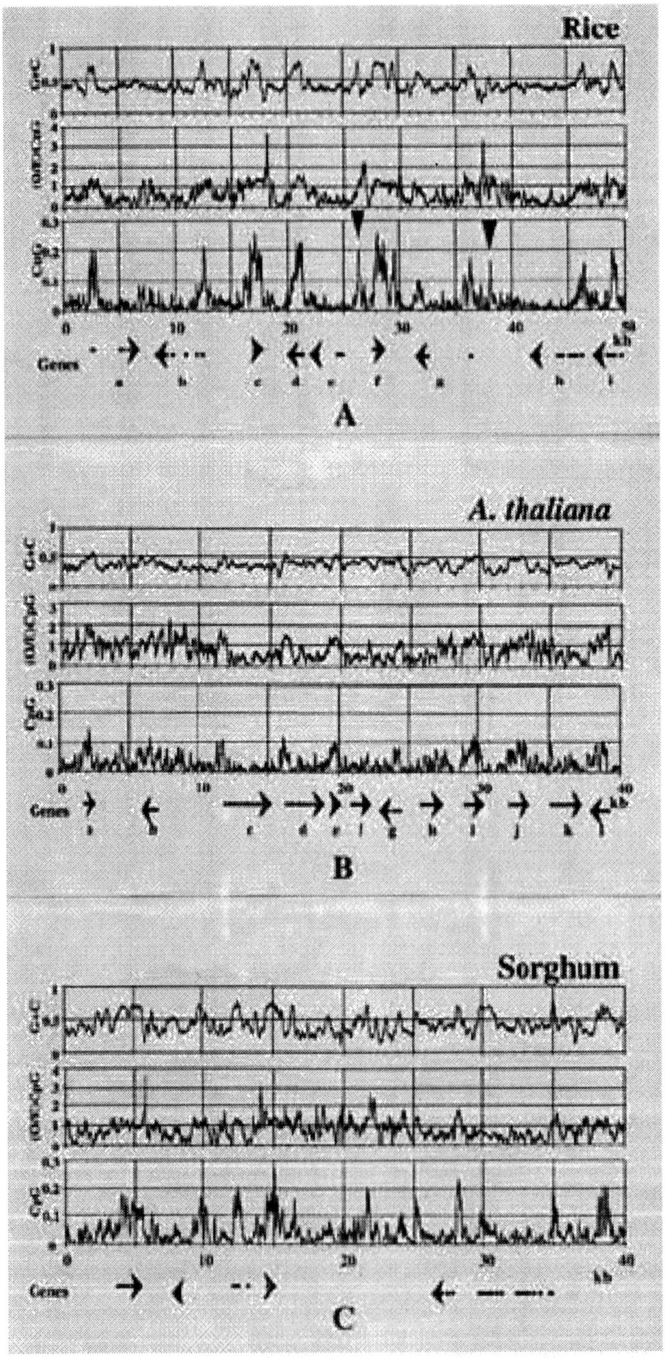

Figure 1. The moving averages of the G+C content, the ratio of observed to expected CpG (O/E CpG), and the CpG frequency plots in (a) a 50 kb DNA segment from the rice genomic sequence, (b) a 40 kb DNA segment from the Arabidopsis genomic sequence, and (c) a 40 kb DNA segment of the sorghum genomic sequence. The horizontal arrows under the CpG frequency graph indicate the position of each gene; each block in the arrow's stem represents an exon in that gene. Two vertical arrowheads in the CpG frequency graph of rice indicate CpG clusters that are not associated with genes. This figure is reproduced from Fig. 1 in [21] with permission from Blackwell Publishing.

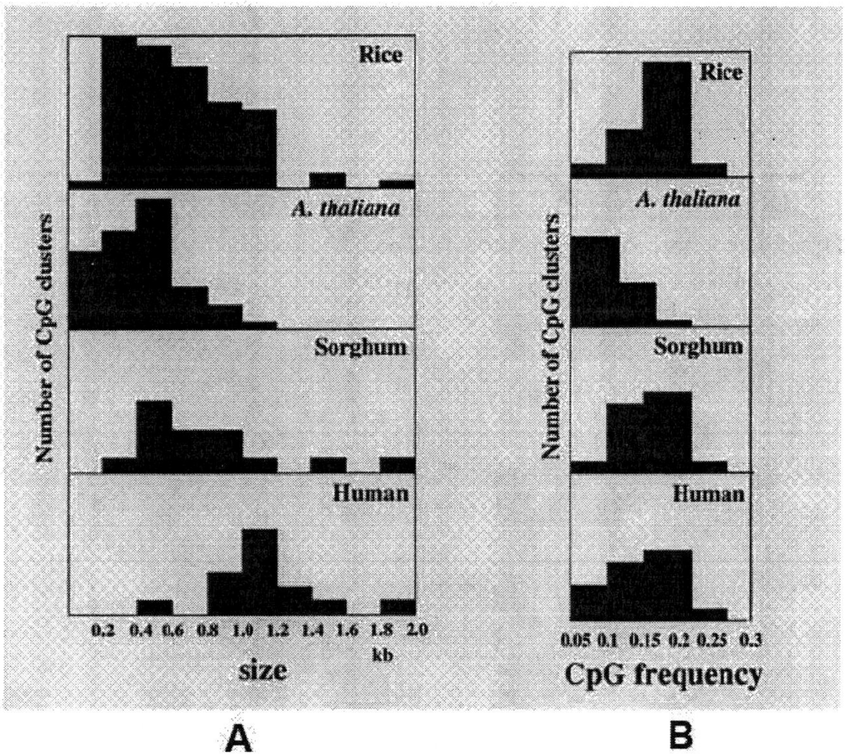

Figure 2. Distribution of (A) the sizes and (B) the CpG frequency of 317 CpG clusters detected in a 1.5 Mb sequence from rice chromosome 1, 297 clusters detected in a 1.2 Mb sequence from Arabidopsis chromosome 2 and 4, 50 clusters detected in a 346 kb sequence from sorghum chromosomes, and 31 clusters detected in a 2.0 Mb sequence from the human 21q22.3 region. This figure is reproduced from Fig. 4 in [21] with permission from Blackwell Publishing.

As described above, it has been known that the majority (ca. 60%) of human CpG islands occur close to genes [10]. Similar to humans, a substantial portion of plant CpG clusters were found to be associated with genes. In plants with small genomes, such as rice and Arabidopsis, almost 80% of the CpG clusters were associated with genes (figure 1). Whereas in plants with larger genomes, such as maize and barley, only 25 to 30 % of CpG clusters were associated with genes [21].

In humans, the position of a CpG island relative to its associated gene is variable. Researchers grouped a human CpG island according to its position within a gene. For example, Bock et al. [23] grouped human CpG islands into the following 4 categories: promoter CpG islands, intragenic CpG islands, gene-terminal CpG islands, and intergenic CpG islands. A large part of gene-associated CpG islands are found at the 5'-end of a gene, as promoter CpG islands. These promoter CpG islands are distinct from CpG islands belonging to other categories in their methylation status and the broadness of gene expression of the associated gene [10, 13, 23]. Likewise, in plants, the position of a CpG-rich cluster relative to its associated gene is variable. We grouped plant genes into five classes according to the position of a CpG-rich cluster in the gene (figure 3) [21]. In class 1 genes, a CpG-rich cluster is located near the 5'-end of the associated gene. In class 2 genes, a CpG-rich cluster covers the whole gene region. In class 3 genes, a CpG-rich cluster occurs considerably downstream of the 5'-end. Class 4 genes contain a CpG-rich cluster near the 5'-end of the associated gene and another cluster downstream of the 5'-end. Class 5 genes lack CpG clusters. Figure 4 shows the results of this grouping of genes from rice and Arabidopsis. These two plants showed an almost similar distribution of genes among the five classes: a large portion of genes belonged to either classes 1, 2, or 5, while classes 3 and 4 genes constituted small minorities. In rice, each of the class 1 and class 2 genes occupied almost one-third of all the genes analyzed, while in Arabidopsis the percentage of class 2 genes was significantly lower than in rice. In contrast to plants, the human genome included class 1 and class 5 genes as major classes and some class 3 genes, but class 2 genes were scarcely detected (figure 4).

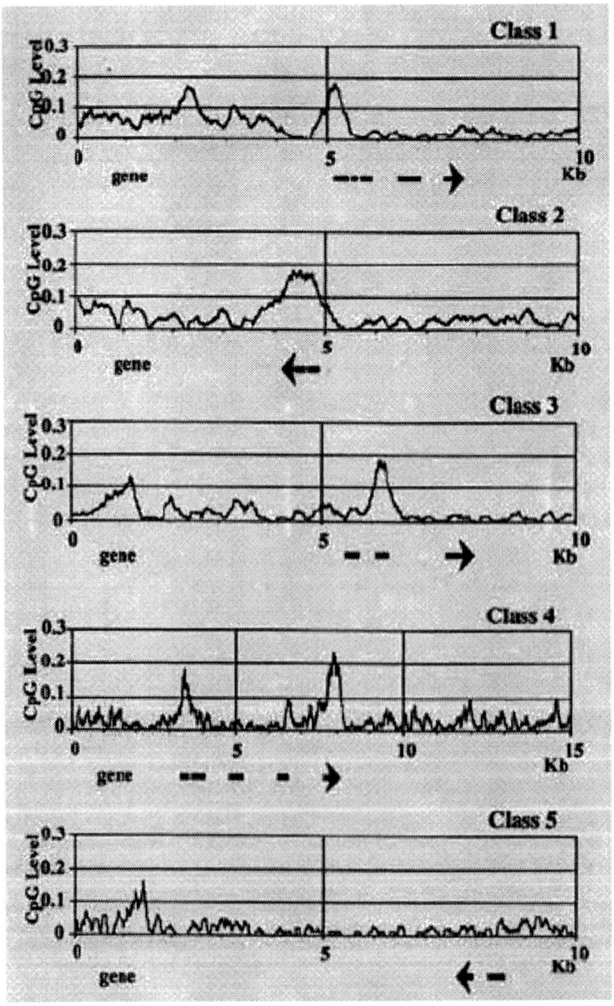

Figure 3. The CpG frequency graphs of five classes of plant genes, grouped according to the position of the associated CpG cluster. In class 1 genes, the CpG cluster is located only at the 5'-end of the gene. The CpG cluster covers the whole gene region of class 2 genes. Class 3 genes comprise those in which a CpG cluster occurs at considerable distance downstream of the 5'-end. Class 4 genes contain a CpG cluster at the 5'-end and another downstream CpG cluster. Class 5 genes do not harbor CpG clusters. This figure is reproduced from figure 3 in [21] with permission from Blackwell Publishing.

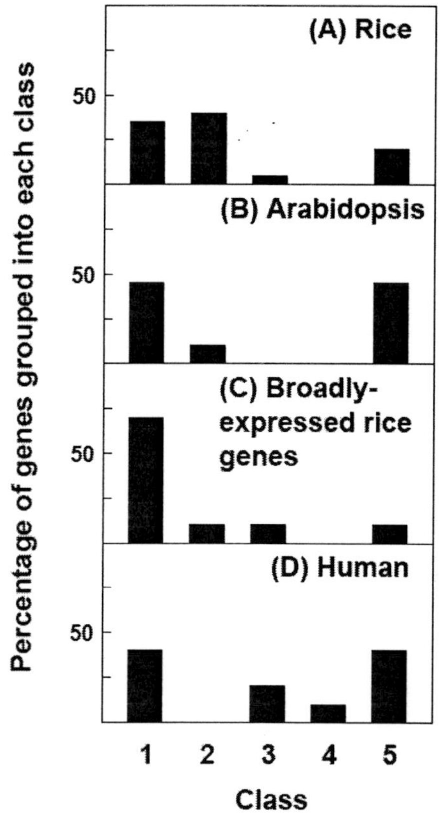

Figure 4. Classification of plant and human genes according to the position of the associated CpG clusters within the genes. (A) Clasification of a total of 1641 rice genes in 8.9 Mb sequences from rice chromosomes 1 and 6. (B) Classification of a total of 875 Arabidopsis genes in 3.4 Mb sequences from Arabidopsis chromosome 1. (C) Classification of 29 rice genes, for which expression was detected by the EST-based analysis in more than two rice tissues. (D) Classification of 25 human genes in the 2.0 Mb sequence from the human 21q22.3 DNA region.

DNA Methylation Status

In mammals, it is believed that CGs in the genome are generally methylated except those located within CpG island regions [1]. A recent chromosome-wide methylation assay using an immunocapturing approach followed by DNA microarray analysis also supported this methylation feature [24]. DNA methylation that has occurred in CpG island regions is considered to regulate (or inactivate) gene expression [5-6]. One of the well known examples is X-chromosome inactivation, i.e., inactivation a set of genes on X-chromosome; the inactive X chromosome has hypermethylated gene-associated CpG islands [25]. Another example is inactivation of genes involved in genomic imprinting [26]. *De novo* DNA methylation has been also found to be associated with cancer [27]. In addition, even in normal tissue, it has been known that a portion of CpG islands were reported to be methylated. Yamada et al. [28], by using the McrBC-PCR technique, examined the methylation status of all of the CpG islands identified on human Chromosome 21q. They showed that most of CpG islands escaped methylation, while some were fully or partially methylated. Methylation status of these CpG islands depended on the location of each CpG island within a gene; about 60% of CpG islands on chromosome 21 were promoter CpG islands and most of them were shown free from methylation. Whereas, a substantial part of intragenic, gene-terminal, and intergenic CpG islands were shown to be methylated, and the number of demethylated and methylated islands were balanced.

The DNA methylation in the plant genomes appeared rather different from that in the mammalian genomes. Rabinowicz et al. [29] compared the cytosine methylation pattern of a set of maize genes to those of mouse genes, by using the McrPCR technique. As expected, both of maize and mouse transposon regions were methylated. In contrast, the methylation profiles of the gene regions of the both organisms were shown to be quite different; the examined mouse exon regions were mostly methylated, whereas most of the maize exons were shown to be undermethylated.

To obtain more detailed knowledge on the plant DNA methylation pattern, we examined the methylation status of rice genes by using the same technique [30]. As described above, the majority of rice genes are grouped into three classes, classes 1, 2, and 5, according to the position of the associated CpG-rich cluster. To seek for any difference in the DNA methylation pattern among these three class genes, we randomly selected 25 class 1, 31 class 2, and 18 class 5 genes from rice chromosome 1, and evaluated the methylation status of their gene regions. The majority of the examined regions were revealed to be undermethylated, whether they were inside or outside the CpG-rich regions; methylation and partial methylation were detected in 1 (4%), 8 (26%), and 4 (22%) of the examined class 1, class 2, and class 5 genes, respectively. This result suggests the following feature for the methylation pattern in the plant genome: in genes having a CpG-rich region around the 5'-end (class 1 genes), both of the gene-associated CpG-rich regions and the remaining gene regions are fairly undermethylated. In contrast, in class 2 and 5 genes, although the majority of the gene regions are also undermethylated, full or partial methylation is detected in almost one fourth of them.

Recently, using a microarray-based technique, Zhang et al. [31] reported a high-resolution genome-wide DNA methylation pattern of the Arabidopsis genome. This map gave a precise view of the distribution of methylated regions in the chromosome. As expected, pericentromeric heterochromatin, repetitive sequences, and regions producing small interfering RNAs were shown to be heavily methylated. Although the gene regions were generally undermethylated, unexpectedly methylation was detected for almost one-third of the genes within their transcribed regions. In addition, DNA methylation around the promoter regions was found for small percent (5%) of genes. Interestingly, this methylation within gene regions was found to be correlated with the expression pattern of genes; genes methylated in the transcribed regions were highly expressed and constitutively active, while genes methylated in their promoter regions showed a greater degree of tissue-specific expression.

CpG Island and Gene Expression

It has been known for a long time that in humans the presence or absence of a CpG island is associated with the tissue distribution of gene expression and that only a promoter CpG island seems to be correlated with the broad gene expression [13].

For the last decade, the advent of the human genome research has produced data about the whole genome DNA sequence, the predicted gene positions in the sequence, and the genome-wide expression pattern of genes. This information has enabled to re-evaluate the above-mentioned notion on the possible association between CpG islands and the broadness of expression of the nearby genes. Yamashita et al. [32] determined the transcription start sites of ca. 9000 human and 4000 mouse genes by aligning full-length cDNA sequences with genome sequence data, and divided the analyzed genes into a group of genes having a promoter CpG island and another group of genes not having it. Then, they analyzed the tissue-specific expression pattern of genes of each group, and finally evaluated whether there is really any correlation between the presence of promoter CpG islands and the broadness of gene expression. Their result supported the presence of the correlation. On the other hand, other researchers suggested another hypothesis regarding the CpG island and gene expression; Ponger et al. [33] noticed that human genes found to be expressed in early embryos contained a promoter CpG island, and suggested a hypothesis that genes having CpG islands were those that were expressed in early embryo rather than those that showed a broad tissue-specific expression pattern. By using a genome-wide Gene Ontology-based approach, Robinson et al. [34] supported this hypothesis. Although at the moment, we do not know which one of these two hypotheses is rational, the above discussion demonstrates that in humans the presence of a promoter CpG island is correlated with the expression pattern of the associated gene.

In plant, Gardiner-Garden and Frommer [14] also suggested the presence of a similar correlation between CpG islands and broad expression of plant genes. Using rice genome sequence data and EST information, we tested whether this suggested correlation actually held. We selected 29 genes, of which expression in more than two tissues were indicated by the EST-based gene expression analysis, from a total of 273 rice genes. We regarded these genes as widely-expressed genes. Then, we grouped these genes into 5 classes according to our classification method. The comparison of the distribution pattern of the selected 29 genes with that of a total of the nonselected 273 genes in each class (figure 4A and C) showed a significant enrichment of genes in the class 1 category thorough the selection. The result confirmed the hypothesis that broadly-expressed genes tend to harbor a CpG cluster at the transcription start site. This plant class 1 CpG cluster structurally corresponds to the promoter CpG island in humans. Thus, the correlation between promoter CpG islands and broad expression of genes in humans seems to hold also in plants [21].

One of the limitations of the above EST-based analysis is that the accuracy of information obtained in this way is highly dependent on the quality of the cDNA libraries used for EST cataloguing and on the number of genes sequenced in each library. To circumvent this possible inaccuracy, we experimentally examined tissue-specific expression of genes and used the obtained result to confirm the correlation [35]. We selected 51 genes, whose expression was discernible by Northern hybridization analysis in at least one of the four rice tissues: shoot, root, callus, and panicle. We then divided these 51 genes into a group of widely-expressed genes and another group of tissue-specific genes on the basis of their gene expression patterns: 25 genes were expressed only in one of the four tissues and grouped into tissue-specific genes, while the remaining 26 genes were expressed in two or more tissues and grouped into widely expressed genes. Classification of genes belonging to these two groups according to the location of a CpG rich cluster indicated a correlation between genes having a CpG cluster at the 5' end and broad gene expression: class 1 genes constituted 81% of the widely expressed genes, while they constituted only 24% of the tissue-specific genes (figure 5A, B).

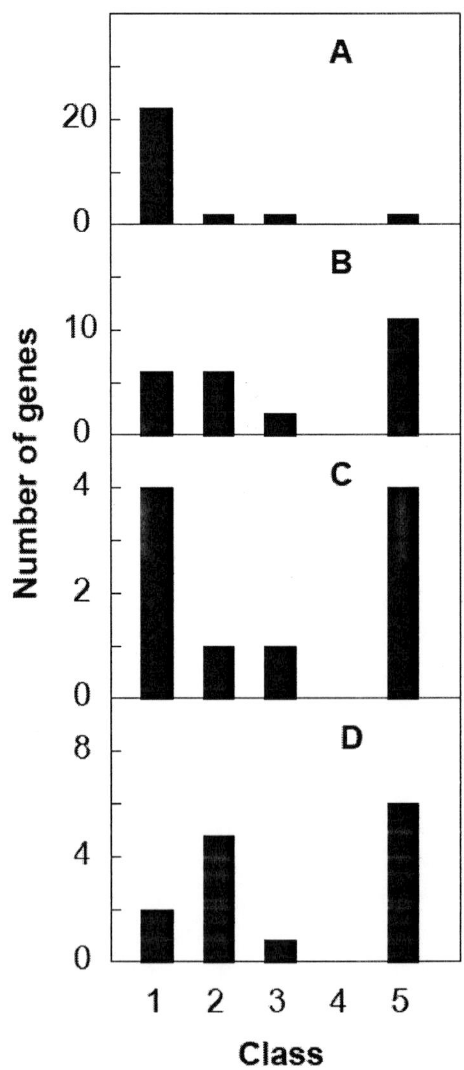

Figure 5. Classification of rice genes, whose expression was examined in the four rice tissues: leaf, root, callus, and panicle, according to the position of their associated CpG clusters. (A) Classification of 26 genes whose expression was detected in two or more tissues. (B) Classification of 25 genes whose expression was detected only in one tissue. (C) Classification of 10 genes whose expression was detected only in callus. (D) Classification of 15 genes whose expression was detected only in one of three tissues: leaf, root, and panicle.

In the genes, of which expression is detected only in one tissue, there seems to exist a relationship between the expressed tissues and the location of their associated CpG clusters; among the genes expressed only in callus, about half belonged to class 1, while most of the genes expressed in one of the three tissues, shoot, root, or panicle, were grouped into class 2 or 5 (figure 5C, D), which suggests that genes expressed in callus are predominantly class 1 genes.

The observed correlation between the CpG-rich cluster and tissue-specific gene expression for rice and humans suggests that the plant CpG cluster functionally corresponds to human CpG islands.

Non-Uniform Distribution of Genes Having a CpG-Rich Cluster

Recent genome-wide gene expression analyses suggested the presence of a chromosomal domain-based regulation of gene expression in humans and plants [36-37]. Such an idea of non-uniform gene activity across a chromosome leads to the hypothesis that a chromosome is not a uniform structure, but rather it is constituted of structurally distinct domains. In light of this hypothesis, it is intriguing to examine whether plant genes having a CpG island-like cluster (class 1 genes) are uniformly or non-uniformly distributed on the chromosome; if class 1 genes are not evenly distributed on a chromosome and exhibit a segment-like distribution, the chromosome would look like non-uniform in terms of gene expression.

Using 8950 kb rice genomic sequences, comprising 5 distinct regions from chromosomes 1 and 6, we investigated the distribution pattern of rice genes having a CpG-rich cluster [30]. The analyzed chromosomal regions contained a total of 1641 predicted genes. We divided these genes into one of the above mentioned 5 classes according to the relative position of the associated-CpG cluster within the genes. This classification divided the predicted genes into 606 class 1 genes, 634 class 2 genes, 44 class 3 genes, 22 class 4 genes, and 335 class 5 genes. Figure 6 shows the distribution of the classified genes in one region, where a total of 331 genes were predicted to be located. This figure reveals that this region was not composed of a random mixture of genes belonging to one of the five classes, but rather it appeared as mosaics comprising mostly chromosomal segments rich in either class 1 (red), 2 (green), or 5 (blue) genes. This segment-like structure was discernible throughout the analyzed regions, although it was more obvious in some regions and less obvious in others (supplementary figures in [30]). These segments typically ranged from 30 to 120 kb in size and contained 5 to 20 genes. The densities of genes embedded in each of the classes 1, 2, and 5 segments (0.18, 0.19, and 0.19 gene/kb, respectively) were almost the same. Comparison of the proportions of class 1, 2, and 5 genes located within the respective segments confirmed that each of the segments showed a significant enrichment of its own class of genes: class 1 gene-enriched segments, which comprised a total of 2605 kb (29.1% of the total analyzed regions), contained 327 (54.0% of the total analyzed genes) class 1 genes. Likewise, class 2 and class 5 gene segments, comprising 23.3% and 6.4% of the total analyzed regions, contained 58.5% and 22.4% of the total class 2 and class 5 genes, respectively.

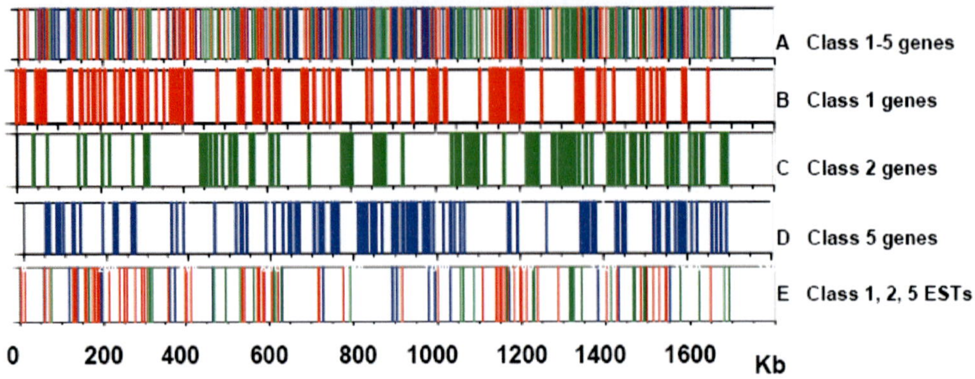

Figure 6. Distributions of predicted genes and mapped ESTs grouped into class 1 (red), 2 (green), 3 (orange), 4 (purple), or 5 (blue) in a 1.7 Mb DNA sequence from rice chromosome 1. Each of the vertical bars in the figure represents a predicted gene or a mapped EST. (A) Distribution of classes 1,2, 3, 4, and 5 genes. Each predicted gene was grouped into one of the five classes according to the relative position of its associated CpG cluster. To show the distribution of genes belonging to each of the three major classes more clearly, groups of genes belonging to classes 1 (B), 2 (C), and 5 (D) are displayed. (E) EST-based expression map generated by locating ESTs from the rice full-length cDNA catalogue in this region. This figure is reproduced from Fig. 1 in [30] with permission from Springer-Verlag.

Co-Localization of Expressed Genes on the Rice Chromosome

As described, class 1 genes (genes having a CpG cluster at the 5'-end) differ from other class genes in that they tend to be expressed in a wider range of tissues. Thus, if these class 1 genes are non-uniformly distributed on a chromosome, the chromosome would look like also non-uniform in light of the expression pattern of genes. To examine whether the segment-like distribution of class 1 rice genes caused a heterogeneity in the expression pattern of genes inside and outside the class 1 gene segments, we investigated the distribution of mapped rice ESTs in the analyzed chromosomal regions and looked for the presence of a non-uniform distribution of mapped ESTs, i.e., any tendency of mapped ESTs to be preferentially distributed within class 1 gene segments [30]. For this, as a source of gene expression, we used rice full-length cDNA cataloguing data [38]. In the analyzed 8950 kb regions, we mapped a total of 615 (37.5% of all the predicted genes) cDNAs. The proportion of class 1-type expressed genes among all the expressed genes was significantly higher (58.7%) than that of class 1-type predicted genes among all the predicted genes (36.9%), indicating that class 1 genes tended to be expressed more broadly and/or intensely, and their cDNAs were more likely to be captured in an EST project. In other words, expression of class 2 and 5 genes tended to be suppressed and their cDNAs were not preferentially captured. The distribution of mapped genes revealed a tendency of preferential localizations of ESTs within the class 1 segments. In contrast, because the number of mapped cDNAs grouped to classes 2 and 5 was decreased, chromosomal regions of classes 2 and 5 genes looked like expressed gene-poor regions. As a result, the mapped expressed genes appeared to exhibit heterogeneity in the local density on the chromosome, and a large proportion of them appeared to be localized as chromosomal segments.

To examine whether the segmental structure detected in the rice chromosome was also present in dicots, we analyzed the distribution of classes 1-5 genes in a total of 3.4 Mb DNA from Arabidopsis chromosome 1 [30]. A total of 875 genes were predicted in the regions. These comprised 367 class 1, 125 class 2, 2 class 3, 2 class 4, and 379 class 5 genes. As compared to rice, the percentage of class 2 genes was significantly lower in Arabidopsis, which may originate from a lower GC level of its genomic DNA. The distribution of the classified genes revealed that although classes 1, 2, and 5 genes tended to cluster to some extent, the segmentation of these 3 class genes was much less obvious as compared to rice (figure 7). This may possibly be due to the much lower proportion of class 2 genes in Arabidopsis than in rice; owning to significant clustering of class 2 genes, a segmental structure was clearly detectable on the rice chromosome, but it was hardly discernible in Arabidopsis. Likewise, although the distribution of mapped expressed genes (we represented by the catalogued Arabidopsis cDNAs) appeared non-uniform in the analyzed region, a segmentation of expressed genes was not so apparent as rice.

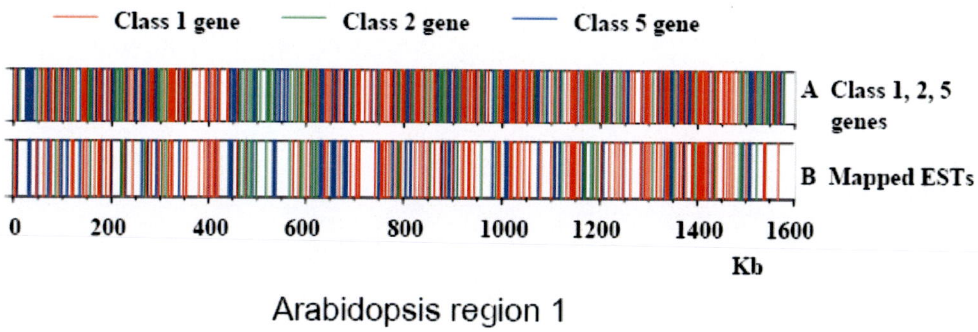

Figure 7. Distributions of predicted genes and mapped ESTs in a 1,583 kb DNA segment from Arabidopsis chromosomal 1. Each of these genes was classified according to its location relative to its associated CpG cluster. (A) 413 genes predicted in this region. (B) ESTs mapped from Arabidopsis full-length cDNA catalogue. This figure is reproduced from Fig. 3 in [30] with permission from Springer-Verlag.

Conclusion

In humans, CpG islands have been extensively studied and a significant amount of knowledge on their role in the epigenetic gene regulation have already been accumulated. In this chapter, I described the summary of our studies on a chromosomal component analogous to mammalian CpG islands in the plant genome. The analyses used the some monocots and Arabidopsis genomes and showed the existence of gene-associated CpG-rich regions. The structural features of these CpG clusters, such as their sizes, CpG densities, and locations in the genome, implied that this chromosomal component is a counterpart of the mammalian CpG island. Especially, the CpG cluster at the 5'-end of a gene, being referred as the class 1 CpG cluster, resemble to the human promoter CpG island in that the presence of these CpG-rich regions is relevant to broad expression of the associated genes. Despite these similarities, plant CpG clusters differ from human CpG islands in the methylation status; in humans only CpG island regions avoid DNA methylation, while in plants most of gene regions, whether it is inside or outside the CpG cluster, are undermethylated.

Recently, genomic DNA sequence data and information on expressed genes of plants other than Arabidopsis and rice have begun to accumulate. In addition, new techniques that enable genome-wide analyses of DNA methylation and chromatin modification have been developed. These information and techniques would be useful for gaining more insights in the role of the CpG cluster in the epigenetic control of plant gene expression. I think that some research, such as ones I describe below, would contribute to understanding the function of the plant CpG cluster. First, the CpG cluster and their involvement in the gene regulation described in this chapter were characterized using data mostly from rice. Thus, obtaining such information also for some other plants is indispensable for concluding that the CpG cluster found in the rice genome is really regarded as a plant counterpart of the human CpG island. Second, a recent genome-wide study [31] revealed that a significant part of Arabidopsis genes constituted distinct groups according to their methylation status in their transcribed or promoter regions. Gene expression was shown to tend to also be distinct between the groups [31]. Because the gene groups classed according to the association of a CpG cluster to a gene also showed distinct tissue-specific expression patterns, it might be intriguing to search for a possible correlation of these gene groups based on methylation pattern with those based on the relative position of CpG clusters, i.e., class 1-5 genes. Third, to understand the mechanism underlying the relationship between CpG clusters and broader gene expression of the associated gene, the comparison of the chromatin structure and/or chromatin modification between class 1 genes and the other class genes would be necessary.

Acknowledgement

I thank Blackwell Publishing and Springer-Verlag for kindly granting permissions of re-using figures 1, 2, and 4 published in the article [21] and figures 1 and 3 in the article [30] to figures in this chapter.

References

[1] Bird, A. P., Taggart, M., Frommer, M., Miller, O. J., and Macleod, D. (1985) A fraction of the mouse genome that is derived from islands of nonmethylated CpG-rich DNA. *Cell, 40*, 91-99.

[2] Bird, A. P. (1987) CpG islands as gene markers in vertebrate nucleus. *Trends Genet., 3*, 342-347.

[3] Gardiner-Garden, M., and Frommer, M. (1987) CpG islands in vertebrate genomes. *J. Mol. Biol., 196*. 261-282.

[4] Cross, S. H., and Bird, A. P. (1995) CpG islands and genes. *Curr. Opin. Genet. Dev., 5*, 309-314.

[5] Bird, A. P. (2002) DNA methylation patterns nd epigenetic memory. *Genes Dev.*, *16*, 6-21.

[6] Hashimshony, T., Zhang, J., Keshet, I., Bustin, M., and Ceder, H. (2003) The role of DNA methylation in setting up chromatin structure during development. *Nat. Genet.*, *34*, 187-192.

[7] Arabidopsis Genome Initiative (2000) Analysis of the genome sequence of the flowering plant *Arabidopsis thaliana*. *Nature*, *408*, 796-815.

[8] International Rice Genome Sequencing Project (2005) The map-based sequence of the rice genome. *Nature*, *436*, 793-800.

[9] International Human Genome Sequencing Consortium (2001) Initial sequencing and analysis of the human genome. *Nature*, *409*, 860-921.

[10] Larsen, F., Gundersen, G., Lopez, R., and Prydz, H. (1992) CpG islands as gene markers in the human genome. *Genomics*, *13*, 1095-1107.

[11] Dunham, I., Shimizu, N., Roe, B. A., et al. (1999) The DNA sequence of human chromosome 22. *Nature*, *402*, 489-495.

[12] Hattori, M., Fujiyama, A., Taylor, T. D., et al. (2000) The DNA sequence of human chromosome 21. *Nature*, *405*, 311-319.

[13] Jones, P. A. (1999) The DNA methylation paradox. Trends Genet., *15*~~15~~, 34-37.

[14] Gardiner-Garden, M., and Frommer, M. (1992) Significant CpG-rich regions in angiosperm genes. *J. Mol. Evol.*, *34*, 231-245.

[15] Langdale, J. A., Taylor, W. C., and Nelson, T. (1991) Cell-specific accumulation of maize phosphoenolpyruvate carboxylase is correlated with demethylation at a specific site >3 kb upstream of the gene. *Mol. Gen. Genet.* *225*, 49-55.

[16] Panstruga, R., Buschges, R., Piffanelli, P., and Schulze-Lefert, P. (1998) A contiguous 60 kb genomic stretch from barley reveals molecular evidence for gene islands in a monocot genome. *Nucl. Acids Res.*, *26*, 1056-1062.

[17] Sorensen, M. B., Muller, M., Skerritt, J., and Simpson, D. (1996) Hordein promoter methylation and transcriptional activity in wild-type and mutant barley endosperm. *Mol. Gen. Genet.*, *250*, 750-760.

[18] Antequera, F., and Bird, A. P. (1988) Unmethylated CpG islands associated with genes in higher plant DNA. *EMBO J.*, *7*, 2295-2299.

[19] Messeguer, R., Ganal, M. W., Steffens, J. C., and Tanksley, S. D. (1991) Characterization of the level, target sites and inheritance of cytosine methylation in tomato nuclear DNA. *Plant Mol. Biol.*, *16*, 753-770.

[20] Moore, G., Abbo, S., Cheung, W., et al. (1993) Key features of cereal genome organization as revealed by the use of cytosine methylation-sensitive restriction endonucleases. *Genomics*, *15*, 472-482.

[21] Ashikawa, I. (2001) Gene-associated CpG islands in plants as revealed by analyses of genomic sequences. *Plant J.*, *26*, 617-625.

[22] Rombauts, S., Florquin, K., Lescot, M., Marchal, K., Rouze, P., and Van de Peer, Y. (2003) Computational approaches to identify promoters and cis-regulatory elements in plant genomes. *Plant Physiol.*, *132*, 1162-1176.

[23] Bock, C., Paulsen, M., Tierling, S., Mikeska, T., Lengauer, T., and Walter, J. (2006) CpG island methylation in human lymphocytes is highly correlated with DNA sequence, repeats, and predicted DNA structure. *PLos Genet.*, *2*, 243-252.

[24] Weber, M., Davies, J. J., Wittig, D., Oakeley, E. J., Haase, M., Lam, W. L., and Schubeler, D. (2005) Chromosome-wide and promoter-specific analyses identify sites of differential DNA methylation in normal and transformed human cells. *Nat. Genet.*, *37*, 853-861.

[25] Heard, E. (2004) Recent advances in X-chromosome inactivation. *Curr. Opin. Cell Biol.*, *16*, 247-255.

[26] Reik, W., and Walter, J. (2001) Genomic imprinting: parental influence on the genome. *Nat. Rev. Genet.*, *2*, 21-32.

[27] Jones, P. A., and Baylin, S. B. (2002) The fundamental role of epigenetic events in cancer. *Nat. Rev. Genet.*, *3*, 415-428.

[28] Yamada, Y., Watanabe, H., Miura, F., Soejima, H., Uchiyama, M., Iwasaka, T., Mukai, T., Sakaki, Y., and Ito, T. (2004) A comprehensive analysis of allelic methylation status of CpG islands on human chromosome 21q. *Genome Res.*, *14*, 247-266.

[29] Rabinowicz, P. D., Palmer, L. E., May, B. P., Hemann, M. T., Lowe, S. W., McCombie, W. R., and Martienssen R. A. (2003) Genes and transposons are differentially methylated in plants, but not in mammals. *Genome Res.*, *13*, 2658-2664.

[30] Ashikawa, I., Numa, H., and Sakata, K. (2006) Segmental distribution of genes harboring a CpG island-like region on rice chromosomes, *Mol. Gen. Genomics*, *275*, 18-25.

[31] Zhang, X., Yazaki, J., Sundaresan, A., Cokus, S., Chan, S. W.-L., Chen, H., Henderson, I. R., Shinn, P., Pellegrini, M., Jacobsen, S. E., and Ecker, J. R. (2006) Genome-wide high-resolution mapping and functional analysis of DNA methylation in Arabidopsis. *Cell*, *126*, 1189-1201.

[32] Yamashita, R., Suzuki, Y., Sugano, S., and Nakai, K. (2005) Genome-wide analysis reveals strong correlation between CpG islands with nearby transcription start sites of genes and their tissue specificity. *Gene*, *350*, 129-136.

[33] Ponger, L., Duret, L., and Mouchiroud, D. (2001) Determinants of CpG islands: expression in early embryo and isochore structure. *Genome Res.*, *11*, 1854-1860.

[34] Robinson, P. N., Bohme, U., Lopez, R., Mundlos, S., and Nurnberg, P. (2004) Gene-Ontology analysis reveals association of tissue-specific 5' CpG island genes with development and embryogenesis. *Hum. Mol. Genet.*, *13*, 1969-1978.

[35] Ashikawa, I. (2002) Gene-associated CpG islands and the expression pattern of genes in rice. *DNA Res.*, *9*, 131-134.

[36] Caron, H., van Schaik, B., van der Mee, M., Baas, F., Riggins, G., van Sluis P., Hermus, M.-C., van Asperen R., Boon, K., Voute, P. A., Heisterkamp, S., van Kampen, A., and Versteeg, R. (2001) The human transcriptome map: clustering of highly expressed genes in chromosomal domains. *Science*, *291*, 1289-1292.

[37] Jiao, Y., Jia, P., Wang, X., Su, N., Yu, S., Zhang, D., Ma, L., Feng, Q., Jin, Z., Li, L., Xue, Y., Cheng, Z., Zhao, H., Han, B., and Deng, X. W. (2005) A tiling microarray expression analysis of rice chromosome 4 suggests a chromosome-level regulation of transcription. *Plant Cell*, *17*, 1641-1657.

[38] Rice Full-Length cDNA Sequencing Consortium (2003) Collection, mapping, and annotation of over 28,000 cDNA clones from *japonica* rice. *Science*, *301*, 376-379.

In: DNA Methylation Research Trends
Editor: Taku B. Kobayashi, pp. 43-66
ISBN: 978-1-60021-721-0

Chapter III

Epigenetic Roles of CpG Methylation at Non-CpG-Islands Revealed by Morphology-Oriented Epigenetic Research

Sohei Kitazawa, Kiyoshi Mori, Takeshi Kondo, Masayo Fujimoto and Riko Kitazawa
Division of Molecular Pathology, Department of Biomedical Informatics, Kobe University Graduate School of Medicine

Abstract

Cytosine methylation plays a major role in the regulation of sequential and tissue-specific expression of genes. *De novo* aberrant DNA methylation and demethylation are also crucial processes in tumorigenesis and tumor progression. The mechanisms of how and when such aberrant methylation and demethylation occur in tumor cells are still obscure, however. In this regard, the utility of archival formalin-fixed, paraffin-embedded pathological specimens for molecular biological analysis has become increasingly important in not only the experimental but also the diagnostic field. In this article, we first introduce technical aspects of morphology-oriented epigenetic research by a modified agarose-bead-mediated bisulfite mapping method with the use of microdissected fragments of paraffin sections. We then present our observation of the unusual upregulation of TrkA, a receptor of nerve growth factor, gene expression by methylation around the negatively regulating elements through the direct interference of c-Jun binding. Recently, we cloned and characterized the TrkA gene promoter region to clarify the molecular mechanism that controls TrkA gene expression during pancreatic cancer progression. By microdissecting a subset of cancer cells showing positive TrkA immunohistochemistry from pancreatic cancer specimens obtained at autopsy and surgery, we noticed that methylation of non-CpG islands around the negatively regulating AP-1 site in the 5'-untranslated region of the TrkA gene was positively related to the expression of the gene. During carcinogenesis and cancer progression, gene silencing by global hypermethylation of CpG islands in the promoter region is observed in many genes related to the cell cycle (p16^{INK4a} and p15^{INK4b}), DNA repair (hMLH1, MGMT, and BRCA1), cell adherence and metastatic processes (CDH1, TIMP3, DAPK), the p53 network (p14^{ARF} and p73), metabolic enzymes (GSTP1), and the APC/β-catenin pathway (APC). Global genomic hypomethylation, which is linked to DNA instability and oncogene activation, has also been found in many types of human cancers including prostate, liver, and uterine tumors. Unlike such global epigenetic alteration at CpG islands that leads to drastic alteration of gene expression, methylation at non-CpG islands has been regarded as a meaningless, random and incidental event and has, therefore, drawn little interest. By selecting a subset of cells based on characteristic morphological features revealed by microdissection, we discovered that methylation at certain CpG sites in non-CpG islands would play a crucial epigenetic role in the versatility and flexibility of gene expression during cancer progression. In the final part of this article, we introduce a novel MeCP2-mediated gene silencing mechanism by the addition of methylation of a single CpG-locus upstream of the TATA-box of the Receptor activator of nuclear factor-κB ligand (RANKL) gene promoter.

1. Introduction

DNA methylation in vertebrate cells is restricted mostly to cytosine nucleotides (C), and 3-5% of all cytosine residues in genomic DNA are present in methylated form [1,2]. Actually, almost all methylated cytosine residues are observed as residues within the context of 5'-cytosine-phosphate- guanine-3' (CpG) dinucleotides. In almost half of human genes, CpG sites are found at a high frequency at the 5'-end of genes, and such clusters are referred to as CpG islands [3]. Although, 60-70% of CpGs in genomic DNA are fully modified by methylation, most CpG sites located within CpG islands escape such methylation; in other word, the majority of CpG methylation in normal individuals occurs within non-CpG islands. While CpG dinucleotides within CpG islands located at the promoter region of transcriptionally active genes tend to be hypomethylated, accumulation of methylated cytosine in CpG islands of specific genes has been attributed to transcriptional inactivity, resulting in suppressed gene expression in a developmental stage-, cell- and tissue-specific manner, and in tumor suppressor gene inactivation in carcinogenesis and tumor progression. Currently, the precise mechanism of the suppression of gene expression through hypermethylation of CpG islands is intensively analyzed and categorized into two ways. First, CpG methylation is capable of directly repressing transcriptional activity by sequestering some of the transcriptional factors that must interact with GC-rich binding sites or CpGs in their DNA recognition elements in the major grooves of the double helix (figure 1A). Second, it is achieved by recruitment of methylcytosine binding proteins (MBPs) that recognize methylated DNA. The MBP-mediated suppressing mechanism consists of two auxiliary machineries: methylated CpGs included in or around the promoter region of genes recruit MBPs, resulting in blocking the access of transcriptional activators to their binding elements (figure 1B); and MBPs binding to methylated CpGs recruit co-repressor molecules to silence gene expression then, by association with large protein complexes (figure 1C), lead to the formation of transcriptionally inactive chromatin (heterochromatin) [4, 5]. Hypermethylation of CpG islands, therefore, involves the controls of gene expression at two distinct levels, transcription and chromatin remodeling. Since this methylation-oriented switching system of gene activity does not depend on the alteration of the DNA sequence, CpG methylation is, consequently, one of the fundamental epigenetic mechanisms that controls gene expression.

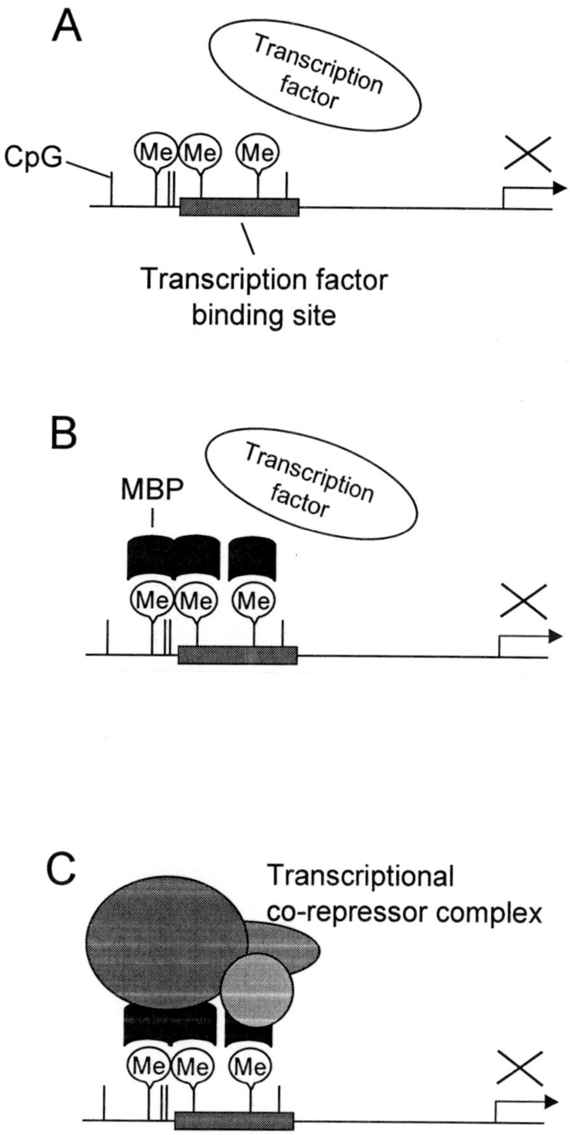

Figure 1. The schema of transcriptional modification by addition of methylation at CpG islands in the promoter region. Direct repression of transcriptional activity by sequestering some of the transcriptional factors that must interact with GC-rich binding sites or CpGs in their DNA recognition elements in the major grooves of the double helix (A). Indirect repression of transcriptional activity by recruitment of methylcytosine binding proteins (MBPs) resulting in blocking access to transcriptional activators by their binding elements (B), or in the formation of transcriptionally inactive chromatin by recruiting a co-repressor consisting of large protein complexes (C).

Since these epigenetic mechanisms play crucial roles in various decision-making events affecting mammalian cell fate and function, aberrant hypermethylation in CpG islands can trigger a considerable number of disorders including cancers. Examples of such genes involved in carcinogenesis and tumor progression by hypermethylation of CpG islands of tumor suppressor genes are the RB1 gene in retinoblastoma, the von Hippel Lindau (VHL) gene in renal carcinoma, the death-associated protein (DAP) in lung cancer, DNA-repair genes like MHL1 in colorectal tumors, and E-cadherin [6-8]. Additionally, aging-dependent change in the CpG methylation status has been reported in various genes [9,10], and therefore, CpG methylation is a flexible machinery that modulates gene expression in a space- and time-dependent fashion.

Although the methylation within CpG islands has been studied intensively, the behavior of methylated CpGs located outside CpG islands, namely non-CpG islands, is still unclear and considered insignificant from the standpoint of genetic function because of their apparent random distribution in genomic DNA. In molecular biology, a crucial point is whether or not tissue samples used in each study are composed of a homogeneous or selected population. Usually, however, tissue samples are composed of different types of cells, that is, heterogeneous populations; therefore, data obtained from mixtures of heterogeneous populations do not always reflect the characteristics of cell populations of interest. In the past decade, through analysis of histomorphometrically distinguishable clusters of cells with the use of tissue samples obtained through laser capture microdissection, however, we have reported that methylation of specific CpG dinucleotides, even those located outside CpG islands, are able to exert their influence on tissue-specific gene expression as well as on cancer progression.

In the next section, theories and methods of epigenetic analyses based on histomorphometric observation, that is, morphology-oriented epigenetic research, are described in detail before presenting actual research data on the regulation of gene expression through direct and indirect mechanisms, independently of hypermethylation of CpG islands.

2. Morphology-Oriented Detection of Methylated Cytosine

With the introduction of the polymerase chain reaction (PCR) technique, genetic alteration as minute as the point mutation of a specific gene can be detected from routine formalin-fixed, paraffin-embedded pathological specimens [11, 12]. Moreover, PCR is sensitive enough to amplify the target DNA from microdissected fragments of paraffin sections, making possible the assignment of morphological alterations to specific genetic events. PCR is also a powerful tool for epigenetic studies. From pathological specimens, methylated cytosine is detectable by PCR in a morphology-oriented manner. Since formalin-fixation and paraffin-embedding cause alterations of DNA, RNA and proteins, fundamental studies are indicated for adequately interpreting the genetic data obtained from archival sources of DNA. In this section, we will first discuss the effect of formalin-fixation on the accessibility of methylated cytosine and then the application of our agarose-bead-mediated bisulfite modification technique to microdissected fragments of paraffin sections.

2.1. Can Formalin-Fixed Paraffin-Embedded Specimens Be Used for Epigenetic Studies?

Formalin-fixation and paraffin-embedding do not alter the accessibility of methylated cytosine.

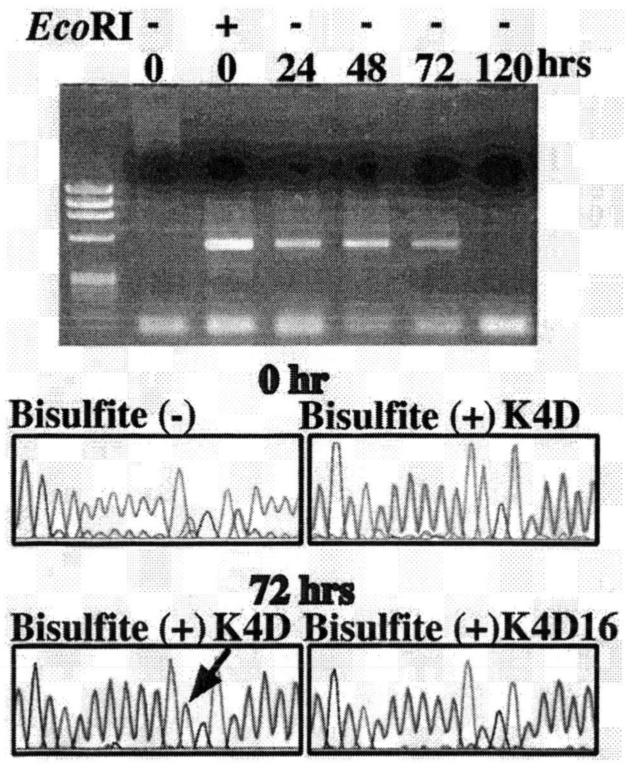

Figure 2. PCR amplification and sequencing after sodium bisulfite treatment. Lane 1, intact high-molecular-weight DNA; lane 2, intact high-molecular DNA predigested with EcoRI restriction enzyme; lanes 3-6, DNA extracted from K4D cell lines fixed with 10% formalin for 24, 48, 72, and 120 hours, without restriction enzyme digestion.

We have analyzed the effect of formalin-fixation on accessing methylated cytosine [13]. Two established cell lines, K4D and K4D16 [14], were used as models for assessing the effects of the fixative condition on the detection of methylated cytosine residues. Freshly prepared 30 μl of 10 mM hydroquinone and 520 μl of 3 M sodium bisulfite at pH 5 were added to the extracted DNA. Each sample was incubated under mineral oil at 50 °C for 16 hours. Modification was completed by NaOH (final concentration, 0.3 M) treatment for 5 minutes at room temperature, then by ethanol precipitation. Bisulfite-modified DNA (100 ng) was amplified with PCR using the primers corresponding to rat cyclin D1 gene [13]. The PCR condition was as previously described [13]. When intact DNA was fragmented, a clear PCR product was formed (figure 2, Lane 2, EcoRI + 0). Cells fixed with 10% formalin for 24, 48, and 72 hours formed a clear PCR product without predigestion with EcoRI restriction endonuclease (figure 2, upper panel, Lanes 3 to 5). No PCR products were formed, however, when the fixation reached 120 hours (figure 2, upper panel, Lane 6). The PCR products obtained from samples after 72 hours of fixation were eluted, purified, and analyzed by sequencing. The methylated cytosine in the 19th CpG locus in the K4D cell line was detectable from the DNA extracted from K4D cells fixed for 72 hours (figure 2, lower panel, arrows). This experiment demonstrates that formalin fixation, for up 72 hours, does not impair access to methylated cytosine.

DNA which is inevitably degraded when DNA extracted from paraffin blocks cannot be used for Southern blot analysis. On the other hand, because of the susceptibility of template DNA to the formation of complex secondary structures after sodium bisulfite modification, PCR is sometimes difficult when intact high-molecular-weight DNA is used as a template for PCR-mediated detection of methylated cytosine. Therefore, DNA degradation, an adverse consequence of formalin-fixation, is apparently an advantage in PCR-mediated detection of methylated cytosine.

From these recent studies, we have shown that almost all duly processed biopsy and surgical specimens can be used for epigenetic studies. Since tissue samples for pathologic diagnosis or from autopsies are routinely processed within a day (or within 72 hours at most), almost all past tissue samples stored in pathology laboratories can be a potential resource for the detection of methylcytosine. This technique allows retrospective and morphology-oriented epigenetic studies as well as clonal analyses with the use of microdissected samples from paraffin-embedded pathological specimens.

2.2. PCR Amplification of Minute Samples for the Detection of Methylated Cytosine

Almost all assays in use today for the detection of methylated cytosine can be roughly divided into two groups: restriction enzyme- and bisulfite treatment-based techniques. The former uses the inhibition of certain restriction enzymes by methylation of their recognition sites for detecting the presence of methylated cytosine. One of the pair of the isoschizomer restriction enzymes can cut the DNA only when its target is unmethylated, whereas the other is not sensitive to methylated cytosine. The most commonly used isoschizomers are the pair of methylation-resistant MspI and methylation-sensitive HpaII. Both enzymes cleave the DNA at the CCGG target, but HpaII cannot cut DNA when the second cytosine is methylated (CmCGG). The results obtained by restriction enzymes are, however, limited only to the specific cleavage sites of the endonuclease. The latter translates the epigenetic information of cytosine methylation in primary sequence differences by converting unmethylated cytosine to uracil (which is subsequently amplified as thymidine), whereas methylated cytosine remains unaltered (figure 3). The bisulfite-converted genomic DNA can be analyzed by a wide variety of PCR-based methods of which direct sequencing of the PCR products or sequencing of individual PCR product clones provides the most detailed information about the methylation pattern of every single CpG site in the given CpG islands. This bisulfite genomic sequencing technique is the most widely used and straightforward method for the study of DNA methylation.

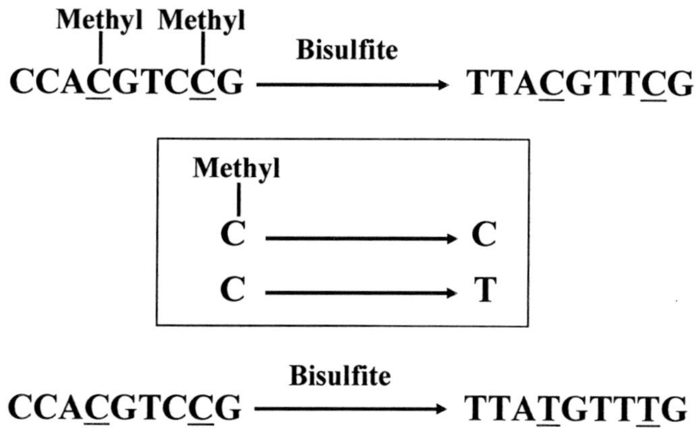

Figure 3. Principle of bisulfite conversion. Cytosine is deaminated in the presence of a high concentration of bisulfite, whereas methylated cytosine remains unaltered. This difference in chemical reactivity translates the epigenetic modification into a difference of the primary sequence.

2.3. Why Did We Adopt the Agarose-Bead-Mediated Method?

Recently, two morphology-oriented methods for the detection of methylated cytosine have been developed. Nuovo *et al.* have developed the methylation-specific PCR *in situ* hybridization (MSP-ISH) technique, which allows the visualization of methylation status of specific DNA sequences in individual cells [15]. The framework of the methodology is the same as conventional solution-phase methylation-specific PCR (MSP). In the hybridization and detection steps of MSP-ISH, however, the selection of the probe, probe concentration, post hybridization stringency conditions, and other factors may affect the final results. Inevitably, there may be non-specific background staining, and the window of signal-to-background noise is narrow; however, further modified methods of MSP-ISH have yet to be reported.

On the other hand, Umetani *et al.* have described the on-slide sodium bisulfite modification (SBM) method, whereby DNA is modified *in situ* before isolation of individual cells by microdissection [16]. The drawbacks of this method are, however, deterioration of microscopic tissue architecture, which is crucial in morphology-oriented and tissue-interpretative epigenetic studies, and increased risk of cross-contamination of DNA during the preparation of slides. Furthermore, total reagent and labor costs do not drop much. We therefore adopted and modified the agarose-bead method to efficiently detect methylated cytosine with the use of bisulfite-converted genomic DNA. We utilized laser capture microdissection [17], the technique for isolating individual cells for molecular analysis, which has the advantage of preserving the original morphology of tissue while avoiding contamination from surrounding tissue. Here, we introduce our detailed protocol for efficiently demonstrating methylated cytosine in microdissected tissue samples by the modified agarose-bead technique.

3. Laboratory Protocol for the Modified Agarose-Bead Method

3.1. Materials

Xylene and 100% ethanol
Proteinase K (10mg/ml)
10% SDS
3.2% low-melting agarose (SeaPlaque, Takara, Kyoto, Japan)
1 x TE (10 mM Tris-HCl (pH 8.0), 10 mM EDTA)
freshly prepared bisulfite solution (3.5 M $NaHSO_3$, 1 mM Hydroquinone, pH 5.0)
0.5N NaOH, 0.3N NaOH, 0.1N NaOH
1N HCl
mineral oil (heavy white oil, Sigma)
centrifuge
incubator
heated dry block bath
rotator
dessicator

3.2. Methods

3.2.1. Microdissection, Bead Formation, and Bisulfite Modification

The tissues are removed and fixed with 10% formaldehyde for 24 hours and embedded in paraffin. Sections (7-μm thick) are cut, and selected areas of the interest are microdissected by a laser capture microdissection system LM200 (Olympus, Tokyo, Japan) and collected in 1.5 ml microcentrifuge tubes. The samples are deparaffinized for a total of 30 minutes in 1 μl xylene before dehydration with 100% ethanol in the microcentrifuge tubes. The samples are further dehydrated in a vacuum for 10 minutes, redissolved, and shredded into minute pieces by pipetting in 10 ml nuclease-free water, then immediately mixed by pipetting with an equal amount of preheated (80 °C) and melted 3.2% low-temperature melting agarose. The agarose beads are made by dropping 5 μl of the mixed samples into a 1.5 ml microcentrifuge tube containing 250 μl mineral oil kept at room temperature. One milliliter of 0.2 mg/ml proteinase K in 150 mM NaCl, 10 mM Tris-HCl (pH 8.0), 20 mM EDTA is added to each tube and maintained at 50 °C in a water bath until the shredded samples become translucent (usually for 24-48 hours in a water bath). After cooling the tubes at room temperature, the agarose beads are treated twice with 1.5 ml of 0.3 N NaOH for 15 minutes each, and once with 1.5 ml of 0.1 N NaOH. To inactivate proteinase K, the samples are heated for 15 minutes at 80 °C and cooled on ice for 5 minutes. One milliliter of freshly prepared bisulfite solution (3.5 M $NaHSO_3$, 1 mM Hydroquinone, pH 5.0) is added to the microcentrifuge tubes, and the samples are incubated for 24 hours at 50 °C under light-protected conditions. The beads are then washed twice with 1 ml of 10 mM Tris-HCl (pH 8.0), 10 mM EDTA for 15 minutes each, and desulfonated by three washes with 1 ml 0.5 N NaOH for 15 minutes each. After neutralization with 200 µl of 1 N HCl for 5 minutes, the beads are washed twice with 1 ml 10 mM Tris-HCl (pH 8.0), 10 mM EDTA, and finally with nuclease-free water. The beads are sliced into several pieces and used directly as a template for PCR amplification (bisulfite sequencing PCR and methylation-specific PCR (MSP)). The schema of our modified agarose-bead method is shown in figure 4.

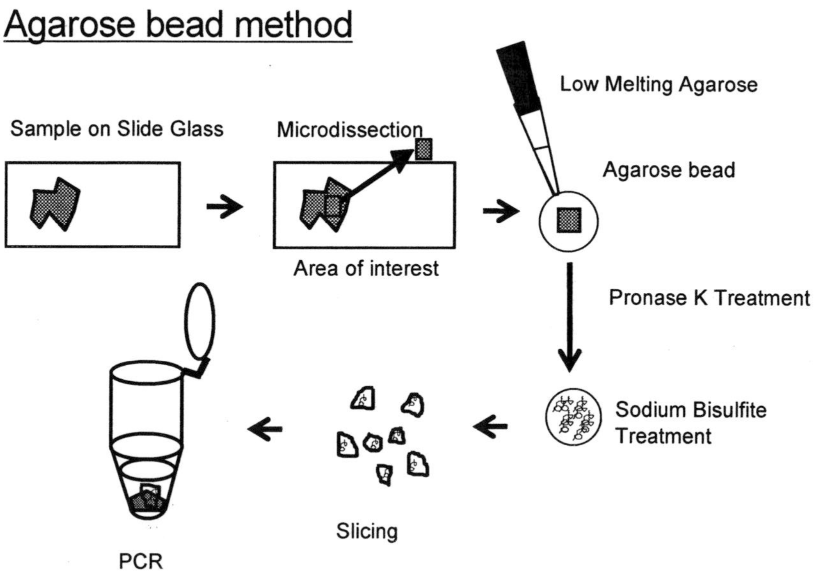

Figure 4. The schema of our newly modified agarose-bead method. The samples are fixed with formaldehyde and embedded in paraffin. Sections are cut, and selected areas of interest are microdissected. The samples are processed and mixed with preheated and melted 3.2% low-temperature melting agarose to make agarose beads. The beads are treated with proteinase K and sodium bisulfite and then sliced into several pieces and used directly as templates for PCR amplification.

3.2.2. Bisulfite Sequencing PCR

Bead fragments are analyzed by nested PCR using sets of converted primers covering the CpG loci. The PCR condition is as follows: 30 cycles of 94 °C for 30 sec, 60 °C for 30 sec, 72 °C for 30 sec, and the final elongation step for 5 min at 72 °C. The PCR mixture contains 1 x buffer (Takara, Tokyo, Japan) with 1.5 mM of $MgCl_2$, 20 pmol of each primer, 0.2 mM of dNTPs, and fragment of bisulfite-treated bead in a final volume of 50 µl. Each PCR product is loaded onto a 3% agarose gel, and the purified DNA from the gel is cloned into the pCR 2.1 plasmid vector (Invitrogen, Groningen, Netherlands). After transforming and culturing the competent bacteria (INVαF'; Invitrogen) overnight on an LB/ agar/ ampicillin plate, colonies (at least 12) are randomly selected, and the recombinant plasmid is recovered for DNA sequencing with an M13F or M13R primer. The sequencing reactions for the cloned PCR products are carried out with a DNA sequencing kit (Applied Biosystems) and the reaction products are analyzed on a 310 Genetic Analyzer (Applied Biosystems). In all molecules examined, all cytosine residues not preceding guanine residues are converted by bisulfite treatment. Bisulfite treatment, PCR, cloning and sequencing analysis of DNA are repeated independently.

3.2.3. Methylation-Specific PCR (MSP)

Methylation-specific PCR (MSP) [18] is one of the most effective choices for investigating the methylation profile of the promoter region of specific genes. The differences between methylated and unmethylated alleles that arise from sodium bisulfite treatment are the basis of methylation-specific PCR. MSP is especially valuable in analyzing CpG islands because of the abundance of CpG sites. Bisulfite-converted DNA is no longer complementary, therefore primer design must be customized for each DNA chain. Primer design is a critical and complex component of the procedure. Methylation patterns of all sequences must be determined in separate reactions with the primers for the methylated and unmethylated alleles (figure 5). A technical merit of the method resides in the use of PCR, which allows the analysis of samples with very low genomic DNA content, such as few haploid cells. Although the application of MSP is limited to known methylated sites and not exploratory, it is a time- and cost-saving technique for molecular diagnoses. MSP can also rapidly assess the methylation status of the specific gene and is suitable for the analysis of large numbers of samples. Comparison between MSP and bisulfite mapping is shown in table 1. We will further discuss the designing of primers for bisulfite-conversion based PCR methods.

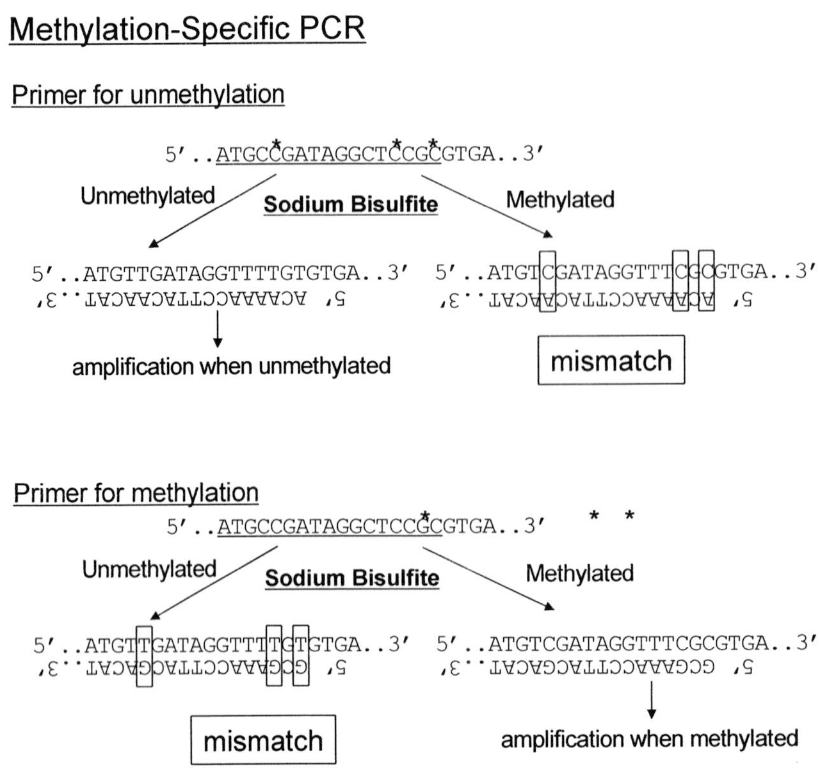

Figure 5. Principle of MSP. For the underlined initial sequence (ATGCCGATAGGCTCCGC), two antisense primers can be designed depending on the methylation status (methylated: 5'-GCGAAACCTTACGACAT-3' and unmethylated: 5'-ACAAAACCTTACAACAT-3'). Each reaction of MSP is carried out separately. For example, using the primer for unmethylation, mismatch occurs for the methylated allele and no PCR product is obtained.

Table 1. Comparison between bisulfite mapping by sequencing and methylation-specific PCR

	False Positive/ Negative	Technique	Information
Bisulfite Mapping	Rare	Multistep and Complicated	Non-CpG, Unexpected Region and Unknown Area Exploratory
MSP	Occasional	Simple and Rapid	Already Known Area Non-exploratory

3.2.4. Primer Design Rules for MSP And Bisulfite Sequencing PCR

3.2.4.A. General Consideration of Primer Selection for Bisulfite-Conversion Based PCR Methods

Modifying DNA with sodium bisulfite as a first step is followed by PCR amplification using primers specific for the modified DNA. Primers are designed to discriminate between methylated and unmethylated alleles after bisulfite treatment and between bisulfite-modified DNA and unmodified DNA. In the latter, primers should have an adequate number of cytosines in their sequence. The challenge resides in the specific amplification of the bisulfite-treated DNA. The high redundancy of the target sequence attributed to the original GC richness creates long stretches of thymines that are often difficult for polymerases to read. These conditions lower the efficiency of the PCR since mispriming occurs often. Therefore, a good primer design is crucial, since the dimer formations are greatly facilitated by the T and A richness of the sense and antisense oligos, respectively. Two pairs of nested primers and a second round of PCR are sometimes requisite.

3.2.4.B. Bisulfite Sequencing PCR

1. Primers should not contain any CpG sites within their sequences in order to avoid confusion between methylated and unmethylated DNA. If CpG sites are not excludable from a primer, degenerated bases should be used at the CpG sites to represent the methylated and unmethylated cytosines.
2. Primers should have an adequate number of non-CpG cytosines in their sequences to allow amplification of only the bisulfite modified DNA. Primers with an ample number of non-CpG cytosines are preferred.
3. To map as many CpG sites as possible, primer pairs should preferably span a CpG-rich region.

3.2.4.C. Methylation-Specific PCR (MSP)

1. Primers should contain at least two CpG sites within their sequences, and the CpG sites should preferably be located at the very 3'-end of their sequences to maximally discriminate between methylated DNA and unmethylated DNA.
2. Primers should have an adequate number of non-CpG cytosines in their sequences to allow amplification of only the bisulfite modified DNA. Primers with an ample number of non-CpG cytosines are preferred.

3.3. Application of the Agarose-Bead-Mediated Method for Epigenetic Studies with the Use of Pathological Specimens

Here we present examples of our recent application of agarose-bead-mediated bisulfite modification and the PCR technique to investigate the methylation status of the 5'-flanking region of the human bone morphogenetic protein (BMP)-6 [19] and the human E-Cadherin [20] gene recovered from microdissected fragments of paraffin sections. The BMP-6 gene was analyzed by bisulfite sequencing PCR and the E-Cadherin gene by MSP in a morphology-oriented manner.

3.3.1. BMP-6

BMPs of the transforming growth β (TGF-β) superfamily are multifunctional molecules that regulate bone induction and organ development [19]. Among these, BMP-6 is overexpressed in prostate cancer and speculated to be associated with bone-forming skeletal metastasis. We investigated the regulatory mechanism of BMP-6 gene expression in prostate cancer cell lines DU-145, LNCaP, PC-3 with regard to the methylation status of the CpG islands in the 5'-flanking region of the human BMP-6 gene. By sequence-specific analysis of methylated cytosines, the methylation status of the CpG loci around the Sp1 site of the BMP-6 promoter was found to be related to its steady-state expression and to an alternative splicing of mRNA in the prostate cancer cell lines [19]. Furthermore, a study of clinical cases of benign and malignant prostate lesions by *in situ* hybridization showed that BMP-6 expression was high at both primary and secondary sites in cases of advanced cancer with metastasis. Demethylation of CpG loci around the Sp1 binding site was demonstrated in cases with high BMP-6 expression by sequence analysis of the methylated cytosine from paraffin-embedded materials (figure 6). Our results suggested that during cancer progression, besides inactivation of tumor suppressor genes by hypermethylation, activation of certain genes like BMP-6 by selective demethylation was a common epigenetic event investing the invading and metastasizing cancer cells with a variable character. In this study, we used surgical and autopsy specimens from 11 cases of prostate cancer that had been routinely formalin-fixed, paraffin-embedded and stored at room temperature (the oldest for as long as 20 years). In the samples, we successfully detected CpG methylation of the promoter region. Here we emphasize that, in principle, once samples are embedded in paraffin, DNA alteration ceases; thus tissues stored in paraffin blocks for a long time are a potential source of DNA for retrospective clinicogenic analyses.

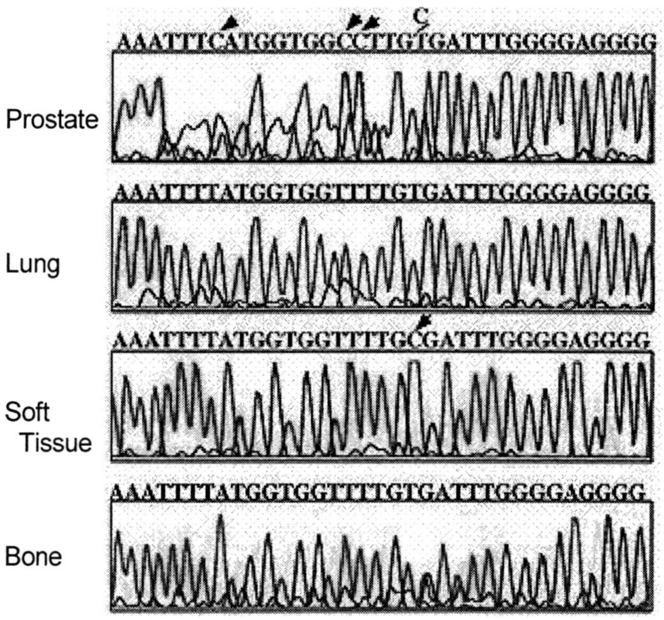

Figure 6. Sequence analysis of methylated cytosine using genomic DNA from the paraffin-embedded pathological specimen of advanced prostatic cancer. Microdissected specimens from paraffin-embedded samples were used for the hypersensitive mapping of methylated cytosine in the promoter region of the human BMP-6 gene. At the primary site (prostate), three methylated and one hemimethylated cytosines including non-CpG loci were identified (arrows). However, at the metastatic sites of lung, soft tissue, and bone most of these methylated cytosines were demethylated.

3.3.2.E. Cadherin

Cancer metastasis and invasion are closely associated with cellular properties including cell-to-cell adhesiveness. The E-cadherin (E-cad) gene encodes a cell-surface adhesion protein that plays a crucial role in homotypic cell-cell adhesion and maintenance of epithelial morphology. It also helps in controlling cell growth and differentiation. The expression of E-cad is lost during the progression of certain cancers, and this loss of expression and function, besides causing loss of cell-cell adhesion, can also convey signals that actively induce tumor-cell invasion and metastasis and promote epithelial cells to a more malignant phenotype. Besides genetic alterations such as loss of chromosome, aberrant hypermethylation of the CpG island in the E-cad promoter has also been postulated to be a major epigenetic mechanism for silencing E-cad gene expression. During tumor progression, loss of E-cad expression can be likened to dedifferentiation of human carcinomas *in vitro* and *in vivo*. Reduced expression of E-cad owing to aberrant 5'CpG island hypermethylation has been regarded as one of the main molecular events involved in the dysfunction of the cell-cell adhesion system. The molecular mechanisms providing diversity and heterogeneity to E-cad expression in colorectal carcinoma were also explored. In 29 cases of colorectal carcinoma in Indonesia, the expression of E-cad was analyzed by immunohistochemical staining, the methylation status of the E-cad promoter was determined by MSP (figure 7), and the expression of methyl-CpG-binding protein (MeCP) 2 was studied by *in situ* hybridization. E-cad expression was strong, and no methylation was observed in normal colon mucosa and in most of the well-differentiated adenocarcinomas. In contrast, both signet-ring cell carcinoma and mucinous adenocarcinoma showed fully methylated patterns and strong MeCP2 expression. In moderately- and poorly-differentiated adenocarcinomas, however, E-cad expression was notably heterogeneous, especially at the border of invasion and in the dissociated areas, where loss of MeCP2 expression correlated with E-cad reexpression even in the presence of E-cad promoter methylation. Thus both CpG methylation of the E-cad promoter and significant MeCP2 expression cooperatively and epigenetically regulate E-cad expression in colorectal cancer [20].

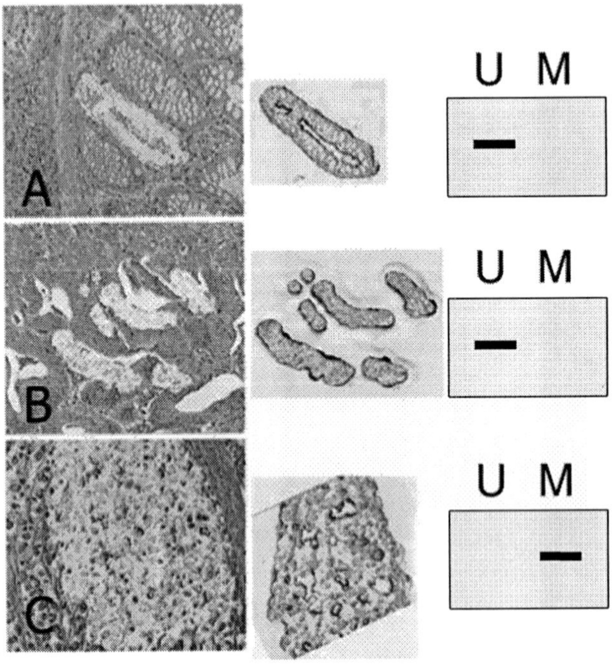

Figure 7. Detection of methylated cytosine by methylation-specific PCR (MSP) analysis of the E-cad CpG island region. Isolation of normal cells, moderately-differentiated adenocarcinoma (MDA) cells, and mucinous adenocarcinoma (MA) cells using laser-assisted microdissection. Tissue samples were stained with hematoxylin and eosin. After bisulfite modification, DNA from each microdissected sample was amplified by MSP. MSP products using primers that specifically amplify only unmethylated DNA are indicated by visible PCR products in lines U while visible PCR products in M indicate those amplified by primers specific for methylated DNA. (A) In normal colon tissues sections, MSP shows an unmethylated pattern (U). (B) Colon tissue section, an example of hemi-methylated pattern in MDA. MSP shows both unmethylated (U) and methylated (M) patterns. (C) An example of methylated pattern (M) in MA.

3.3.3. *Clonal Analysis*

Although the possibility of *de novo* alteration of the methylation status and contamination of the surrounding minor cell component is inevitable, methylation analysis of microdissected sections can be a useful alternative technique for clonal analysis. Our study showed the practicability and usefulness of routine formalin-fixed, paraffin-embedded pathological specimens as a valuable source for the morphology-oriented analysis of methylated cytosine.

4. Functional Significance of Methylated Cytosine at Non-CpG islands

From our morphology-oriented epigenetic studies, we noticed that the apparent randomness of methylated cytosine outside conventional CpG islands, when analyzed as a selected population by microdissection, sometimes showed specific distribution around major transcription factor binding sites, especially around the transcription start site. In the succeeding part of this chapter, we introduce two distinct regulatory mechanisms of gene expression by the addition of methylation at non-CpG islands.

4.1. Modulation of Gene Expression by a Direct Sequestering Mechanism–An Exceptional Event that the Addition of Methylation at Non-CpG Islands Increases Transcription of TrkA Gene

As already mentioned in this chapter, methylation at cytosine residues can directly control the binding ability of transcription factors to DNA. Basically, the target base in DNA is originally accessible to the amino-acid side chain of transcription factors by interacting in the major groove structure of the double helical DNA without methylation. Since the addition of methylation to DNA results in the intrusion of obstructive structures into the groove, methylation can directly disrupt the binding of some of the transcription factors to double helical DNA [21]. Actually, transcription factors are broadly divided into two main categories: "methylation sensitive" and "methylation insensitive" depending on whether or not DNA binding of transcription factors is affected by the addition of the methylation to the DNA binding site. The former includes transcription factors such as AP-2, c-Myc, CREB, and NF-κB that do not bind to DNA when their respective recognition sequences are methylated at cytosine residues; the latter, on the other hand, includes AP-1, Sp1, CTF, and TCR-ATF that can bind equally to both methylated and unmethylated binding elements [22]. Even among the latter group of transcription factors, the addition of methylation not inside but adjacent to the recognition sequences, albeit not a universal phenomenon, can change the affinity between transcription factors and DNA. Zhu *et al.* [23] and our group [14,24] have shown examples where, although methylation of the central CpG dinucleotides on either or both strands has no effect on the efficiency of Sp1 binding, methylation of the outer cytosines of the consensus Sp1-binding site significantly inhibits Sp1 binding. Therefore, addition of methylation to even a single dinucleotide at critical transcription binding sites can greatly affect gene expression. At the same time, unlike global hypermethylation at CpG islands that inevitably leads to the silencing of genes, depending on the degree of the contribution of transcription factors on the expression of target genes, the effect of adding methylation that interferes with the binding of transcription factors of DNA binding may range from total gene silencing to subtle changes. Furthermore, since some transcription factors function as positive regulators and others as negative regulators, methylation may yield two opposite results. While methylation inhibits the function of positively regulating transcription factors and leads to the repression of the transcription of genes, the same mechanism inhibits the function of negatively regulating transcription factors and, consequently, gene transcription is promoted. Thus, methylation of non-CpG islands can be associated with the control of gene transcription and may lead to opposing effects. Moreover, these events may be transient and plastic, and may have no effect on the global chromatin structure. To illustrate in greater detail this direct modulation mechanism induced by methylation, various patterns of DNA modification of the AP-1 binding site are presented here.

The AP-1 transcription factor mediates gene regulation under various physiological and pathological conditions, such as proliferation, differentiation, apoptosis and transformation. Dimeric AP-1 is composed mainly of Jun, Fos or ATF subunits and binds to an AP-1 site (consensus sequence 5'-TGAG/CTCA-3'). Since many subunit proteins are recognized as of the AP-1 family and can function as homodimers or heterodimers, various combinations of subunits are possible; moreover, it is known that the individual Jun and Fos proteins have significantly different transactivation potentials. In some cases, relatively weak transactivators, such as JunB, JunD, Fra-1 and Fra-2, might act as repressors of transcription by competing for binding to the site through the formation of inactive heterodimers with stronger transactivators, such as Jun, Fos or FosB [25]. As a first mechanism that blocks AP-1 binding to DNA by methylation interference that leads to the suppression of the gene (figure 8), Jones *et al.* [26] have reported that during T helper cell development, rapid methylation of the AP-1 site suppresses IFN-γ transcription in T helper type 2 cells directly; also, methylation prevents positive transcription factors from binding to the promoter, while the binding is not prevented in T helper type 1 cells whose AP-1 site is not methylated. Consequently, IFN-γ is expressed not in T helper type 2 cells but in T helper type 1 cells, and the polarization of naive CD4 T cells into T helper 2 cells or T helper type 1 cells becomes distinct. The remaining CpG loci in the promoter of polarized T helper type 2 cells are then methylated after a certain period of time, which contributes to stable and conventional chromatin remodeling and the permanent suppression of the loci. In another somewhat contradictory example, site-specific methylation directly interferes with the reaction of the negatively regulating AP-1-like site and activates TrkA (a nerve growth factor (NGF) receptor) gene expression during cancer progression (figure 9) [24]. With the use of microdissected surgical specimens of pancreatic cancer, we have shown that methylation tends to accumulate in CpG loci adjacent to the AP-1-like site of the TrkA gene in cases with perineural invasion sites where strong TrkA expression is evinced by immunohistochemistry. As some methylation-sensitive transcription factors like E2F can be both activating and repressing [27], selective methylation at specific transcription factor-binding sites can also be both repressing and activating. Interestingly, the induction of methylation of a single CpG dinucleotide within the AP-1-binding site per se did not affect the DNA binding of the transcription factors, indicating that induction of methylation not within but adjacent to the AP-1-binding site directly blocks the binding of AP-1 complexes. Because methylation adjacent to the transcription factor-binding site can block the binding of some methylation-insensitive transcription factors like Sp1 [23] and AP-1 [24], we speculate that sporadic methylation at or adjacent to the transcription factor-binding site located in non-CpG islands can modulate gene expression, both negatively and positively, without affecting the global chromatin structure. The final unique and rousing example of the modulation of the AP-1

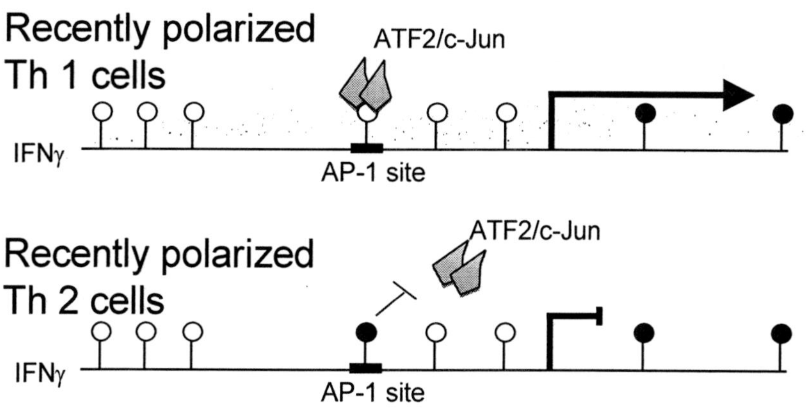

Figure 8. The schema of the direct mechanism that blocks AP-1 binding to DNA by methylation interference that leads to the suppression of the gene. During T helper cell development, while the binding is not prevented in T helper type 1 cells whose AP-1 site is not methylated, rapid methylation of the AP-1 site prevents positive transcription factors from binding to the promoter, and directly suppresses IFN-γ transcription in T helper type 2 cells.

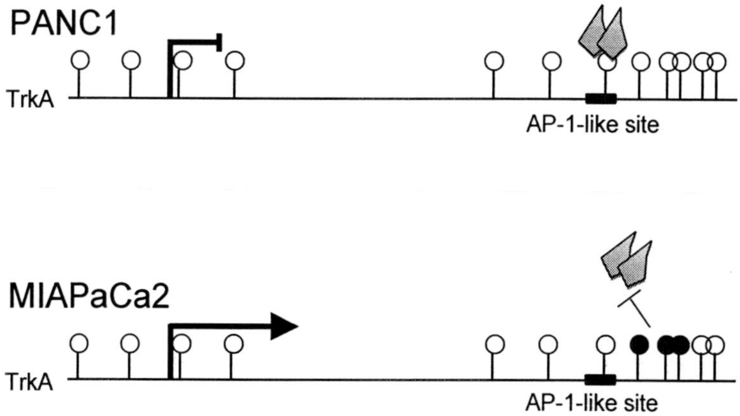

Figure 9. Site-specific methylation directly interferes with the reaction of the negatively regulating AP-1-like site and activates TrkA gene expression. Sporadic methylation at or adjacent to the transcription factor-binding site located in non-CpG islands can modulate gene expression, both negatively and positively, without affecting the global chromatin structure. Selective methylation at specific transcription factor-binding sites can also be both repressing and activating. Induction of methylation not within but adjacent to the AP-1-binding site directly blocks the binding of AP-1 complexes and upregulates TrkA gene expression in cases with perineural invasion sites, where strong TrkA expression is evinced by immunohistochemistry.

binding site by DNA methylation is the creation of a novel AP-1 binding site by the addition of methylation within the AP-1binding site-like sequence [28], demonstrating that AP-1 binding in the first intron of the metastasis-associated mts1 gene is dependent on DNA methylation within the AP-1 binding site-like sequence (figure 10). From these three examples of the modulation of AP-1 binding by DNA methylation, unlike hypermethylation at CpG islands at promoter regions that leads to condensation and typically irreversible inactivation of the gene at the chromatin level, direct interference of transcription factors binding to target sequences by methylation can be reversible, subtle and diverse. Therefore such a mechanism may account for the heterogeneity of cancer cells that occurs during invasion from the primary site to surrounding tissues in order to adapt to an altered microenvironment, and may be stabilized by subsequent subclonal selection according to subtle epigenetic differences. Furthermore, methylation of non-CpG islands may be a prerequisite to the occurrence of global methylation of CpG islands and/or genetic alteration.

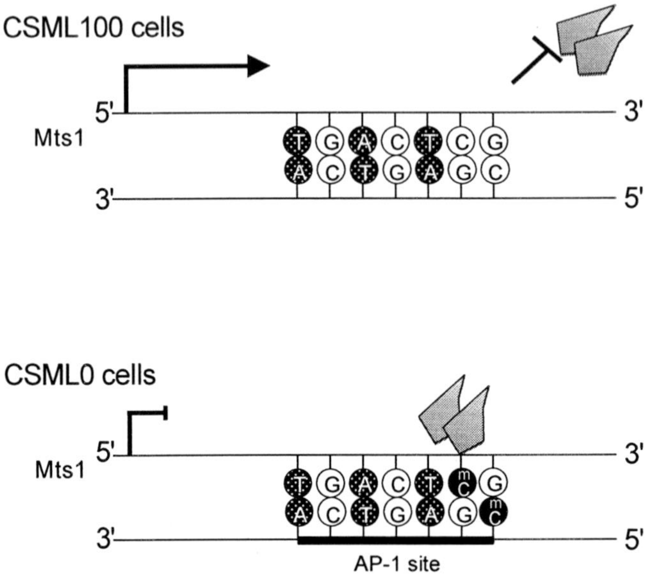

Figure 10. Creation of a novel AP-1 binding site by the addition of methylation within the AP-1 binding site-like sequence. A unique and rousing example of the modulation of the AP-1 binding site by DNA methylation is the creation of a novel AP-1 binding site by the addition of methylation within the AP-1binding site-like sequence, demonstrating that AP-1 binding in the first intron of the metastasis-associated mts1 gene is dependent on DNA methylation within the AP-1 binding site-like sequence

4.2. Modulation of Gene Expression by an Indirect Mechanism- MeCP2 Mediated Silencing of the Gene by the Addition of Methylation at the ZZCpG-Locus Upstream of the TATA-Box

Receptor activator of nuclear factor-κB ligand (RANKL) is a membrane-bound signal transducer essential for osteoclastogenesis [29-31]. RANKL expressed on stromal cells and osteoblasts and its cognate receptor, RANK [31, 32], expressed on osteoclast precursors are critical regulators of osteoclastogenesis and bone resorption in pathological conditions such as autoimmune arthritis [33] and osteolytic bone metastasis of cancer [34]. To clarify the regulatory mechanism of RANKL expression, we have characterized the 5'-flanking region of the mouse RANKL gene that contains Vitamin D responsive element (VDRE) overlapped with c-AMP responsive element (CRE), the glucocorticoid responsive element (GRE) half site, AP-1, Runx2 binding sites, inverted CCAAT- and inverted TATA-boxes [35, 36]. RANKL gene expression is, on the other hand, tissue-specific and limited to certain subsets of T-lymphocytes and osteoblastic cells [37]. Even among osteoblastic cells, 20% of those lining the bone surface express RANKL in the normal mouse [34]. To explain cell- and tissue-specific expression of the RANKL gene, we assessed epigenetic alterations during osteoblastic cell differentiation [38]. In light of a recent observation that DNA-enriched fragments containing A/T-rich bases (more than four continuous A/T sequences) adjacent to methyl-CpG are required for MeCP2, a most abundant methyl-binding domain protein, to selectively bind to the target DNA [39], we noticed that methylation at CpG locus #1 located just 3 nucleotides upstream of the inverted TATA-box (CGaggTTTATAA) was a typical MeCP2 target region and could, therefore, be a critical site for silencing the RANKL gene.

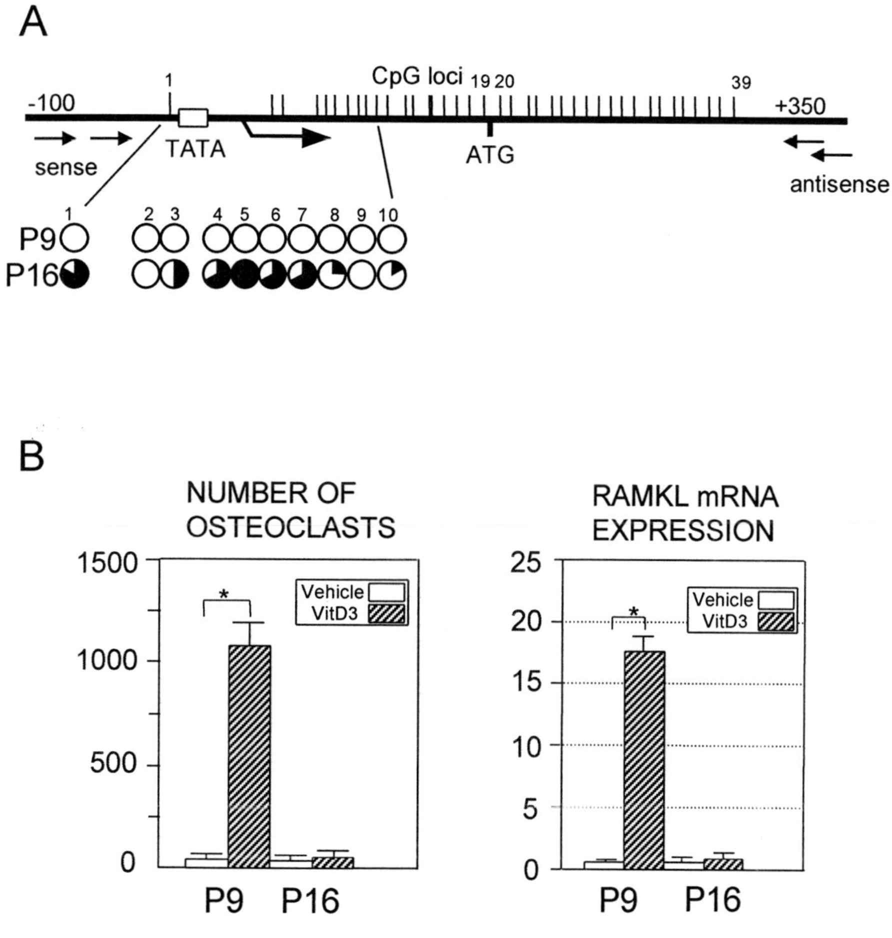

Figure 11. (A) Bisulfite mapping of -100/+120 of CpG loci #1 to #10 of the RANKL gene promoter in P9 and P16 ST2 cells. The black slices in the pie charts represent the ratio of methylated clones of the 12 clones examined. Bisulfite mapping showed a higher rate of cytosine methylation in cytosines 1, 3, 4-8, 10 in P16 than in P9 ST2 cells. (B) Comparison between ST2 cell subpopulations of osteoclastogenesis and RANKL mRNA expression. Osteoclastogenesis by ST2 cells in the coculture with mouse bone marrow macrophages for 7 days (left panel). In the presence of 10 nM vitamin D3, ST2 cells of passages 9 (P9) generated a large number of osteoclasts, but those of passage 16 (P16) did not. Results are expressed as the means ± SD of 4 culture wells. RANKL mRNA expression by quantitative RT-PCR (right panel). In the presence of 10 nM vitamin D3, RANKL mRNA increased significantly in P9, but not in P16. Results are expressed as the means ± SD of the relative mRNA amount standardized by GAPDH from 4 cultures.

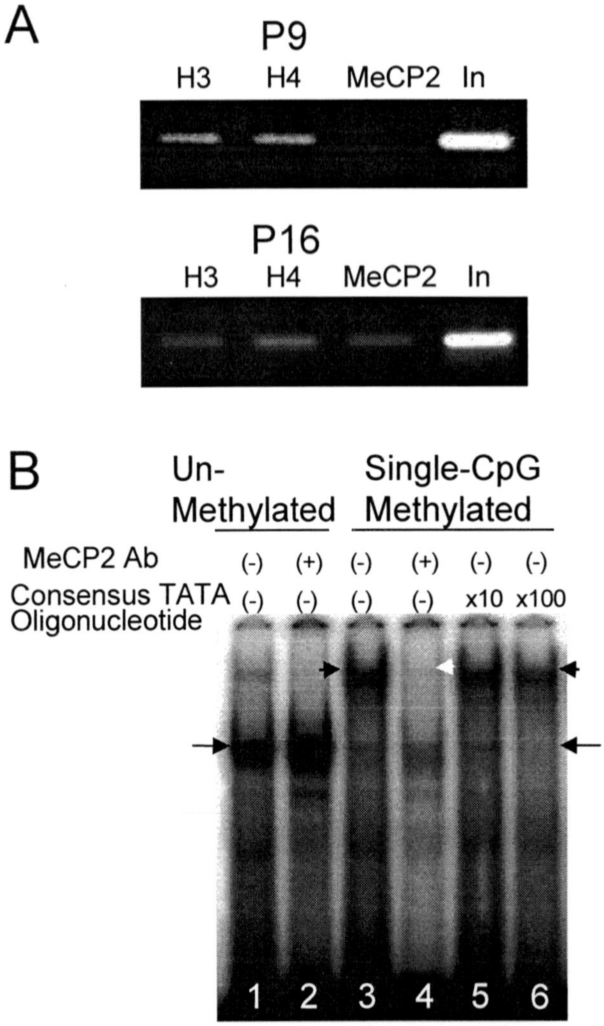

Figure 12. (A) ChIP assay of MeCP2 on the mouse RANKL gene promoter. P9 and P16 ST2 cells were subjected to immunoprecipitation. Input and immunoprecipitated DNA with either anti-acetylated histone H3, H4 or anti-MeCP2 antibodies were assessed by PCR using the sets of primers designed to amplify region -96/+12 containing an inverted TATA-box. A reactive PCR product reflecting endogenous MeCP2 binding within –96/+12 was found only in P16. H4 and H3 histone acetylation, on the other hand, was detected predominantly in P9. (B) EMSA. Double-stranded oligonucleotides spanning the TATA-box (-42/-13), an unmethylated one and a single-CpG methylated 5'-GCTCTCTCCA[5Me-dC]GAGGTTTATAAGAGTTAGG-3', were ^{32}P-labeled and subjected to the binding reaction with nuclear protein extracts from ST2 cells. The unmethylated oligonucleotides containing an inverted TATA-box (-42/-13) showed a clear protein DNA binding complex (arrow in lane 1). A single-CpG mehtylated probe at CpG locus #1 showed a strong protein DNA binding complex at a high position (arrowhead in lane 3), which was selectively and completely blockshifted by the addition of anti-MeCP2 antibody (white arrowhead in lane 4). A single-CpG methylated probe also showed weak binding at the same position as in the unmethylated probe; the binding was washed out with a 10- to 100-fold excess amount of the oligonucleotide containing the consensus TATA-box sequence (lanes 5, 6).

By sodium bisulfite mapping (figure 11A), P16 exhibited accumulation of methylation at the CpG loci (-10/+120) of the RANKL gene promoter, especially around #1 and #3-7 (downstream of the transcription start site). A CpG at #1 located three nucleotides upstream of the TATA-box was highly methylated in P16 ST2 cells that did not express RANKL (figure 11B). Both ChIP assay and EMSA showed that CpG locus #1 was a target of MeCP2 binding and that the recruitment of MeCP2 at the TATA-box interferes with the binding of the TATA-box binding protein (TBP) to the TATA-box (figure 12A). Moreover, the methylated oligonucleotide (-42/-13) showed TBP binding at the TATA-box sequence, which was increased by blocking the DNA-MeCP2 binding through the addition of an interfering anti-MeCP2 antibody (figure 12B, Lane 4). From these ChIP and gel shift assay studies, we speculated that the TATA-box of the mouse RANKL gene promoter cannot be co-occupied by TBP and MeCP2 and that the occupancy of the TATA-box by MeCP2 probably shuts down RANKL gene transcription (figure 13) [8]. This hypothesis would explain the difference in osteoclastogenesis and RANKL gene expression between P9 and P16 ST2 cells (figure 11B). In fact, in drug-resistant human breast cancer cells, CpG methylation close to the TATA-box is also associated with the silencing of the WTH3 gene [40]. Besides such site-specific methylation critical for TBP binding, generalized accumulation of methylation at a CpG island in the mouse RANKL gene promoter may result in the recruitment of histone deacetylase [41] and chromatin condensation, leading to the stable epigenetic silencing of the RANKL gene. Again, the addition of methylation at non-CpG islands may be important as the earliest epigenetic event, in both direct and indirect MeCP2 binding, that precedes global or generalized hypermethylation of CpG islands.

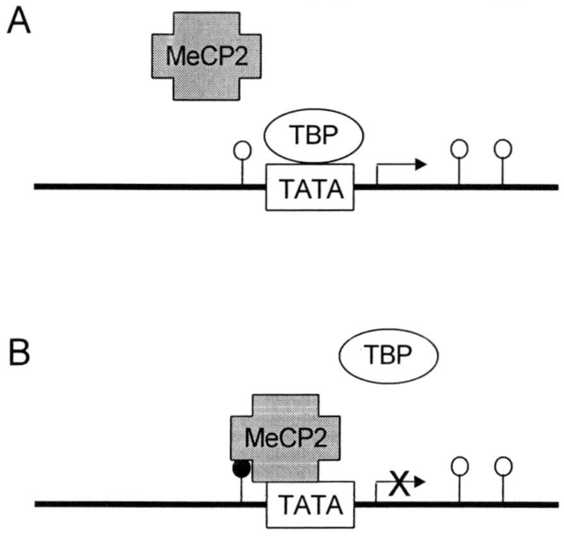

Figure 13. A model by which a single-CpG methylation adjacent to TATA-box can suppress transcription of mouse RANKL gene by MeCP2 binding. (A) When CpG locus #1 (3 bases upstream of the TATA-box) is unmethylated, TBP binds to the TATA-box, then transcription of RANKL gene occurs. (B) When CpG locus #1 is methylated, MeCP2 binds to its target region (CGaggTTTATAA) and suppresses RANKL transcription by blocking the TATA-box and TBP binding.

Conclusion

To evaluate subtle epigenetic alteration of non-CpG islands among minor subclonal populations, morphology-oriented epigenetic analysis is requisite, especially where heterogeneity and flexibility are as notable as in the process of cancer progression and cellular differentiation at critical stages.

References

[1] Ehrlich, M; Gama-Sosa, MA; Huang, L-H; Midgett, RM; Kuo, KC; McCune, RA. (1982). Amount and distribution of 5-methycytosine in human DNA from different types of tissue or cells. *Nucleic Acids Res,* 10, 2709-2721.

[2] Ng, H-H; Bird, AP. (1999). DNA methylation and chromatin modification. *Curr. Opin. Genet. Dev,* 9, 158-163.

[3] Cross, SH; Bird, AP. (1995). CpG islands and genes. *Curr. Opin. Genet. Dev,* 5, 309-314.

[4] Wade, PA. (2001). Methyl CpG-binding proteins and transcriptional repression. *BioEssays,* 23, 1131-1137.

[5] Klose, RJ; Bird, AP. (2006). Genomic DNA methylation: the mark and its mediators. *Trends Biochem. Sci,* 31, 89-97.

[6] Baylin, SB. (2005). DNA methylation and gene silencing in cancer. *Oncology,* 2, S4-S11.

[7] Baylin, S; Bestor, TH. (2002). Altered methylation patterns in cancer cell genomes: cause or consequence? *Cancer Cell,* 1, 229-305.

[8] Kitazawa, R; Kitazawa, S. (2006). Methylation status of a single CpG locus 3 bases upstream of TATA-box of RANKL gene promoter modulates cell- and tissue-specific RANKL expression and osteoclastogenesis. *Mol. Endocrinol,* (in press).

[9] Richardson, BC. (2002). Role of DNA methylation in the regulation of cell function: autoimmunity, aging and cancer. *J. Nutr,* 132, 2401S-2405S.

[10] Agrelo, R. (2006). A new molecular model of cellular aging based on Werner syndrome. *Medical Hypothesis,* in press.

[11] Weirich, G; Hornauer, MA; Bruning, T; Hofler, H; Brauch, H. (1997). Fixed archival tissue. Purify DNA and primers for good PCR yield! *Mol. Biotechnol,* 8, 299-301.

[12] Going, JJ; Lamb, RF. (1996). Practical histological microdissection for PCR analysis. *J. Pathol,* 179, 121-124.

[13] Kitazawa, S; Kitazawa, R; Maeda, S. (2000). Identification of methylated cytosine from archival formalin-fixed paraffin-embedded specimens. *Lab. Invest,* 80, 275-276.

[14] Kitazawa, S; Kitazawa, R; Maeda, S. (1999). Transcriptional regulation of rat cyclin D1 gene by CpG methylation status in promoter region. *J. Biol. Chem,* 274, 28787-28793.

[15] Nuovo, GJ; Plaia, TW; Belinsky, SA; Baylin, SB; Herman JG. (1999). In situ detection of the hypermethylation-induced inactivation of the p16 gene as an early event in oncogenesis. *Proc. Natl. Acad. Sci. U S A,* 96, 12754-12759.

[16] Umetani, N; de Maat, FG; Sunami, E; Hiramatsu, S; Martinez, S; Hoon, DSB. (2006). Methylation of p16 and Ras Association Domain Family Protein 1a during colorectal malignant transformation. *Mol. Cancer Res,* 4, 303-309.

[17] Emmert-Buck, MR; Bonner, RF; Smith, PD; Chuaqui, RF; Zhuang, Z; Goldstein, SR; Weiss, RA; Liotta, LA. (1996). Laser capture microdissection. *Science,* 274, 998-1001.

[18] Herman, JG; Graff, JR; Myohanen, S; Nelkin, BD; Baylin, SB. (1996). Methylation-specific PCR: a novel PCR assay for methylation status of CpG islands. *Proc. Natl. Acad. Sci. U S A,* 93, 9821-9826.

[19] Tamada, H; Kitazawa, R; Gohji, K; Kitazawa, S. (2001). Epigenetic regulation of human bone morphogenetic protein 6 gene expression in prostate cancer, *J. Bone Miner Res,* 16, 487-496.

[20] Darwanto, A; Kitazawa, R; Maeda, S; Kitazawa, S. (2003). MeCP2 and promoter methylation cooperatively regulate E-cadherin gene expression in colorectal carcinoma. *Cancer Sci,* 94, 442-447.

[21] Razin, A; Cedar, H. (1991). DNA methylation and gene expression. *Microbiol. Rev,* 55, 451-458.

[22] Tate, PH; Bird, AP. (1993). Effects of DNA methylation on DNA-binding proteins and gene expression. *Curr. Opin. Genet. Dev,* 3, 226-231.

[23] Zhu, WG; Srinivasan, K; Dai, Z; Duan, W; Druhan, LJ; Ding, H; Yee, L; Villalona-Calero, MA; Plass, C; Otterson, GA. (2003). Methylation of adjacent CpG sites affects Sp1/Sp3 binding and activity in the p21(Cip1) promoter. *Mol.Cell. Biol.,* 23, 4056-4065.

[24] Fujimoto, M; Kitazawa, R; Maeda, S; Kitazawa, S. (2005). Methylation adjacent to negatively regulating AP-1 site reactivates TrkA gene expression during cancer progression. *Oncogene,* 24, 5108-5118.

[25] Hess, J; Angel, P; Schorpp-Kistner, M. (2004). AP-1 subunits: quarrel and harmony among siblings. *J. Cell Sci,* 117, 5965-5973.

[26] Jones, B; Chen, J. (2006). Inhibition of IFN-γ transcription by site-specific methylation during T helper cell development. *EMBO J,* 25, 2443-2452.

[27] Attwooll, C; Lazzerini-Denchi, E; Helin, K. (2004). The E2F family: specific functions and overlapping interests. *EMBO J,* 23, 4709-4716.

[28] Tulchinsky, EM; Georgiev, GP; Lukanidin, EM. (1996). Novel AP-1 binding site created by DNA-methylation. *Oncogene,* 18, 1737-1745.

[29] Yasuda, H; Shima, N; Nakagawa, N; Yamaguchi, K; Kinosaki, M; Mochizuki, S; Tomoyasu, A; Yano, K; Goto, M; Murakami, A; Tsuda, E; Morinaga, T; Higashio, K; Udagawa, N; Takahashi, N; Suda, T. (1998). Osteoclast differentiation factor is a ligand for osteoprotegerin/osteoclastogenesis-inhibitory factor and is identical to TRANCE/RANKL. *Proc. Natl. Acad. Sci. U S A,* 95,3597-3602.

[30] Lacey, DL; Timms, E; Tan, HL; Kelley, MJ; Dunstan, CR; Burgess, T; Elliott, R; Colombero, A; Elliott, G; Scully, S; Hsu, H; Sullivan, J; Hawkins, N; Davy, E; Capparelli, C; Eli, A; Qian, YX; Kaufman, S; Sarosi, I; Shalhoub, V; Senaldi, G; Guo, J; Delaney, J; Boyle, WJ. (1998). Osteoprotegerin ligand is a cytokine that regulates osteoclast differentiation and activation. *Cell,* 93, 165-176.

[31] Anderson, DM; Maraskovsky, E; Billingsley, WL; Dougall, WC; Tometsko, ME; Roux, ER; Teepe, MC; DuBose, RF; Cosman, D; Galibert, L. (1997). A homologue of the TNF receptor and its ligand enhance T-cell growth and dendritic-cell function. *Nature,* 390, 175-179.

[32] Nakagawa, N; Kinosaki, M; Yamaguchi, K; Shima, N; Yasuda, H; Yano, K; Morinaga, T; Higashio, K. (1998). RANK is the essential signaling receptor for osteoclast differentiation factor in osteoclastogenesis. *Biochem. Biophys. Res. Commun,* 253, 395-400.

[33] Mori, H; Kitazawa, R; Mizuki, S; Nose, M; Maeda, S; Kitazawa, S. (2002). RANK ligand, RANK, and OPG expression in type II collagen-induced arthritis mouse. *Histochem. Cell Biol,* 117, 283-292.

[34] Kitazawa, S; Kitazawa, R. (2002). RANK ligand is a prerequisite for cancer-associated osteolytic lesions. *J. Pathol,* 198, 228-236.

[35] Kitazawa, R; Kitazawa, S. (2002). Vitamin D_3 augments osteoclastogenesis via vitamin D-responsive element of mouse RANKL gene promoter. *Biochem. Biophys. Res. Commun,* 290, 650-655.

[36] Kitazawa, R; Kitazawa, S; Maeda, S. (1999) Promoter structure of mouse RANKL/ TRANCE/ OPGL/ ODF gene. *Biochim. Biophys. Acta,* 1445,134-141.

[37] Kartsogiannis, V; Zhou, H; Horwood, NJ; Thomas, RJ; Hards, DK; Quinn, JM; Niforas, P; Ng, KW; Martin, TJ; Gillespie, MT. (1999). Localization of RANKL (receptor activator of NF-κB ligand) mRNA and protein in skeletal and extraskeletal tissues. *Bone,* 25, 525-534.

[38] Kitazawa, S; Kitazawa, R. (2002). Epigenetic control of mouse receptor activator of NF-κB ligand gene expression. *Biochem. Biophys. Res. Commun,* 293, 126-131.

[39] Klose, RJ; Sarraf, SA; Schmiedeberg, L; McDermott, SM; Stancheva, I;Bird, AP. (2005). DNA binding selectivity of MeCP2 due to a requirement for A/T sequences adjacent to methyl-CpG. *Mol. Cell,* 19, 667-678.

[40] Tian, K; Jurukovski, V; Wang, XP; Kaplan, MH; Xu, H. (2005). Epigenetic regulation of WTH3 in primary and cultured drug-resistant breast cancer cells. *Cancer Res,* 65, 10024-10031.

[41] Jones, PL; Veenstra, GJ; Wada, PA; Vermaak, L; Kass, SU; Landsberger, N; Strouboulis, J; Wolffe AP. (1998). Methylated DNA and MeCP2 recruit histone deacetylase to repress transcription. *Nat. Genet,* 19, 187-191.

In: DNA Methylation Research Trends
Editor: Taku B. Kobayashi, pp. 67-85
ISBN: 978-1-60021-721-0

Chapter IV

DNA Methyltransferase Inhibitors as Therapeutic Agents in Hematological Malignancies

***Takashi Kumagai*[1,2] *and H.Phillip Koeffler*[2]**
[1] Department of Hematology, Ohme Municipal General Hospital
[2] Hematology/Oncology, Cedars-Sinai Medical Center, UCLA School of Medicine

Abstract

Aberrant DNA methylation is an epigenetic event, which has an important role in carcinogenesis by silencing tumor suppressor genes in many cancers. Therefore, DNA methyltransferase (DNMT) inhibitors have been investigated as anti-cancer agents. These inhibitors transiently induce DNA hypomethylation of tumor suppressor genes, such as p15INK4B, as observed in clinical studies, as well as, in vitro and in vivo studies. Furthermore, two DNMTs, azacytidine (5-azacytidine) and decitabine (5-aza-2'-deoxycytidine), have shown promising clinical results in hematological malignancies, including myelodysplastic syndrome (MDS), acute leukemia, and chronic myelogenous leukemia. Recently, these drugs have been approved by the FDA for the treatment of MDS. Because of their toxicity, they have been used at low doses; however, the optimization of dose and duration has improved the epigenetic modulation as well as clinical response. Moreover, some studies have demonstrated the improved clinical responses of these drugs in combination with other clinically useful drugs, especially histone deacetylation inhibitors. Therefore, DNA methyltransferase inhibitors, alone or in combination with other drugs, is a promising molecular targeted therapy in hematological malignancies.

Introduction

Epigenetic changes such as DNA methylations have important roles in regulating the expressions of genes during normal cell maturation, but such changes are also thought to contribute to carcinogenesis [1]. Aberrant DNA methylation can transform normal cells especially by silencing tumor suppressor genes [2-4]. This phenomenon has been extensively examined in myelodysplastic syndrome and leukemia [5, 6]. An epigenetic therapy which can reverse the silencing of genes has become a promising approach for cancer therapy because these silenced genes are genes associated with guarding against inappropriate growth [7].

Myelodysplastic syndromes (MDS) are considered to be a heterogeneous group of bone-marrow diseases that are characterized by ineffective hematopoiesis of blood cells due to a clonal genomic abnormality. Survival ranges between months to years [5, 8]. Typical disease symptoms include fatigue from anemia and infection; and these patients are at high risk of progression to acute myeloid leukemia (AML) [5, 8]. MDS occurs usually in individuals over the age of 65. Standard therapy consists of supportive treatments such as blood transfusions, haematopoietic growth factors and antibiotics to treat opportunistic infections [5, 8]. Stem-cell transplantation can be a potentially curative therapy, but it is often not appropriate for MDS patients because of either the patients' advanced age or accompanying diseases [5, 8]. Recently advances in understanding of the pathogenesis of MDS have encouraged us to test several compounds that effect epigenetic modification such as those that decrease DNA methylation [5]. One such drug is 5- AZA-cytidine (Vidaza; azacytidine, Pharmion). It was approved by the FDA for the treatment of MDS in May 2004 [6].

A related drug, 5-Aza-2'-deoxycytidine (MGI Pharma; decitabine, Dacogen) also inhibits DNA methylation by suppressing the ability of DNA methyltransferase to reactivate silenced genes [9]. It also has been approved for the therapy for MDS by FDA in May, 2006. The drug has been examined in clinical trials for hematological malignancies including MDS, chronic myeloid leukemia (CML) and acute myeloid leukemia (AML) [10-17]. The drug has been well reviewed [18-23].

In this issue, we will provide an overview of the mechanism of action of DNA methyltransferase (DNMT) inhibitors as therapeutic agents and their efficiency in clinical trials for hematological malignancies. We will also explore future directions.

DNA Methylation and Cancer

DNA methylation plays an important role in carcinogenesis. It is induced by DNA methyltransferases (DNMTs) which catalyze the addition of the covalent methyl group to the 5' position of cytosine in the CpG dinucleotide from donor S-adenosylmethionine [24]. DNA methylation often occurs in the promoter regions of the gene repressing transcription of these target genes.

Methylation frequently occurs in CpG islands [25, 26]. They range in size from 200 to 5000 bp (basepairs), are GC rich (60-70%), and occur every 100 kb on average [26, 27]. They show very high density of CpG dinucleotides, which is in contrast to other regions of the gene [7]. The CpG dinucleotide in the mammalian genome can be spontaneously deaminased to uracil of the relatively unstable base of 5-methylcytosine. CpG islands usually occur within either regulatory regions of the genome such as promoters or in repetitive DNA regions including Alu repeats [28].

When the CpG island is methylated, it induces changes in chromatin structure and represses gene transcription [29]. In normal cells, promoter methylation is usually found in selectively silenced genes such as imprinted genes [30] and genes located on the inactivated X chromosome in females [31]. Why transcription is repressed by methylation is not totally clear, but probably involves a direct interactions of methylated DNA with certain regulatory proteins such as methylated DNA-binding proteins, and the recruitment of silencing complexes [32]. For example, methyl-CpG-binding protein 2, containing methyl-CpG-binding and transcriptional-repression domains [33] forms a complex with corepressors including histone deacetylases. DNA methyltransferase, methylated DNA-binding proteins, and histone deacetylase interact synergistically with methylated DNA and deacetylated histones to repress transcription [34, 35, 36].

Cancers frequently have aberrant DNA methylations resulting in the aberrant silencing of genes including CDK inhibitors which are normally associated with inhibiting cell cycle progression (e.g., p15INK4B, p16CDKN2A) [37, 38], inhibitors of tumor metastasis (E-cadherin, TIMP3) [39], differentiation (RARβ2) [40], apoptosis (caspase 8, TMS-1) [41, 42], angiogenesis (TSP-1) [43], DNA repair genes (hMLH-1) [44], hormone-related transcription factors (ERα) [45] , and tumor antigens (MAGE) [46, 47]. These genes are hallmarks of cancer [48, 49].

For instances, p15INK4B and p16CDKN2A genes are often methylated in hematological malignancies and other cancers, contributing to carcinogenesis [50]. Disease progression of CML can be associated with increased of DNA methylations [51,]. Gene methylation in acute lymphoblastic leukemia (ALL) has been associated with poor prognosis [52]. Both examples emphasize the association of abnormal DNA methylation and either the development or progression of hematological malignancies.

Drug Pharmacology of DNA Methyltransferase Inhibitors

5-AZA-cytidine and 5-AZA-CdR are nucleoside analogue inhibitors in which the carbon at the 5th position of the pyrimidine ring is replaced by a nitrogen (figure 1). This can prevent their enzymatic methylation to form the modified base 5-methylcytosine. The presence of ribose in 5-azacytidine can allow it to be incorporation into RNA and subsequently also DNA, while the deoxyribose of decitabine (5-aza-2'-deoxycytidine) directs the compound into DNA. Decitabine is activated via phosphorylation by deoxycytidine kinase. Other kinases in the cell rapidly convert the monophosphate (5-AZA-dCMP) to its triphosphate form (5-AZA-dCTP), the active form of the drug in the cell. 5-AZA-dCTP is a good substrate for DNA polymerarse and is rapidly incorporated into DNA [53] (figure 1). Cytidine deaminase causes deamination of 5-AZA-dCR resulting in a loss of its antineoplastic activity. The incorporation of 5-AZA-dCR into the 5-methylcytosine position during the replication of DNA results in a potent inhibition of DNA methyltransferase due to the formation of a covalent bond between this enzyme and the 5-azacytosine ring [54]. Decrease of the DNA methyltransferase activities in the cell results in the global hypomethylation of the genome associated with reactivation of tumor suppressor genes silenced by aberrant methylations (figure 2). Understanding this process is important for the development and testing of novel methyltransferase inhibitors as anti-cancer agents.

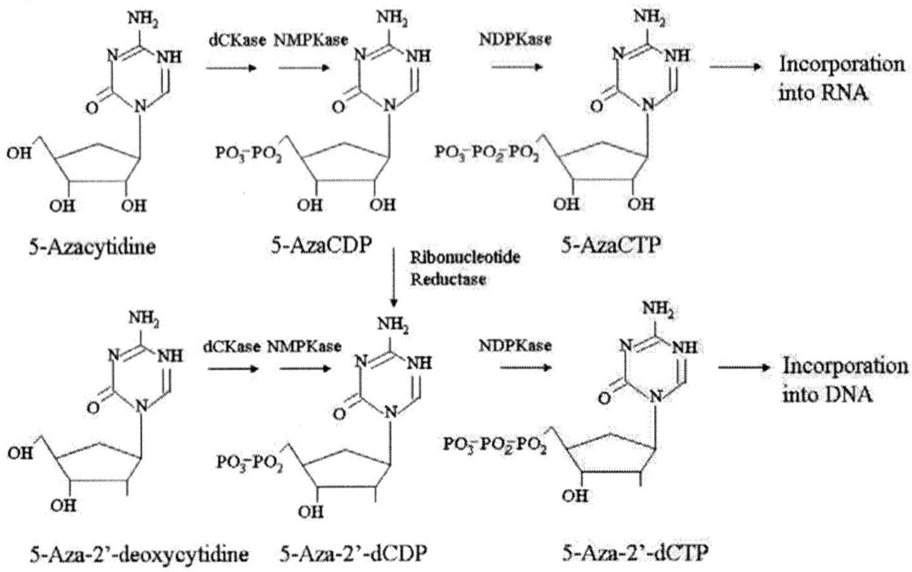

Figure 1. Metabolism of DNA methyltransferase inhibitors which are incorporated into DNA or RNA. Sequential phosphorylation at the 5'-sugar position by several enzymes; deoxycytidine kinase (dCKase), nucleotide monophosphate kinase and nucleotide diphosphate kinase (NAPKase) allows formations of the RNA precursor, 5-AzaCTP and the DNA precursor, A-Aza-2'-dCTP.

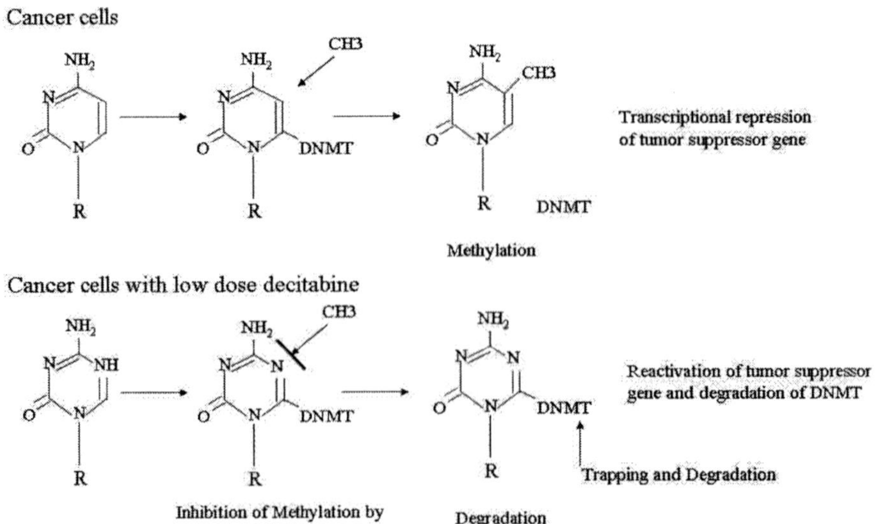

Figure 2. Mechanism by which DNA methyltransferase inhibitors reverse the methylation in cells. DNMT normally binds to a cytosine ring in DNA and transfers a methyl group from S-adenosylmethionine to the cytosine ring releasing DNMT. If DNMT binds to decitabine incorporated into DNA, the methyl group can not bind to the N group at 5' position. Consequently, DNMT is trapped and degraded.

Clinical Studies of DNMT Inhibitors for MDS and other Hematological Malignancies

Initially, DNMT inhibitors were considered to be similer to conventional anti-cancer drugs, and often given at or near to the maximum tolerable dose.

For example, in a study by EORTC (European Organization for Research and Treatment of Cancer), decitabine was used at various doses (20 mg - 660 mg/m^2/course) but was not very effective against solid tumors including melanoma, colorectal, head and neck, and ovarian cancers [55]. In contrast, the agent showed significant anti-tumor activity against hematological malignancies [13, 56]. Monparker et al. demonstrated an overall response (OR) rate of single-agent decitabine (31-81 mg/kg over 40-60 hrs) of 37% and complete response (CR) rate of 22% in relapsed pediatric leukemia in an early study [13]. As this study showed a dose-dependent effect of decitabine, subsequent leukemia studies used more than 400 mg/m^2 per course.

A 25% CR rate was achieved in a pilot study of decitabine (90-120 mg/m^2 every 8hr for 3 days) for 12 untreated patients with MDS with poor prognosis [14]. Kantarjian et al. reported that 90 mg/m^2 of decitabine every 8 hrs for 3 days resulted in only one CR among eight patients with relapsed and refractory AML [15]. EORTEC conducted studies of decitabine in combination with other agents against either refractory or relapsed AML, comprising 125 mg/m^2 of decitabine every 12hr for 6 days (total 1500 mg) with amsacrine (120 mg/m^2 on day 6 and 7) in 30 patients, or with idarubicin (12 mg/m^2 on day 5-7) in 33 patients [57]. The CR rates were 26% and 45%, respectively. The median disease-free survival interval was 8 months.

This high dose regimen of decitabine (50-100 mg/m^2 twice a day for 5 days) was tested in a total of 130 patients with CML (chronic myelogenous leukemia) [17]. In blast crisis, 28% of patients (18 of 64 individuals) achieved an objective response. They include six complete hematologic responses (CHR). Three patients return to the chronic phase (second CP) and five patients (8%) had a cytogenetic response. In the accelerated phase, 55% of patients (28 of 51 individuals) had an objective response (12 in CHR and 3 returned to their second CP). Seven patients (14%) had a cytogenetic response. Among 8 patients treated in the chronic phase, five (63%) had an objective response. In Ph1-negative CML (7 patients), four (57%) had an objective response. No evidence of dose-response effect was observed in this study.

Side-effects were frequent, especially severe myelosuppression which was often delayed, prolonged and dose-dependent. Because of the toxicity and lack of clear superiority over conventional anti-cancer agent such as cytarabine, its use as an anti-cancer agent at high dose has faded into obscurity.

Interestingly, at the same time, in vitro studies found that hypomethylation of DNA of cancer cells occurred in the presence of low concentrations of the DNMT inhibitors [58, 59], prompting to clinical trials at lower doses in the hopes of reducing toxicity including clinical trials in patients with MDS.

A randomized, controlled Phase III study of 191 patients with MDS treated with 5-AzaC (75 mg/m^2/day), given subcutaneously for 7 days every 28 days compared with supportive care was performed [60]. Patients of both groups received transfusions and antibiotics, and patients in the supportive care group were permitted to cross over to 5-AzaC when their disease worsened. Clinical responses were observed in 60% of the 5-AzaC treatment group compared to 5% improvement found in the supportive care group ($p < 0.001$). Although the mean number of red cell transfusions increased in patients treated with 5-AzaC, all patients who attained a complete or partial remission no longer required transfusions. Therefore, 45% of 65 patients who needed transfusions at the beginning of trial, were transfusion-free after treatment with 5-Aza, and 6% had a 50% reduction in transfusions. Median times for initial and best response were 64 and 93 days, respectively, with median duration of response of 15 months in the responders. Tranformation of MDS to acute myeloid leukemic (AML) as the first event occurred in 15% of 5-AzaC treated patients which was significantly lower than 38% of those in the supportive care cohort ($p = 0.001$). Median time to transformation or death was 20 months for 5-AzaC while it was 14 months for supportive care ($p = 0.1$). The major side effect was myelosuppression [grade III/IV leukopenia (59%), granulocytopenia (81%), and thrombocytopenia (70%)] in 5-AzaC treated group [60].

The increase in survival of the patients treated with 5-Aza was reported though it was not significant [60]. The patients receiving 5-Aza treatment had improvement in physical function, symptoms (particularly for fatigue and dyspnea), and psychological state [61].

These results showed that the patients receiving 5-Aza had significantly higher response rates, improved quality of life, reduced risk of leukemic transformation, and improved survival compared with the patients receiving supportive care [60, 61], suggesting that 5-AzaC therapy was superior to supportive care for patients with MDS. This study was the basis of FDA approval of 5-AzaC as monotherapy for MDS and CMML (chronic myelomonocytic leukemia) [60, 61, 62].

Low dose decitabine was examined in several small studies [59, 63, 64, 65], which was followed by a large phase II study of 66 MDS patients who received 45 mg/m^2/day for 3 days every 6 weeks [10]. Overall response and CR rate was 49 % and 20%, respectively. Platelet counts increased after even one cycle of treatment in most patients who responded to the treatment. Of the 61 patients who had abnormal cytogenetics, 19 had a major cytogenetic response (31%) after a median 3 courses of decitabine. According to the International Prognostic system (IPSS) of classification [66], 5, 30 and 26 patients were in either low, intermediate or high risk groups, respectively; and 3, 6 and 10 of these individuals achieved a cytogenetic response, respectively.

This success was the harbinger of a Phase III clinical trial of 170 MDS patients [12]. Eighty-nine individuals received intravenously decitabine (15 mg/m^2 every 8 hr for 3 days every 6 week) and 81 patients received only supportive care including transfusions, growth factors and antibiotics. Median age was 70 years, and about 14 % of individuals in both groups had secondary MDS. In 70 patients, the MDS was IPSS class intermediate-2 or higher, and the distribution pattern was similar in both groups. Decitabine was repeated every 6 week according to the clinical response, recovery and toxicities. The primary endpoints were the overall frequency of complete and partial response in the intent-to-treat population, and the time to progression or death. Responses were classified using the International Working Group (IWG) criteria [67]. Of the 81 patients, the overall response rate was 17% including 9% CR, compared to 0% in the supportive care cohort. Hematological improvement occur in an additional 13 percent of decitabine-treated patients compared to 7 percents of those in the supportive care group. Median time to response was 93 days, about 3 cycles of therapies. Most patients who responded to decitabine therapy achieved either a complete or partial response after 4 cycles of therapies. Median duration of complete and partial responses was 288 days in the treated patients. The drug did not significantly delay the median time to acute leukemia or death (12.1 in treatment group vs 7.8 months in supportive cohort, $p<0.16$). But either outcome was significantly prolonged in the high risk IPSS (2.8 vs 9.3 months, $p<0.010$) [12]. Quality of life as evaluated according to the EORTC scale was superior in the decitabine group compared to the supportive care cohort (global health status, physical functioning, fatigue and dyspnea).

Two additional open-label, single-arm studies evaluated the safety and efficiency of decitabine in 66 and 89 MDS patients with various FAB subtypes [68]. Decitabine was administered intravenously for 4 hrs at a dose of 15 mg/m3, every 8 hrs on 3 successive days every 6 week. The overall response rate in each study was 26% and 24%, respectively which paralleled the results of the above mentioned Phase III study.

Further studies have attempted to optimize dose and dose schedules for AML, CML and MDS patients. Decitabine (15 mg/m^2/day for 10 days) was administered to imanitib-refractory or –intolerant CML patients (12 chronic phase, 17 accelerated phase and 6 patients blastic phase) [69]. Hematological CR and PR occurred in 34% and 20% of patients, for an overall hematological response rate of 54% (83 % in chronic phase, 41% accelerated phase and 34 % blast phase). A total of 17% and 29% of patients had either a major or minor cytogenetic response, respectively. The median response duration was 3.5 months (range 2->13 months).

A randomized phase II study of different dose schedule of low dose decitabine for MDS patients (100 mg/m^2/course) showed that 20 mg/m^2 IV for 5 days induced a better clinical responses (CR 39%, 25 of 64 patients) than 10 mg/m^2 IV for 10 days (CR 4 of 17 patients, 24 %, $p=0.2$) or 20 mg/m^2 subcutaneously (SC) for 5 days (CR in 3 of 14 patients, 21%, $p=0.2$) [11]. Previous studies also showed that continuous infusion schedules of decitabine were not as effective as i.v infusion for 1 hr [10, 56]. These results suggest that low-dose decitabine (100-150 mg/m^2/course) at high dose intensity (short infusion times rather than prolonged infusion) might be optimal therapy.

Although the hypomethylating potency of azatidine is considered to be less than that of decitabine [58], currently available data suggest that clinical efficacy of the two drugs is comparable for treatment of MDS [60, 62]. Randomized clinical trials will be necessary to asses if either is more efficacious. The key clinical features of clinical studies of these two agents are shown in table 1.

Table 1. Clinical Trials of either 5-Azacytidine (Vizada) or 5-aza-2'-deoxycytidine (Decitabine) For Hematological Malignancies

Drug	Disease	Number of Patients	Dosage (course)	Interval between courses (weeks)	Response Rate	Source
Decitabine	MDS	66	45 mg/m^2/day x 3days	6	CR(20) PR(5) CI(25)	(10) Wijermans P. et al. 2000
Decitabine	MDS	170	15 mg/m^2 TID x 3days	6	CR(9) PR (8) CI (13)	(12) Kantarjian H et al. 2006
Decitabine	MDS	98	15 mg/m^2 TID x 3days	6	CR+PR (24)	(68)
Decitabine	higher risk MDS	115	20mg iv or s.c x 5 days or 10mg iv x 10 days	4	CR(35) PR(2) CI(10)	(96) Kantajian H et al. 2007
Decitabine	CML	130	50 or 75 or 100 mg/m^2 BID x 5 days	4 - 8	OR(42) CHR(16) PHR(13) PCP(5)	(17) Kantarjian H et al. 2003
Decitabine	Imanitib Resistent CML	35	15 mg iv x 5 days x 2 weeks		OR(65) CHR(34) PHR(20)	(69) Issa JP et al. 2006
5-Azacytidine	MDS	191	75 mg/m^2 s.c x 7days	4	CR(7) PR(16) CI(37)	(60) Silverman LR et al. 2002
5-Azacytidine	MDS	43	75 mg/m^2 s.c x 7days	4	CR(12) PR(25) CI(5)	(97) Silverman LR et al. 1993

Abbreviations: MDS, myelodysplastic syndrome; CML, chronic myelogenous leukemia; CR, complete ressponse; PR, partial responese; CI clinical improvement;
OR, overall response; CHR, complete hematological response; PHR, partial hematological response; sc, subcutaneous; iv intravenous;
BID, twice a day; TID, three times a day.

In vitro data are accumulating suggesting that inhibitors of both DNMT and histone deacetylase may have synergistic anti-proliferative activity against cancer cells [70], and clinical trials of this approach have commenced. For example, combination of decitabine and valproic acid possessed activity for patients with AML and MDS, with a response rate of 22% [71]. Also, azacytidine combined with phenylbutyrate (histone deacetylase inhibitor) showed promising efficacy in myeloid malignancies [72]. Also, the laboratory data suggested that DNMT inhibitors may increase the sensitivity of the cancer cells for chemotherapeutic agents partly by regulating expression of genes associated with drug resistance [73]. Taken together, the combination of drugs having different epigenetic targets deserves further clinical attention.

Effect of Decitabine on Methylation of Target Genes

Several of the above described clinical trials explored the effect of decitabine on DNA methylation.

Global gene methylation analysis focused on the repetitive sequences called long interspersed nucleotide element (LINE) and Alu sequences which are normally methylated and can be quantified by bisulfite-pyrosequenceing or several other techniques [74]. Changes of LINE methylation can be used as a surrogate marker of global changes [74]. Global and gene-specific DNA methylation was examined in the peripheral blood of patients with leukemia treated with decitabine [74] either during a high dose (phase II study) [17], or a low dose (phase I study) [16]. Dose-dependent linear decrease of methylation was noted by day 5 of low dose (5-20 mg/m^2/day) treatment. Maximal hypomethylation occurred at 20 mg/m^2/day, suggesting a plateau effect. In contrast, assessment by liquid chromatography did not show a plateau. The difference in results requires further studies [74].

Several low dose studies including a randomized phase II investigation showed that the induction of hypomethylation and clinical response paralleled each other [75]; decitabine induced significant LINE hypomethylation as early as the 5th day of beginning of therapy and the degree of hypomethylation was most prominent in patients receiving 20 mg/m^2 IV x 5 days compared to either 20 mg/m2 SC x 5 days or 10 mg/m^2 IV x 10 days. Likewise, the best clinical response was observed in those who recieved 20 mg/m^2 IV x 5 days. Hypomethylation reached a plateau after repeated cycles of decitabine, and the complete loss of methylation in LINE was observed. But the degree of LINE hypomethylation did not correlate with clinical response, suggesting that the LINE hypomethylation can reflect the pharmacological activity of decitabine but it is not necessarily a good biological marker of clinical response. These data suggest that either hypomethylation of target genes, not LINE repetitive elements, is the important measurement or decitabine may mediate anti-cancer activity independent of DNA methylation.

The degree of global hypomethylation in patients with MDS receiving decitabine was also analyzed by capillary electrophoresis [76]. Hypomethylation in the bone marrow appeared most prominently after karyotypic normalization, presumably normal hematopoietic cells. Multiple cycles of decitabine treatment induced significant (up to 70%) demethylation in five of seven patients. In this study, no clear description is provided of when the sample was obtained in the relationship to the decitabine treatment.

Line methylation was also analyzed in a phase II study of decitabine in patients with imanitib-resistant CML [69]. Methylation rapidly decreased, averaging 16% decrease at day 12 of treatment; methylation recovered by 4-6 weeks after stopping decitabine. On day 5 of therapy, LINE1 methylation was not correlated with clinical responses. But the absolute decrease in methylation was less in responders than in non-responders (14% versus 27%, p=0,007) on day 12 of therapy. This inverse correlation between methylation on day 12 and clinical response is not explained, but could suggest that the hypomethylated cells died and the normal or resistant cells were able to withstand greater hypomethylation. Similar results were reported in another study of treatment of CML with decitabine [77].

In addition to changes in global methylation, specific gene methylation has been examined including p15INK4B and p21WAF1 genes. The promoter region of p15INK4B is frequently hypermethylated in hematological malignancies, including acute lymphoblastic leukemia, AML and MDS [78,79,80]. Daskalakis et al. showed that samples from 9 out of 12 MDS patients with p15INK4B hypermethylation [identified by methylation-sensitive single-nucleotide primer extension (Ms-SNuPE)] had decreased p15INK4B methylation levels after more than one course of decitabine treatment [81]. The decrease was correlated with a clinical response, suggesting that hypomethylation of p15INK4B among other genes, may be functionally important in mediating the clinical response to decitabine in patients with MDS. Some of the hypomethylation could represent replacement of the MDS clone with a normal haematopoietic clone. But the investigators showed gradual demethylation in the MDS clone rather than replacement of the MDS clone with a normal unmethylated clone.

The effect of decitabine on the methylation of the p15INK4B promoter was investigated by pyrosequencing in several studies including phase I and II high dose drug trials in CML [82], phase II low dose decitabine in CML [69] and a randomized phase II trial in MDS [75]. In this latter trial, promoter hypermethylation of p15INK4B was observed in samples from 32 of 124 patients. The individuals with p15INK4B hypermethylation tended to have a lower response rate than those without hypermethylation (34% versus 45%, p=0.31). Rapid induction of p15INK4B hypomethylation (i.e. days 5-10 after treatment) occured in all the studies, but was not correlated with clinical responses.

MDS patients who achieved CR sustained p15INK4B demethylation compared to those who did not achieved CR. The latter group, those achieved who achieved various degrees of hypomethylation, had rapid re-methylation on day 10 of the therapy.

In the multivariate analysis of a randomized phase III study of decitabine in MDS, the baseline hypermethylation of multiple genes (including p15INK4B, E-cadherin, etc.) was significantly correlated with overall survival [83]. Methylation status was followed for 3-6 months of therapy. The results suggested that patients with MDS generally tended to gain methylation in the abnormal clone over time unless decitabine treatment was given and hypomethylation could not be completely separated from clonal replacement by unmethylated normal progenitors.

Other non-monitored genes are also likely targets of hypomethylation by decitabine. Furthermore, decitabine may also induce responses by mechanisms other than demethylation.

p15INK4B expression levels were also analyzed as a correlative investigation during a phase II study of decitabine in patients with MDS [75]. Basal p15INK4B levels were similar in patients in the three treated groups. But expression increased more in either 20mg/m^2 IV dosing schedule than in either the 20 mg/m^2 SC or 10mg/m^2 IV. Furthermore, levels of p15INK4B were higher in the responders than non-responders at baseline, days 5 and12 therapy, as well as, at recovery of hematopoiesis, suggesting that p15INK4B levels may be a better predictor of clinical response than amount of p15INK4B methylation. The level of expression increased after treatment in both responders and non-responders, but more significant in the responders.

Expression levels of p21WAF1 have been examined in several decitabine studies. Scott et al. found that p21WAF1 mRNA was undetectable in 6 of 24 AML patient samples, irrespective of the levels of p21WAF1 promoter methylation [84]. Decitabine induced p21WAF1 expression in the AML cell line, KG-1 and subline KG-1a even though the p21WAF1 promoter was not methylated; the investigators explained the results by showing that the drug blocked HDAC1 (histone deacetylase 1) and increased acetylated histone H3 at the unmethylated p21WAF1 promoter. Our study also showed a similar result (Kumagai, IJC, in press).

Several human tumor antigens have been identified since the initial discovery of the MAGE-A antigen expressed on melanoma cells [85,86]. These antigens are categorized into various classes including differentiation antigens (*e.g.*, tyrosinase and MelanA/MART-1), antigens derived from overexpressioned/amplificationed genes (*e.g.*, HER-2/*neu*), or mutation (*e.g.*, p53), cancer/testis (CT) antigens (*e.g.*, MAGE, GAGE and NY-ESO-1), and viral antigens (*e.g.*, human papillomavirus and EBV) [86]. Among them, the CT antigens have received particular attention as potential vaccine targets because of their unique tissue expression [87, 88]. We have shown that 5-aza-dC and suberoylanilide hydroxamic acid (SAHA) up-regulated the expression of several CT antigens on cancer cells [89]. CT antigens on the tumor cells are often silenced in cancer cells and may be reactivated by epigenetic therapy [90, 91]. Perhaps a demethylating agent like 5-aza-CdR and a histone deacetylase inhibitor may induce a T-cell-mediated antitumor activity against tumor cells.

A recent study demonstrated global changes in gene expression using microarray array analysis before and after exposure of an AML cell line, as well as, primary AML and MDS cells to decitabine both in vitro and in vivo [92]. Decitabine induced expression of about 80 of a possible 22,000 genes. The enhanced expression could be direct effect on target genes or secondary to inducing expression of previously methylated transcription factors which can then inactivate target genes. Also decitabine may have methylation-independent effects on gene expression.

Decitabine might exert a cytoreductive effect on the MDS clone by initiating growth inhibition, apoptosis and/or terminal differentiation. Recently, Lubbert et al. showed that re-established granulopoiesis in patients with MDS treated with low-dose decitabine was mainly due to the emergence of normal, non-clonal granulocytes rather than differentiation of the abnormal MDS clone [93].

Clinical Use of DNMT Inhibitors

As mentioned earlier, two DNMT inhibitors are FDA approved for use in the treatment of MDS patients. The formulations of azacitidine (Vizada, Pharmion) and decitabine (Dacogen, MGI Pharma) are 100 mg and 50 mg single-use vials, respectively. Dosage of azacitidine is 75 mg/m^2 SC once/day x 7days every 4 week administered for 4 or more cycles. Dose of decitabine is 15 mg/m^2 continuous iv over 3 hrs q8hr x3 /day every 6 week for at least 4 cycles. However, because of the inconvenience of administration, other approaches are being explored such as 20 mg/m^2 i.v over 1 hr for 5 days [11]. Costs of azacitidine and decitabine are $424.53 and $1,250 /vial, respectively. So the costs for one cycle for an individual with a body surface area of 1.7 m^2 are $5936 and $11,250, respectively [22].

Main adverse effects of these drugs include myelosuppression, nausea and vomiting. Diarrhea, constipation and fever are also common [22]. Administering of either drug to patients with cytopenias requires very close monitoring. Also, the cytopenia can be delayed as compared to typical cytotoxic effect. The safety of decitabine in patients with either hepatic or renal impairment requires further study.

Drug Resistance

Drug resistance to deoxycytidine analogues, such as decitabine (5-AZA-dCR), can occur by several different mechanisms [94] including deficiency of deoxycytidine kinase in the cells. Individuals with increase in cytidine deaminase will also have resistance to these drugs [95]. Another cause of resistance is an increase in the intracellular pool of dCTP which competes with 5-AZA-dCTP blocking the incorporation of the azo-compounds into DNA.

Conclusion

Two DNA hypomethylating agents (DNMT inhibitors) (5-AzaC and decitabine), have recently been approved for patients with MDS. Besides MDS, these compounds have activity against AML and CML cells. Decitabine may have milder adverse effect than 5-AzaC. The mechanism of action of these drugs in MDS is not fully understood, but they appear to be more useful at low rather than high doses, suggesting that inhibition of DNMT rather than a direct cytotoxic activity, may be clinically important. Optimization of drug dosage, duration of usage and delivering are being investigated. Inhibition of DNMT may activate tumor-suppressor genes through hypomethylation of their promoters leading to inhibition of tumor cell growth and cell repair, as well as, induction of terminal differentiation and/or apoptosis. Recently, clinical trials have begun exploring the combined use of a DNA hypomethylating agent and either a chemotherapeutic agent or an additional epigenetic therapeutic agent, especially histone deacetylase inhibitors. In summary, hypomethylating agents are novel therapies that are promising new therapeutic options for cancers especially hematological malignancies.

References

[1] Ducasse M, Brown MA. Epigenetic aberrations and cancer. *Mol. Cancer.* 2006;5:60. Review.

[2] Jones PA, Baylin SB. The fundamental role of epigenetic events in cancer. *Nat. Rev. Genet.* 2002 Jun;3(6):415-28. Review.

[3] Bird A. DNA methylation patterns and epigenetic memory. *Genes Dev.* 2002 Jan 1;16(1):6-21. Review.

[4] Momparler RL. Cancer epigenetics. *Oncogene.* 2003;22(42):6479-83. Review.

[5] Silverman LR, Mufti GJ. Methylation inhibitor therapy in the treatment of myelodysplastic syndrome. Nat Clin Pract Oncol. 2005 Dec;2 Suppl 1:S12-23. Review.

[6] Issa JP, Kantarjian H. Azacitidine. *Nat. Rev. Drug Discov.* 2005;Suppl:S6-7. Review.

[7] Bhalla KN. Epigenetic and chromatin modifiers as targeted therapy of hematologic malignancies. *J. Clin. Oncol.* 2005 ;23(17):3971-93. Epub 2005 May 16. Review.

[8] Heaney ML, Golde DW. Myelodysplasia. *N. Engl. J. Med.* 1999 May 27;340(21):1649-60. Review. No abstract available.

[9] Issa JP. Decitabine. *Curr. Opin. Oncol.* 2003;15(6):446-51. Review.

[10] Wijermans P, Lubbert M, Verhoef G, Bosly A, Ravoet C, Andre M, Ferrant A. Low-dose 5-aza-2'-deoxycytidine, a DNA hypomethylating agent, for the treatment of high-risk myelodysplastic syndrome: a multicenter phase II study in elderly patients. *J. Clin. Oncol.* 2000 ;18(5):956-62.

[11] Kantarjian H, Oki Y, Garcia-Manero G, Huang X, O'brien S, Cortes J, Faderl S, Bueso-Ramos C, Ravandi F, Estrov Z, Ferrajoli A, Wierda W, Shan J, Davis J, Giles F, Saba HI, Issa JP. Results of a randomized study of three schedules of low-dose decitabine in higher risk myelodysplastic syndrome and chronic myelomonocytic leukemia. *Blood.* 2006 Aug 1; [Epub ahead of print], in press

[12] Kantarjian H, Issa JP, Rosenfeld CS, Bennett JM, Albitar M, DiPersio J, Klimek V, Slack J, de Castro C, Ravandi F, Helmer R 3rd, Shen L, Nimer SD, Leavitt R, Raza A, Saba H. Decitabine improves patient outcomes in myelodysplastic syndromes: results of a phase III randomized study. *Cancer.* 2006;106(8):1794-803.

[13] Momparler RL, Rivard GE, Gyger M. Clinical trial on 5-aza-2'-deoxycytidine in patients with acute leukemia. *Pharmacol. Ther.* 1985;30(3):277-86.

[14] Petti MC, Mandelli F, Zagonel V, De Gregoris C, Merola MC, Latagliata R, Gattei V, Fazi P, Monfardini S, Pinto A. Pilot study of 5-aza-2'-deoxycytidine (Decitabine) in the treatment of poor prognosis acute myelogenous leukemia patients: preliminary results. *Leukemia.* 1993 ;7 Suppl 1:36-41.

[15] Kantarjian HM, O'Brien SM, Estey E, Giralt S, Beran M, Rios MB, Keating M, de Vos D, Talpaz M. Decitabine studies in chronic and acute myelogenous leukemia. *Leukemia.* 1997;11 Suppl 1:S35-6.

[16] Issa JP, Garcia-Manero G, Giles FJ, Mannari R, Thomas D, Faderl S, Bayar E, Lyons J, Rosenfeld CS, Cortes J, Kantarjian HM. Phase 1 study of low-dose prolonged exposure schedules of the hypomethylating agent 5-aza-2'-deoxycytidine (decitabine) in hematopoietic malignancies. *Blood.* 2004;103(5):1635-40.

[17] Kantarjian HM, O'Brien S, Cortes J, Giles FJ, Faderl S, Issa JP, Garcia-Manero G, Rios MB, Shan J, Andreeff M, Keating M, Talpaz M. Results of decitabine (5-aza-2'deoxycytidine) therapy in 130 patients with chronic myelogenous leukemia. *Cancer.* 2003 ;98(3):522-8.

[18] Momparler RL. Epigenetic therapy of cancer with 5-aza-2'-deoxycytidine (decitabine). *Semin. Oncol.* 2005 ;32(5):443-51. Review.

[19] Oki Y, Aoki E, Issa JP. Decitabine-Bedside to bench. *Crit. Rev. Oncol. Hematol.* 2007;61(2):140-52.

[20] Kuykendall JR. 5-azacytidine and decitabine monotherapies of myelodysplastic disorders. *Ann. Pharmacother.* 2005;39(10):1700-9. Review.

[21] Ruter B, Wijermans PW, Lubbert M. DNA methylation as a therapeutic target in hematologic disorders: recent results in older patients with myelodysplasia and acute myeloid leukemia. *Int. J. Hematol.* 2004;80(2):128-35. Review.

[22] [No authors listed] Decitabine (Dacogen) for myelodysplastic syndromes. *Med. Lett. Drugs Ther.* 2006;48(1247):91-2.

[23] Gore SD, Jones C, Kirkpatrick P. Decitabine. *Nat. Rev. Drug Discov.* 2006; 5(11):891-2.

[24] Herman JG, Baylin SB. Gene silencing in cancer in association with promoter hypermethylation. *N. Engl. J. Med.* 2003;349(21):2042-54. Review.

[25] Bird AP. CpG-rich islands and the function of DNA methylation. *Nature.* 1986 May 15-21;321(6067):209-13. Review.

[26] Cross SH, Bird AP. CpG islands and genes. *Curr. Opin. Genet. Dev.* 1995 Jun;5(3):309-14. Review

[27] Gardiner-Garden M, Frommer M. CpG islands in vertebrate genomes. *J. Mol. Biol.* 1987;196(2):261-82.

[28] Graff JR, Herman JG, Myohanen S, Baylin SB, Vertino PM. Mapping patterns of CpG island methylation in normal and neoplastic cells implicates both upstream and downstream regions in de novo methylation. *J. Biol. Chem.* 1997;272(35):22322-9.

[29] Baylin SB, Herman JG, Graff JR, Vertino PM, Issa JP. Alterations in DNA methylation: a fundamental aspect of neoplasia. *Adv. Cancer Res.* 1998;72:141-96. Review.

[30] Barlow DP. Gametic imprinting in mammals. *Science.* 1995;270(5242):1610-3. Review.

[31] Goto T, Monk M. Regulation of X-chromosome inactivation in development in mice and humans. *Microbiol. Mol. Biol. Rev.* 1998 ;62(2):362-78. Review.:

[32] Bird AP, Wolffe AP. Methylation-induced repression--belts, braces, and chromatin. *Cell.* 1999 ;99(5):451-4. Review.

[33] Jones PL, Veenstra GJ, Wade PA, Vermaak D, Kass SU, Landsberger N, Strouboulis J, Wolffe AP. Methylated DNA and MeCP2 recruit histone deacetylase to repress transcription. *Nat. Genet.* 1998;19(2):187-91.

[34] Jones PL, Wolffe AP. Relationships between chromatin organization and DNA methylation in determining gene expression. *Semin. Cancer Biol.* 1999;9(5):339-47. Review.

[35] Jaenisch R, Bird A. Epigenetic regulation of gene expression: how the genome integrates intrinsic and environmental signals. *Nat. Genet.* 2003 Mar;33 Suppl:245-54. Review.

[36] Kouzarides T. Histone methylation in transcriptional control. *Curr. Opin. Genet. Dev.* 2002 Apr;12(2):198-209. Review. Erratum in: *Curr. Opin. Genet. Dev.* 2002;12(3):371.

[37] Herman JG, Jen J, Merlo A, Baylin SB. Hypermethylation-associated inactivation indicates a tumor suppressor role for p15INK4B. *Cancer Res.* 1996;56(4):722-7.

[38] Otterson GA, Khleif SN, Chen W, Coxon AB, Kaye FJ. CDKN2 gene silencing in lung cancer by DNA hypermethylation and kinetics of p16INK4 protein induction by 5-aza 2'deoxycytidine. *Oncogene.* 1995;11(6):1211-6.

[39] Graff JR, Herman JG, Lapidus RG, Chopra H, Xu R, Jarrard DF, Isaacs WB, Pitha PM, Davidson NE, Baylin SB. E-cadherin expression is silenced by DNA hypermethylation in human breast and prostate carcinomas. *Cancer Res.* 1995 ;55(22):5195-9.

[40] Cote S, Sinnett D, Momparler RL. Demethylation by 5-aza-2'-deoxycytidine of specific 5-methylcytosine sites in the promoter region of the retinoic acid receptor beta gene in human colon carcinoma cells. *Anticancer Drugs.* 1998;9(9):743-50.

[41] Teitz T, Wei T, Valentine MB, Vanin EF, Grenet J, Valentine VA, Behm FG, Look AT, Lahti JM, Kidd VJ. Caspase 8 is deleted or silenced preferentially in childhood neuroblastomas with amplification of MYCN. *Nat. Med.* 2000;6(5):529-35.

[42] Conway KE, McConnell BB, Bowring CE, Donald CD, Warren ST, Vertino PM. TMS1, a novel proapoptotic caspase recruitment domain protein, is a target of methylation-induced gene silencing in human breast cancers. *Cancer Res.* 2000;60(22):6236-42.

[43] Li Q, Ahuja N, Burger PC, Issa JP. Methylation and silencing of the Thrombospondin-1 promoter in human cancer. *Oncogene.* 1999 May 27;18(21):3284-9.

[44] Herman JG, Umar A, Polyak K, Graff JR, Ahuja N, Issa JP, Markowitz S, Willson JK, Hamilton SR, Kinzler KW, Kane MF, Kolodner RD, Vogelstein B, Kunkel TA, Baylin SB. Incidence and functional consequences of hMLH1 promoter hypermethylation in colorectal carcinoma. *Proc. Natl. Acad. Sci. U S A.* 1998 ;95(12):6870-5.

[45] Ferguson AT, Lapidus RG, Baylin SB, Davidson NE. Demethylation of the estrogen receptor gene in estrogen receptor-negative breast cancer cells can reactivate estrogen receptor gene expression. *Cancer Res.* 1995;55(11):2279-83.

[46] Weber J, Salgaller M, Samid D, Johnson B, Herlyn M, Lassam N, Treisman J, Rosenberg SA. Expression of the MAGE-1 tumor antigen is up-regulated by the demethylating agent 5-aza-2'-deoxycytidine. *Cancer Res.* 1994;54(7):1766-71.

[47] Sigalotti L, Coral S, Nardi G, Spessotto A, Cortini E, Cattarossi I, Colizzi F, Altomonte M, Maio M. Promoter methylation controls the expression of MAGE2, 3 and 4 genes in human cutaneous melanoma. *J. Immunother.* 2002 Jan-Feb;25(1):16-26.

[48] Balch C, Montgomery JS, Paik HI, Kim S, Kim S, Huang TH, Nephew KP. New anti-cancer strategies: epigenetic therapies and biomarkers. *Front. Biosci.* 2005;10:1897-931. Review.

[49] Hanahan D, Weinberg RA. The hallmarks of cancer. *Cell.* 2000 Jan 7;100(1):57-70. Review. No abstract available.

[50] Santini V, Kantarjian HM, Issa JP. Changes in DNA methylation in neoplasia: pathophysiology and therapeutic implications. *Ann. Intern. Med.* 2001;134(7):573-86. Review.

[51] Nguyen TT, Mohrbacher AF, Tsai YC, Groffen J, Heisterkamp N, Nichols PW, Yu MC, Lubbert M, Jones PA. Quantitative measure of c-abl and p15 methylation in chronic myelogenous leukemia: biological implications. *Blood.* 2000;95(9):2990-2.

[52] Roman-Gomez J, Jimenez-Velasco A, Castillejo JA, Agirre X, Barrios M, Navarro G, Molina FJ, Calasanz MJ, Prosper F, Heiniger A, Torres A. Promoter hypermethylation of cancer-related genes: a strong independent prognostic factor in acute lymphoblastic leukemia. *Blood.* 2004;104(8):2492-8.

[53] Bouchard J, Momparler RL. Incorporation of 5-Aza-2'-deoxycytidine-5'-triphosphate into DNA. Interactions with mammalian DNA polymerase alpha and DNA methylase. *Mol. Pharmacol.* 1983;24(1):109-14.

[54] Juttermann R, Li E, Jaenisch R. Toxicity of 5-aza-2'-deoxycytidine to mammalian cells is mediated primarily by covalent trapping of DNA methyltransferase rather than DNA demethylation. *Proc. Natl. Acad. Sci. U S A.* 1994;91(25):11797-801.

[55] Aparicio A, Weber JS. Review of the clinical experience with 5-azacytidine and 5-aza-2'-deoxycytidine in solid tumors. *Curr. Opin. Investig Drugs.* 2002;3(4):627-33. Review.

[56] Rivard GE, Momparler RL, Demers J, Benoit P, Raymond R, Lin K, Momparler LF. Phase I study on 5-aza-2'-deoxycytidine in children with acute leukemia. *Leuk. Res.* 1981;5(6):453-62.

[57] Willemze R, Suciu S, Archimbaud E, Muus P, Stryckmans P, Louwagie EA, Berneman Z, Tjean M, Wijermans P, Dohner H, Jehn U, Labar B, Jaksic B, Dardenne M, Zittoun R. A randomized phase II study on the effects of 5-Aza-2'-deoxycytidine combined with either amsacrine or idarubicin in patients with relapsed acute leukemia: an EORTC Leukemia Cooperative Group phase II study (06893). *Leukemia.* 1997 Mar;11 Suppl 1:S24-7.

[58] Jones PA, Taylor SM. Cellular differentiation, cytidine analogs and DNA methylation. *Cell.* 1980;20(1):85-93.

[59] Pinto A, Attadia V, Fusco A, Ferrara F, Spada OA, Di Fiore PP. 5-Aza-2'-deoxycytidine induces terminal differentiation of leukemic blasts from patients with acute myeloid leukemias. *Blood.* 1984;64(4):922-9.

[60] Silverman LR, Demakos EP, Peterson BL, Kornblith AB, Holland JC, Odchimar-Reissig R, Stone RM, Nelson D, Powell BL, DeCastro CM, Ellerton J, Larson RA, Schiffer CA, Holland JF. Randomized controlled trial of azacitidine in patients with the myelodysplastic syndrome: a study of the cancer and leukemia group B. *J. Clin. Oncol.* 2002;20(10):2429-40.

[61] Kornblith AB, Herndon JE 2nd, Silverman LR, Demakos EP, Odchimar-Reissig R, Holland JF, Powell BL, DeCastro C, Ellerton J, Larson RA, Schiffer CA, Holland JC. Impact of azacytidine on the quality of life of patients with myelodysplastic syndrome treated in a randomized phase III trial: a Cancer and Leukemia Group B study. *J. Clin. Oncol.* 2002 May 15;20(10):2441-52.

[62] Kaminskas E, Farrell A, Abraham S, Baird A, Hsieh LS, Lee SL, Leighton JK, Patel H, Rahman A, Sridhara R, Wang YC, Pazdur R; FDA. Approval summary: azacitidine for treatment of myelodysplastic syndrome subtypes. *Clin. Cancer Res.* 2005;11(10):3604-8.

[63] Pinto A, Zagonel V. 5-Aza-2'-deoxycytidine (Decitabine) and 5-azacytidine in the treatment of acute myeloid leukemias and myelodysplastic syndromes: past, present and future trends. *Leukemia.* 1993;7 Suppl 1:51-60. Review.

[64] Zagonel V, Lo Re G, Marotta G, Babare R, Sardeo G, Gattei V, De Angelis V, Monfardini S, Pinto A. 5-Aza-2'-deoxycytidine (Decitabine) induces trilineage response in unfavourable myelodysplastic syndromes. *Leukemia.* 1993;7 Suppl 1:30-5.

[65] Wijermans PW, Krulder JW, Huijgens PC, Neve P. Continuous infusion of low-dose 5-Aza-2'-deoxycytidine in elderly patients with high-risk myelodysplastic syndrome. *Leukemia.* 1997;11(1):1-5.

[66] Greenberg P, Cox C, LeBeau MM, Fenaux P, Morel P, Sanz G, Sanz M, Vallespi T, Hamblin T, Oscier D, Ohyashiki K, Toyama K, Aul C, Mufti G, Bennett J. International scoring system for evaluating prognosis in myelodysplastic syndromes. Blood. 1997;89(6):2079-88. Erratum in: *Blood.* 1998 ;91(3):1100.

[67] Cheson BD, Greenberg PL, Bennett JM, Lowenberg B, Wijermans PW, Nimer SD, Pinto A, Beran M, de Witte TM, Stone RM, Mittelman M, Sanz GF, Gore SD, Schiffer CA, Kantarjian H. Clinical application and proposal for modification of the International Working Group (IWG) response criteria in myelodysplasia. *Blood.* 2006;108(2):419-25.

[68] http://www.fda.gov/cder/Offices/OODP/whatsnew/decitabine.htm

[69] Issa JP, Gharibyan V, Cortes J, Jelinek J, Morris G, Verstovsek S, Talpaz M, Garcia-Manero G, Kantarjian HM. Phase II study of low-dose decitabine in patients with chronic myelogenous leukemia resistant to imatinib mesylate. *J. Clin. Oncol.* 2005;23(17):3948-56.

[70] Cameron EE, Bachman KE, Myohanen S, Herman JG, Baylin SB. Synergy of demethylation and histone deacetylase inhibition in the re-expression of genes silenced in cancer. *Nat. Genet.* 1999;21(1):103-7.

[71] Garcia-Manero G, Kantarjian HM, Sanchez-Gonzalez B, Yang H, Rosner G, Verstovsek S, Rytting M, Wierda WG, Ravandi F, Koller C, Xiao L, Faderl S, Estrov Z, Cortes J, O'brien S, Estey E, Bueso-Ramos C, Fiorentino J, Jabbour E, Issa JP. Phase 1/2 study of the combination of 5-aza-2'-deoxycytidine with valproic acid in patients with leukemia. *Blood.* 2006;108(10):3271-9.

[72] Gore SD, Baylin S, Sugar E, Carraway H, Miller CB, Carducci M, Grever M, Galm O, Dauses T, Karp JE, Rudek MA, Zhao M, Smith BD, Manning J, Jiemjit A, Dover G, Mays A, Zwiebel J, Murgo A, Weng LJ, Herman JG. Combined DNA methyltransferase and histone deacetylase inhibition in the treatment of myeloid neoplasms. *Cancer Res.* 2006;66(12):6361-9.

[73] Balch C, Montgomery JS, Paik HI, Kim S, Kim S, Huang TH, Nephew KP. New anti-cancer strategies: epigenetic therapies and biomarkers. *Front. Biosci.* 2005 May 1;10:1897-931. Review.

[74] Yang AS, Estecio MR, Doshi K, Kondo Y, Tajara EH, Issa JP. A simple method for estimating global DNA methylation using bisulfite PCR of repetitive DNA elements. *Nucleic Acids Res.* 2004 Feb 18;32(3):e38.

[75] Oki Y, Kantajian HM, Davis J, Ravandi-Kashani F, Verstovsek S, Cortes JE et al. Hypomethylation induction in MDS after treatment with decitabine at three different doses. *J. Clin. Oncol.* 2005; 23(16S) Abs.6546

[76] Mund C, Hackanson B, Stresemann C, Lubbert M, Lyko F. Characterization of DNA demethylation effects induced by 5-Aza-2'-deoxycytidine in patients with myelodysplastic syndrome. *Cancer Res.* 2005;65(16):7086-90.

[77] Oki Y, Kantajian HM, O'Bean S, Verstovsek S, Gharibyan V, Cortes JE, et al. Phase II study of decitabine in combination with imanitib mesylate in patients with accelelated (AP) or blastic phase (BP) of chronic myeloid leukemia. *Blood.* 2005; 106(11) Abs. 1099

[78] Batova A, Diccianni MB, Yu JC, Nobori T, Link MP, Pullen J, Yu AL. Frequent and selective methylation of p15 and deletion of both p15 and p16 in T-cell acute lymphoblastic leukemia. *Cancer Res.* 1997;57(5):832-6.

[79] Uchida T, Kinoshita T, Nagai H, Nakahara Y, Saito H, Hotta T, Murate T. Hypermethylation of the p15INK4B gene in myelodysplastic syndromes. *Blood.* 1997 ;90(4):1403-9.

[80] Aoki E, Uchida T, Ohashi H, Nagai H, Murase T, Ichikawa A, Yamao K, Hotta T, Kinoshita T, Saito H, Murate T. Methylation status of the p15INK4B gene in hematopoietic progenitors and peripheral blood cells in myelodysplastic syndromes. *Leukemia.* 2000;14(4):586-93.

[81] Daskalakis M, Nguyen TT, Nguyen C, Guldberg P, Kohler G, Wijermans P, Jones PA, Lubbert M. Demethylation of a hypermethylated P15/INK4B gene in patients with myelodysplastic syndrome by 5-Aza-2'-deoxycytidine (decitabine) treatment. *Blood.* 2002;100(8):2957-64.

[82] Yang AS, Doshi KD, Choi SW, Mason JB, Mannari RK, Gharybian V, Luna R, Rashid A, Shen L, Estecio MR, Kantarjian HM, Garcia-Manero G, Issa JP. DNA methylation changes after 5-aza-2'-deoxycytidine therapy in patients with leukemia. *Cancer Res.* 2006;66(10):5495-503.

[83] Shen L, Kantajian H, Saba H, Lin E, Berry D, Ahmed S, Jelimec J, Issa J. CpG island methylation is a poor prognostic factors in myelodysplastic syndrome patients and is reversed by decitabine therapy – results of phase III randomized study. *Blood.* 2005 106(11.) Abs 790

[84] Scott SA, Dong WF, Ichinohasama R, Hirsch C, Sheridan D, Sanche SE, Geyer CR, Decoteau JF. 5-Aza-2'-deoxycytidine (decitabine) can relieve p21WAF1 repression in human acute myeloid leukemia by a mechanism involving release of histone deacetylase 1 (HDAC1) without requiring p21WAF1 promoter demethylation. *Leuk. Res.* 2006;30(1):69-76.

[85] van der Bruggen P, Traversari C, Chomez P, Lurquin C, De Plaen E, Van den Eynde B, Knuth A, Boon T. A gene encoding an antigen recognized by cytolytic T lymphocytes on a human melanoma. *Science.* 1991;254(5038):1643-7.

[86] Jager E, Jager D, Knuth A. Clinical cancer vaccine trials. *Curr. Opin. Immunol.* 2002;14(2):178-82. Review.

[87] Chen YT, Old LJ. Cancer-testis antigens: targets for cancer immunotherapy. *Cancer J. Sci. Am.* 1999;5(1):16-7. Review. No abstract available.

[88] Jager E, Stockert E, Zidianakis Z, Chen YT, Karbach J, Jager D, Arand M, Ritter G, Old LJ, Knuth A. Humoral immune responses of cancer patients against "Cancer-Testis" antigen NY-ESO-1: correlation with clinical events. *Int. J. Cancer.* 1999 Oct 22;84(5):506-10.

[89] Kumagai T, Takai N, Chumakov A, Desmond J.C, Kawamata N, Gery S, Miller C.W, Kitano A, Suzuki T, Imai Y, Koeffler H.P C/EBPα; a new candidate tumor suppressor gene as a molecular target in pancreatic cancer. *American Association of Cancer Research.* 2006, Abs 3318

[90] Wischnewski F, Pantel K, Schwarzenbach H. Promoter Demethylation and Histone Acetylation Mediate Gene Expression of MAGE-A1, -A2, -A3, and -A12 in Human Cancer Cells. *Mol. Cancer Res.* 2006; 4(5): 339-49.

[91] Claus R, Almstedt M, Lubbert M. Epigenetic treatment of hematopoietic malignancies: in vivo targets of demethylating agents. *Semin. Oncol.* 2005;32(5):511-20. Review.

[92] Tamm I, Sattler N, Wagner M, Lubbert M, leCourtre P, Dorken B, et al. Decitabine: where is the target? *Blood.* 2005; 106(11) Abs 495

[93] Lubbert M, Daskalakis M, Kunzmann R, Engelhardt M, Guo Y, Wijermans P. Nonclonal neutrophil responses after successful treatment of myelodysplasia with low-dose 5-aza-2'-deoxycytidine (decitabine). *Leuk. Res.* 2004;28(12):1267-71.

[94] Momparler RL. Pharmacology of 5-Aza-2'-deoxycytidine (decitabine). *Semin. Hematol.* 2005;42(3 Suppl 2):S9-16. Review.

[95] Eliopoulos N, Cournoyer D, Momparler RL. Drug resistance to 5-aza-2'-deoxycytidine, 2',2'-difluorodeoxycytidine, and cytosine arabinoside conferred by retroviral-mediated transfer of human cytidine deaminase cDNA into murine cells. *Cancer Chemother. Pharmacol.* 1998;42(5):373-8.

[96] Kantarjian HM, O'brien S, Shan J, Aribi A, Garcia-Manero G, Jabbour E, Ravandi F, Cortes J, Davisson J, Issa JP. Update of the decitabine experience in higher risk myelodysplastic syndrome and analysis of prognostic factors associated with outcome.

[97] Silverman LR, Holland JF, Weinberg RS, Alter BP, Davis RB, Ellison RR, Demakos EP, Cornell CJ Jr, Carey RW, Schiffer C, et al. Effects of treatment with 5-azacytidine on the in vivo and in vitro hematopoiesis in patients with myelodysplastic syndromes. *Leukemia.* 1993 May;7 Suppl 1:21-9.

In: DNA Methylation Research Trends
Editor: Taku B. Kobayashi, pp. 87-133
ISBN: 978-1-60021-721-0

Chapter V

Analysis of DNA Methylation Patterns for the Early Diagnosis, Classification and Therapy of Human Cancers

Jörg Tost*
Laboratoire d'Epigénétique, Centre National de Génotypage,
Bâtiment G2, 2 rue Gaston Crémieux, CP5721, 91057 Evry, France

Abstract

A cumulative series of genetic and epigenetic alterations leading to unregulated cell growth is the foundation of tumorigenesis. Changes in DNA methylation patterns occur early in the progression process and often precede malignancy. Global hypomethylation of the genome is observed and has been suggested to initiate and propagate oncogenesis by inducing chromosome instabilities and transcriptional activation of oncogenes. The overall decrease in DNA methylation is accompanied by a region- and gene-specific hypermethylation of CpG islands. Genes involved in DNA repair, detoxification, cell cycle regulation, and apoptosis are often inappropriately inactivated. It has been suggested that epigenetic lesions in normal tissue set the stage for neoplasia. Recent reports of DNA methylation analyses carried out on tumor biopsies and matched DNA samples extracted from body fluids such as serum, plasma, urine and sputum have demonstrated a high level of concordance. Sensitive technologies detecting low levels of aberrant promoter methylation in easily accessible biological material are therefore of great interest for the early detection of cancer, which is the most efficient way to improve survival of patients, as well as for the detection of cancer recurrence. Methylation patterns can be shared by different types of tumors as well as being tumor type-specific and the extent of hypo- and hypermethylation often correlates with the malignancy grade and/or the disease stage. Classifications based on the methylation profile are able to predict the metastatic potential of a cancer as well as the response to a specific anticancer drug or therapy. As DNA methylation is a non-mutational and therefore reversible modification, it can be used as point of departure for anticancer treatment by chemically

* Jörg Tost : tost@cng.fr

or antisense oligonucleotides induced demethylation. Inhibitors of DNA methyltransferases such as the FDA approved 5-azacytidine or 5-aza-2'deoxycytidine and their analogues are incorporated into *de novo* synthesized DNA strands during replication. In promising clinical trials, they have been shown to induce stable re-expression of some previously epigenetic silenced tumor suppressor genes in hematological malignancies and to sensitize solid tumors to conventional cytotoxic drugs. DNA methylation based biomarkers promise to have a large impact on cancer diagnostics and treatment in the near future.

1. Introduction

A cumulative series of genetic and epigenetic alterations leading to unregulated cell growth is the foundation of tumorigenesis and gives rise to a cancer-specific gene expression signature. Genetic changes encompass point mutations, gene deletions as well as amplifications, chromosomal rearrangements (translocations), loss of heterozygosity and aneuploidy. These changes can be either inherited or acquired during the multistep progression from a normal to a cancerous cell. Epigenetic changes are stable modifications of gene expression in a certain cell that are heritable during somatic cell divisions (and sometimes germ line transmissions) but do not involve changes of the DNA sequence itself. Epigenetic phenomena are mediated by several molecular mechanisms comprising histone modifications, polycomb/trithorax protein complexes, small non-coding or antisense RNAs and DNA methylation. These different modifications are closely interconnected, but in this review I will focus only on the most stable and best studied epigenetic modification in cancer: DNA methylation.

1.1. DNA Methylation

DNA methylation is the only genetically programmed DNA modification in mammals. This post-replication modification is almost exclusively found on the 5 position of cytosines in the context of the dinucleotide sequence CpG (Bird, 2002). 5-methylcytosine accounts for ~1 % of all bases, varying slightly in different tissue types (Ehrlich et al., 1982). The CpG dinucleotide is underrepresented in the genome, probably because it is a mutation hotspot resulting in depletion during evolution (Holliday and Grigg, 1993) and is generally methylated. Relatively CpG rich clusters of approximately 1-2 kb length - CpG islands - are found in the promoter region and first exons of many genes, and are mostly not methylated (Bird, 2002). However, it should be noted that a growing number of CpG islands is identified that are methylated in normal tissues. It is hypothesized that 60 % of human genes are regulated by DNA methylation. This regulation is controlled in a tissue- and developmental-stage specific manner and is maintained through the life on an individual with some changes with age and in response to environmental effects such as diet. So far 5 DNA methyltransferases have been identified (DNMT1, DNMT2, DNMT3A, DNMT3B and DNMT3L) (Bestor, 2000). They catalyze the transfer of a methyl group from S-adenosyl-L-methionine to the cytosine base. Simplified, DNMT1 acts as maintenance methyltransferase thus passing the epigenetic information through cell generations as it prefers hemimethylated templates. *De novo* methylation is carried out by the methyltransferases DNMT3A and DNMT3B. Possible trigger mechanisms to initiate *de novo* methylation include preferred target DNA sequences, RNA interference and certain chromatin structures induced by histone modifications (Bird, 2002).

Cytosine methylation is essential for mammalian embryogenesis and development, where methylation levels change dynamically (Reik et al., 2001). Cytosine methylation is critical for imprinting, i.e. the asymmetric expression of either the maternal or paternal allele in a parent-of-origin-specific manner in somatic cells of the offspring (Reik and Walter, 2001) and is implicated in X chromosome inactivation (Chang et al., 2006). It plays a key role in the maintenance of genome integrity by transcriptional silencing of repetitive DNA sequences and endogenous transposons. Aberrant methylation patterns have been reported in various neurodevelopmental disorders and imprinting anomalies lead to disorders such as Prader-Willi, Angelman and Beckwith-Wiedemann syndrome (Robertson, 2005). A possible role for epigenetic events in complex, non Mendelian disorders like type 2 diabetes has been postulated (Schumacher and Petronis, 2006).

1.2. DNA Methylation Changes in Cancer

Cancer is probably the best studied disease with a strong epigenetic component (Jones and Baylin, 2002; Laird, 2005). In tumors a global loss of DNA methylation (hypomethylation) of the genome is observed (Feinberg and Vogelstein, 1983) and has been suggested to initiate and propagate oncogenesis by inducing chromosome instabilities and transcriptional activation of oncogenes and pro-metastatic genes (Ehrlich, 2002). The overall decrease in DNA methylation is accompanied by a region- and gene-specific increase of methylation (hypermethylation) of multiple CpG islands (figure 1, Jones and Baylin, 2002; Laird, 2005). Hypermethylation of CpG islands in the promoter region of a tumor suppressor or otherwise cancer-related gene is often associated with transcriptional silencing of the associated gene. The number of gene-associated promoters that are known to become hypermethylated during carcinogenesis is rapidly growing. Genes of numerous pathways involved in signal transduction (*APC*), DNA repair (*MGMT, MLH1, BRCA1*), detoxification (*GSTP1*), cell cycle regulation (*p15, p16, RB*), differentiation (*MYOD1*), angiogenesis (*THBS1, VHL*) and apoptosis (*Caspases, p14, DAPK*) are often inappropriately inactivated by DNA methylation. It should be noted that so far no single gene has been identified that is always methylated in a certain type of cancer. Both hypo- and hypermethylation are found in the same tumor, but the underlying mechanisms for both phenomena have not yet been elucidated. While the contribution of genetic factors to carcinogenesis such as the high-penetrance germ line mutations in genes such as *BRCA1* and *p53* in familial cancers has long been recognized, it has become evident that epigenetic changes leading to transcriptional silencing of tumor suppressor genes constitute an at least equally contributing mechanism. For example, microarray expression profiles of breast tumors with *BRCA1* mutations are very similar to those of sporadic breast cancer cases with *BRCA1* promoter hypermethylation demonstrating that disruption of *BRCA1* function by either genetic or epigenetic pathways leads to the same perturbations (Hedenfalk et al., 2001). Epigenetic changes occur at higher frequency compared to genetic changes and maybe especially important in early-stage human neoplasia. They often precede malignancy as extensive CpG island hypermethylation can be detected in benign polyps of the colon, in low as well as in high grade tumors (Costello et al., 2000; Goelz et al., 1985). It has therefore been suggested that epigenetic lesions in normal tissue set the stage for neoplasia. DNA hypermethylation could for example not only be detected in dysplastic epithelium of patients with ulcerative colitis, a condition associated with an increased risk for the development of colon cancer, but already in pathological normal epithelium (Issa et al., 2001). Aberrant DNA methylation patterns are therefore probably not a consequence or by-product of malignancy. It has been extrapolated that aberrant promoter methylation is initiated at ~1 % of all CpG islands and as much as 10 % become methylated during the multistep process of tumorigenesis (Costello et al., 2000). With the exception of haploinsufficient genes, 'two hits' are necessary to inactivate the two alleles of a gatekeeper tumor suppressor gene to enable oncogenic progression according to Knudson's two hit hypothesis (Knudson, 2001). DNA methylation can act as a hit having the same functional effect as a genetic mutation or deletion as proven by numerous experiments in which re-establishing expression of tumor suppressor genes could be achieved through drugs inducing demethylation. Epimutations can inactivate one of the two alleles while the other is lost by genetic mechanisms or silence both alleles (Balmain et al., 2003).

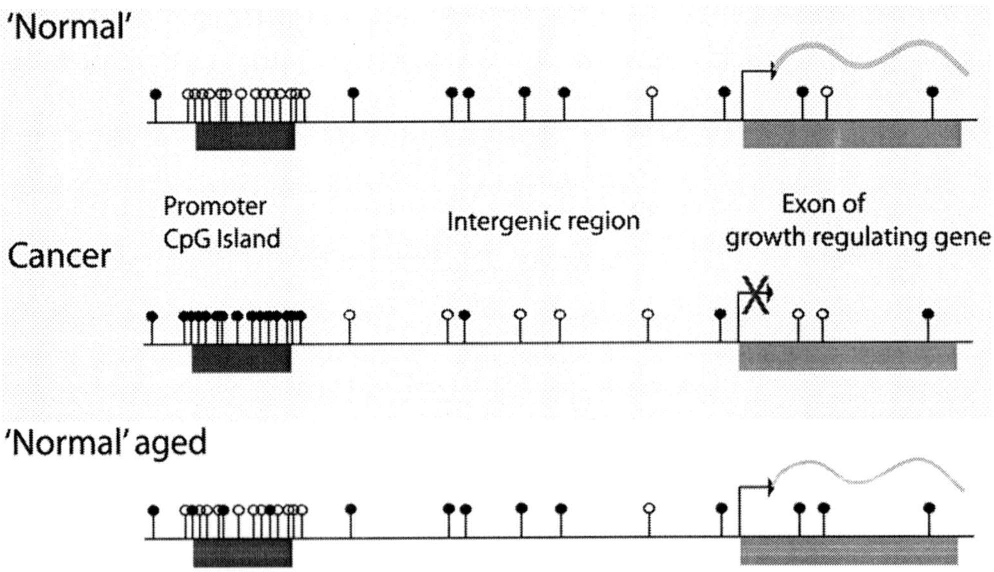

Figure 1. Distribution of DNA methylation in normal tissue, tumors and normal ageing tissue. Unmethylated CpGs are represented by empty lollipops, methylated ones by filled lollipops. Exons are shown as lightly shaded boxes, the preceding CpG island regulating its expression as darkly shaded box. Distances are not drawn to scale and in many cases the CpG island overlaps with the first exon. In normal tissues CpG islands are mostly unmethylated while exonic and intergenic CpG positions are methylated. During carcinogenesis the promoter CpG island gets methylated while methylation in the other regions is lost. Methylation patterns are dynamic and change during age and in response to environmental factors. Individual CpGs positions change their methylation status, which has stimulated the hypothesis that these age-related changes might predispose to cancer.

1.3. Where to Look at DNA Methylation Changes

Analysis of DNA methylation patterns in the field of cancer research aims at the identification of (tumor suppressor) genes that are aberrantly silenced by epigenetic mechanisms in human cancer and their correlation to clinical parameters such as response to chemotherapy or survival. Most of the DNA methylation changes during carcinogenesis occur within regions that do not seem to be directly implicated in the regulation of gene expression or influence genomic stability. Research has therefore mainly focused on the hypermethylation of promoter associated CpG islands where hypermethylation is inversely correlated to their transcriptional activity. Epigenetic inactivation is non-random but promoters of genes with similar function can display a different susceptibility to become methylated. While the mismatch repair protein *MLH1* gets frequently hypermethylated in various cancers leading to genetic instability, the repair gene *MSH2* is rarely epigenetically inactivated. This differential susceptibility could be due to intrinsic sequence properties (Feltus et al., 2006). The presence of methylation, however, does not necessarily induce silencing of nearby genes. Only when a specific core region of the promoter, that is often but not necessarily spanning the transcription start site, gets hypermethylated expression of the associated gene is modified. The methylation status at specific CpG dinucleotides in the core region might therefore better correlate with the expression of the gene than the overall methylation level of the entire CpG island. CpG islands are also found outside promoter regions and these appear generally to be more susceptible to aberrant methylation than the respective promoter sequences in various cancers as well as during cell culture (Nguyen et al., 2001; Ushijima et al., 2003). However, methylation of these CpG islands does usually not diminish transcription. One exception is the *MCJ* gene, where transcription depends on the methylation status of a CpG island starting in the first exon of the gene rather than on CpGs in the promoter region (Strathdee et al., 2004). It has been proposed that methylation begins in exonic regions and then progressively spreads to CpG islands in other locations including promoter regions (Nguyen et al., 2001). The exact mechanism is still to be elucidated but it could be that the protection from methylation is lost through the absence of transcription facilitating the access of DNMTs to the DNA as transcription factors and/or transcription/initiation complexes are absent. A second possibility would be the attraction of methylation through either recognition sequences or the association of proteins with methylated cytosines interfering with transcription factor binding sites (Bird, 2002). In some cases the methylation density in a promoter core region seems to be crucial to induce transcriptional silencing (Cameron et al., 1999b) while in other cases the demethylation of specific CpG sites is sufficient for transcriptional reactivation (Pao et al., 2001).

A new dimension has recently been added to epigenetic cancer research with the demonstration of long-range gene silencing by epigenetic modifications in various cancers. Epigenetic inactivation has previously been thought to be a gene-specific event. However, in colorectal cancer a large chromosomal region of 4 Mb on chromosome 2q14.2 is transcriptionally repressed by epigenetic remodeling including histone methylation and DNA methylation, whereas DNA methylation is concentrated in regions containing multiple CpG islands and spreads over regions of up to 1 Mb in size (Frigola et al., 2006). A similar situation is found in breast cancer where the entire *HOXA* gene cluster spanning ~100 kb is epigenetically inactivated probably through a mechanism dominated by DNA methylation (Novak et al., 2006). Long range epigenetic silencing seems to be a prevalent phenomenon during carcinogenesis as a recent survey identified 28 regions of copy-number independent transcriptional deregulation in bladder cancer that are potentially regulated through epigenetic mechanisms (Stransky et al., 2006).

Correlation between DNA methylation and gene inactivation is a prerequisite for the identification and validation of novel tumor suppressor genes. However, a large number of promoters get hypermethylated during carcinogenesis where there is no evidence that the corresponding gene acts a tumor suppressor gene. In this case DNA methylation might still be a useful biomarker for tumor diagnosis or risk assessment if the methylation pattern is specific for a certain tumor type and/or correlates with clinically important parameters. The same is true for DNA methylation in regions without a direct correlation between hypermethylation and gene silencing in which therefore methylation might not have direct functional consequences. A good example is the classic panel for the detection of the CpG island methylator phenotype defining a subtype of colorectal cancers with a distinct phenotype that comprises three MINT (Methylated IN Tumors) fragments (Issa, 2004). DNA methylation at these loci was identified through genome-wide scanning for tumor-specific methylation differences and subsequent cloning of differentially methylated sequences but their exact function and/or location in the genome is not always known.

The analysis of DNA methylation patterns is complicated by the fact that some changes are due to exposure to environmental influences as well as accumulation of DNA methylation at some promoters during aging (Feil, 2006). To be useful as biomarker for early diagnosis age-associated DNA methylation changes have therefore to be distinguished from cancer predisposing alterations.

1.4. How to Analyze DNA Methylation

For the analysis of DNA methylation sensitive and quantitative methods are needed to detect even subtle changes in the degree of methylation as biological samples often represent a heterogeneous mixture of different cells, especially tumor and non-tumor cells from tissue biopsies. Realizing the importance of epigenetic changes in development and disease a variety of techniques for the study of DNA methylation has been developed in the last years (Brena et al., 2006b; Fraga and Esteller, 2002; Laird, 2003). No single method has emerged as the "gold" standard technique unifying quantitative accuracy and high sensitivity, possibility for whole genome analysis and precise investigation of individual CpG positions. So the choice of the method mainly depends on the desired application.

Traditionally methylation patterns have been analyzed by digestion of genomic DNA with methylation sensitive restriction endonucleases and subsequent detection by Southern blotting (Bird and Southern, 1978) or PCR amplification (Singer-Sam et al., 1990). Restriction landmark genomic scanning (RLGS) permits the genome-wide quantitative assessment of epigenetic alterations between samples by digestion with a methylation sensitive enzyme and subsequent radiolabeling of the created cleavage sites (Rush and Plass, 2002). Differential methylation is then analyzed by comparing spot intensities on a two dimensional gel. Methods like methylated CpG island amplification (Toyota et al., 1999), methylation sensitive arbitrarily primed PCR (Gonzalgo et al., 1997) and differential methylation hybridization (DMH, Huang et al., 1999) compare genome-wide methylation patterns between two samples (or two pools of samples). They are based on the restriction digest of DNA with methylation sensitive enzymes, followed by size fragmentation, PCR amplification and comparative analysis of patterns on a gel or hybridization patterns to a genomic library of CpG islands. However, due to the dependence on restriction sites for methylation sensitive restriction enzymes only CpG sites found within these recognition sequences can be analyzed. A more recently developed approach isolates the methylated fraction of a genome by immunoprecipitation of methylated DNA with an antibody specific for 5-methylcytosine (Weber et al., 2005). Analysis can then either be performed specifically for a gene by quantitative real-time PCR or genome-wide on various microarrays. Tiling arrays enable the DNA methylation map of an entire genome at high resolution as recently demonstrated on the plant *Arabidopsis thaliana* with a 35 base pair resolution (Zhang et al., 2006).

Another approach to analyze DNA methylation patterns besides methylation sensitive restriction endonucleases relies on the conversion of genomic DNA extracted from a tissue or a body fluid with sodium bisulfite. This treatment permits to freeze the methylation status of a sample as methylation information would not be retained during PCR amplification. Sodium bisulfite deaminates unmethylated cytosines to uracils while methylated cytosines are resistant to conversion under the employed reaction conditions (Frommer et al., 1992). Following PCR amplification in which uracils are replaced by their DNA analogue thymine, the methylation status at a given position is manifested in the ratio C (former methylated cytosine) to T (former non-methylated cytosine) and can be analyzed as a virtual C/T polymorphism in the bisulfite treated DNA. A main advantage of bisulfite treatment is the possibility to amplify any target sequence by PCR thus requiring only a very small amount of input DNA. A wide range of technologies that are available of the quantitative analysis of single nucleotide polymorphisms (SNPs) have been adapted to the analysis of gene-specific investigation of DNA methylation patterns and levels.

Direct Sanger sequencing of PCR products amplified from bisulfite treated DNA can be used for detailed analysis of DNA methylation in long CpG stretches from small amounts of starting material (Clark et al., 1994; Frommer et al., 1992). However, direct sequencing has proved to be technically demanding and quantitative accuracy is limited in regions with heterogeneous methylation levels. Cloning and sequencing of multiple colonies is currently regarded as the reference method for the analysis of DNA methylation due to the richness of information provided and most novel methods are validated by comparison to the results obtained by sequencing and cloning. However, this approach gets rapidly expensive and time-consuming when applied in high-throughput. Pyrosequencing or MALDI mass spectrometry based fragmentation methods represent novel approaches for the simultaneous analysis and quantification of the methylation degree of several CpG positions in close proximity (Colella et al., 2003; Ehrich et al., 2005; Tost et al., 2003a; Tost et al., 2006). Methylation levels of small regions can be semi-quantitatively analyzed by techniques like denaturing high-performance liquid chromatography (dHPLC, (Deng et al., 2002)), methylation sensitive single-strand conformation analysis (Burri and Chaubert, 1999) or analysis of melting curves monitored in real time (Akey et al., 2002; Worm et al., 2001) making all use of the differential retention or migration behavior due to the sequence difference between former methylated and unmethylated molecules induced by bisulfite treatment. However, heterogeneous methylation levels might result in patterns difficult to interpret. Quantitative information on single CpG positions can be obtained by a variety of methods most of which have originally been developed for quantitative genotyping applications such as determination of allele frequencies in pooled samples. Combined bisulfite restriction analysis (COBRA) makes use of the sequence difference that are caused by bisulfite treatment between methylated and unmethylated positions and that might create or abolish recognition sites for restriction enzymes (Xiong and Laird, 1997). The original protocol for quantitation including polyacrylamide gel electrophoresis with subsequent electroblotting, hybridization with labeled oligonucleotides and phosphoimager quantitation, has recently been facilitated by the transfer of the assay to the BioAnalyzer platform (Brena et al., 2006a). Methylation sensitive single nucleotide primer extension (Ms-SNuPE) uses internal primers annealing to the PCR generated template and terminating immediately 5' of the CpG position to be assayed. Extension is carried out with (labeled) terminating dideoxynucleotides. Ms-SNuPE has been combined with radioactive (Gonzalgo and Jones, 1997), ion-pair reverse-phase HPLC (El-Maarri et al., 2002), MALDI mass spectrometry (Tost et al., 2003b), capillary electrophoresis (SNaPShot, Kaminsky et al., 2005; Uhlmann et al., 2002) or real-time allele-specific PCR (Thomassin et al., 2004) detection. As these methods interrogate only single or very few CpG sites in a multiplex reaction, they are well suited for the confirmation of a limited number of epigenetic markers in large cohorts but less well for the discovery of these markers. Recently an assay for the simultaneous analysis of 1536 CpGs based on Illumina's bead array platform was demonstrated permitting rapid screening for aberrant hypermethylation of over 800 genes in 96 samples in parallel (Bibikova et al., 2006).

However, for the below described analyses of DNA methylation patterns in body fluids extremely sensitive detection is required to specifically pick up methylated molecules derived from tumors in the presence of a large excess of unmethylated molecules. Analysis is in most studies performed using methylation-specific PCR (MSP), a technique permitting the amplification of virtually any CpG sites following bisulfite treatment with three pairs of primers for amplification, complementary to either the methylated or unmethylated allele as well as a control for complete bisulfite conversion (Herman et al., 1996). The detection limit can be further decreased while simultaneously the quantitative accuracy increased through the use of fluorescent primers. The measurement of amplification in real-time permits with the help of a reference gene the absolute quantification of the number of methylated molecules in a sample. Methods such as MethyLight based on the TaqMan^{TM} technology are able to detect a single hypermethylated allele against a background of 10.000 unmethylated alleles (Eads et al., 2000a). A probe labeled with a fluorescent and a quencher dye is hybridized to the template. If the probe is fully complementary, the probe is cleaved by the intrinsic 5'→3' exonucleolytic activity of Taq polymerase during amplification resulting in spatial separation of the two dyes and a fluorescent signal proportional to the amount of PCR product generated. By using an internal reference such as the β-actin gene sample inputs can be normalized and the percentage of methylation in the samples can be calculated (Eads et al., 2001). This avoids false positive results due to minimal levels of methylation that are probably without biological relevance and that might constitute a problem for conventional MSP. Several variants for the selective amplification of molecules with a specific methylation pattern have been devised such as QAMA, which uses a TaqMan probe conjugated to a minor groove binder improving specificity of the probes (Zeschnigk et al., 2004), or HeavyMethyl which increases still further the sensitivity and specificity of real-time PCR based assays using methylation dependent blocking oligonucleotides (Cottrell et al., 2004). Conventional gel-based methylation-specific PCR is probably still the most frequently used technology for DNA methylation detection due to its simplicity and the fact that it does not rely on any special instrumentation. However, the need for quantitative measurements becomes more and more acknowledged and methods like MethyLight or pyrosequencing become more widespread.

1.5. Epigenetic Changes at the DNA Level Are an Attractive Target for Diagnostics

Biomarkers capable of distinguishing cancerous cells from normal ones must be specific, sensitive and detectable in specimens obtained through minimally invasive procedures to be clinically applicable. Many biomarkers on the protein, RNA or DNA level fulfilling these criteria have been discovered. In clinical routine practice, most tumor diagnostic is carried out by biochemical assays determining the presence and/or quantity of enzymes, receptors, growth factors or hormones. Despite the wide-spread use of (microarray-based) RNA detection techniques in research facilities, there are some potential pitfalls associated with its potential use in routine clinical diagnostics such as the required preservation of mRNA from the tissue, tissue heterogeneity and the need for normalization. Attention to numerous details of sample extraction, storage and handling has to be paid to ensure intra- and interlaboratory reproducibility (Simon, 2005). DNA based molecular biomarkers can be more easily transferred from a research laboratory setting into routine diagnostics in a clinic due to the amplifiable and stable nature of DNA. Methylgroups on cytosines are part of the covalent structure of the DNA in contrast to other epigenetic marks such as chromatin. Once methylation is acquired it is in most cases chemically and biologically stable over time while expression of mRNA and/or proteins can also be modified by non-disease related environmental conditions and vary over the cell cycle. As by most methods the ratio between methylated and unmethylated CpGs is determined, DNA methylation analysis is independent of the total amount of starting material. It provides a binary and positive signal that can be detected independent of expression levels. It is therefore also easier to detect than negative signals like the analysis of loss of heterozygosity. If the core region of a CpG island in a promoter that is controlling transcriptional activity is defined, the stable DNA based analyte can be used as a proxy to monitor gene expression during treatment. DNA methylation can be analyzed with a growing number of methods that are amenable to high throughput and quantitative assays eliminate the need for normalization. They are applicable to formalin fixed paraffin embedded clinical specimens and other archived material. Epigenetic changes - similar to genetic alterations - lead to an altered phenotype of a certain cell conferring a selective advantage to those. However, in contrast to DNA based alterations such as point mutations that can theoretically occur at any position in the coding regions of a gene, DNA methylation changes are always confined to the same small regions of a gene (usually the promoter associated CpG island). Also, the assessment of the methylation status examines the lesion itself which is the epigenetic inactivation of the promoter rather than the effect of this alteration such as loss of protein expression or modified enzyme activity. A further advantage is the potential reversal of epigenetic changes by treatment with pharmacological agents, while genetic changes are irrevocable. One of the most important criteria for a clinically useful biomarker enabling screening of individuals at potential risk and monitoring of therapy response or disease recurrence is the analysis of the reliable biomarker in surrogate tissues such as blood or body fluids that can be obtained through minimal invasive procedures. Similar approaches based on the detection of RNA have been complicated by the inherent instability of these molecules and difficulties in detecting changes in the level of tumor derived RNA in the background of a large number of molecules derived from normal cells. The sensitive and specific detection of tumor-specific DNA methylation patterns at distal sites makes DNA methylation a biomarker of choice for the clinical management of cancer patients.

In the following I will briefly summarize the potential of DNA methylation as biomarker for early diagnosis and classification of cancer, its use for the detection and assessment of high-risk individuals and the possible use of DNA methylation profiles for the choice of treatment. Current knowledge on epigenetic therapy either with direct effects in hematological malignancies or as sensitizer for conventional therapy with cytotoxic drugs will be described.

2. Early Detection

Comparison of mortality rates for cancers for which population wide screening programs for early detection have been established such as colon and breast cancer with cancers for which so far no screening exists due to the absence of suitable biomarkers demonstrates clearly the efficacy and benefit of such programs. Survival rates are greatly improved if the disease is detected at an early stage with a locally limited disease and most common cancers can be cured at this stage. However, most traditional detection methods based on - for example cytological screening - do not provide the required sensitivity and modern methods for early detection such as spiral computed tomography recommended for lung and other cancers or endoscopy for gastric cancers are too expensive to be used on a population-wide scale. Tremendous effort has therefore been put into research to identify suitable molecular biomarkers for the detection of preneoplastic lesions. Quantitative analyses of DNA methylation changes demonstrated that epigenetic changes are early events in carcinogenesis and that the methylation degree or the frequency of methylation correlates with the stage of the disease. Methylation changes occurring at an early stage during tumorigenesis can thus be detected at a time point when the patient can be cured or at least the outcome can be improved. Novel epigenetic biomarkers are in most cases discovered in retrospective studies on samples of patients who have developed a certain type of cancer or that have a strong predisposition to develop a cancer in the future due to bacterial or viral infection using the above described (fluorescent real-time) MSP approaches which enable detection of methylated alleles with high sensitivity. One of the first examples was the methylation of the promoter of the cell cycle regulator *p16* in non-small cell lung cancer that was already detected in adenomas and hyperplastic lesions and could also be detected in the sputum of lung cancer patients and people that are at high risk (cancer-free smokers) to develop cancer opening possibilities of DNA methylation based analysis for risk assessment and early diagnosis (Belinsky et al., 1998). Methylation changes are also already present in preneoplastic lesions in breast tissue as demonstrated by the presence of methylation at the *14-3-3σ* promoter, a gene involved in the G2 cell cycle checkpoint control, in atypical hyperplasia and histological normal adjacent breast epithelium of breast cancer patients (Umbricht et al., 2001). Further, epigenetic field changes are already present in morphologically normal tissue and might therefore predispose to tumor development. For example, patches of cells with methylation of the *p16* promoter can be detected in normal mammary epithelial tissue of cancer-free women which suggests that this constitutes a very early event in the multi-step progression that might eventually lead to sporadic breast cancer (Holst et al., 2003). Women at calculated high-risk to develop breast cancer (Gail score ≥ 2) showed more frequent methylation in benign breast tissue, especially at *APC* or *RASSF1A* suggesting that DNA methylation changes predispose breast epithelium to progress ultimately to cancer (Lewis et al., 2005). In a recent study DNA methylation analysis of the *RASSF1A* CpG island and three newly identified loci (*KCNAB1*, *CYP26A1*, and *SNCA*) was used to delineate the zones of histological normal breast tissue into which potentially malignant cells have spread from a tumor (Yan et al., 2006). The identification of the area in which premalignant cells, that could give rise to a secondary tumor, are hidden between normal cells would be useful to ensure that surgical margins are cancer-free. DNA methylation changes seemed to diffuse out of the tumor center as shown by the gradual diminution of methylation levels and were found up to a distance of 4 cm from the primary tumor potentially explaining the high risk of local recurrence in breast cancer after lumpectomy of the primary tumor. In colorectal cancer patients methylation of the *MGMT* promoter was frequently detected at a distance of 1 cm and also 10 cm from the tumor (Shen et al., 2005) and DNA methylation was routinely found in normal colonic mucosa in sites distant from the tumor whereby methylation levels are higher in patients displaying the methylator phenotype (see below) in their respective cancers (Kawakami et al., 2006). The same field effect was observed in esophageal adenocarcinomas patients where methylation at *CALCA*, *MGMT* and *TIMP3* was also observed in normal esophageal mucosa distant of the tumor or stomach tissue (Eads et al., 2001). Methylation analysis of different stages in the development of esophageal adenocarcinoma showed *ESR1* was already present in inflammatory reflux esophagitis and all subsequent stages of tumor progression while *APC* and *p16* were detectable from the metaplastic stage on (Bian et al., 2002; Eads et al., 2000b). The same phenomenon was confirmed in non-small cell lung cancer where the bronchial margins of resected tumors were positive for aberrant methylation supporting the idea of an epigenetic field defect (Guo et al., 2004).

The 3' part of the CpG island spanning the transcription start and the first exon of *p16* was found to be frequently methylated in the colonic mucosa of patients with ulcerative colitis, a condition conferring a substantial risk to the development of colon cancer (Issa et al., 2001). The same locus becomes methylated in patients with cirrhosis or chronic hepatitis, which might promote liver carcinogenesis (Kondo et al., 2000) and precedes genetic changes such as loss of heterozygosity (Kanai et al., 2000). In patients with liver adenomas, methylation of *RASSF1A*, *APC* and/or *cyclinD2* is more frequently and to larger extent found in adjacent tissue than in patients without any cancer suspicion (Lehmann et al., 2005). These results demonstrate that aberrant DNA methylation is a very early event in carcinogenesis or in conditions that might be precursors to neoplastic lesions. A potential confounding factor are age-dependent changes of methylation that might confer a certain predisposition to the development of cancer but do not necessarily implicate the presence of malignant cell in the tissue investigated (Yuasa, 2002). In gastric cancer patients methylation of *E-cadherin*, *MLH1* and *p16* was found in tumors and adjacent tissue, however compared to cancer free stomachs from individuals older than 45 years only *MLH1* remained as potential biomarker for early detection associated with carcinogenesis and present in both neoplastic and corresponding non-neoplastic gastric epithelia (Waki et al., 2002).

Bacterial infection is another possible mechanism for the induction of field defects. Levels of aberrant DNA methylation were found to be up to 300-fold increased in the gastric mucosa of healthy volunteers infected with *Heliobacter pylori* compared to non-infected volunteers demonstrating that promoter hypermethylation is a very early event in gastric carcinogenesis induced by the bacterial infection (Maekita et al., 2006). When compared to *H. pylori* positive gastric cancer cases, no further increase in the methylation patterns was observed in the eight genes analyzed, while methylation patterns were elevated in *H. pylori* negative gastric cancers compared to the normal mucosa without bacterial infection.

In numerous studies epigenetic changes that are present in the tumor have also been detected in the adjacent histological normal tissue. The number of genes that have been found to be methylated in different tumor types and their adjacent tissue might be slightly overestimated as age-matched control tissues are not always included in the study design and many of the studies utilized conventional methylation-specific PCR for detection. This technique might detect also very low levels of normal DNA methylation and quantitative methods should be applied to confirm changes in the frequency or the degree of DNA methylation between normal and cancerous tissue.

2.1. Early Detection in Body Fluids

However, access to the primary tissue may be complicated, inconvenient and does certainly not meet the needs for population-based screening for risk assessment, which requires the biological material to be obtained by cost effective and minimal invasive procedures. Recent reports of DNA methylation analyses carried out on tumor biopsies and matched DNA samples extracted from body fluids such as serum, plasma, urine and sputum have demonstrated a high level of concordance. DNA methylation based markers are therefore promising tools for noninvasive detection of different tumor types and sensitive assays for the detection of specific methylation patterns in tumor derived DNA extracted from body fluids such as serum and plasma or urine have been developed. As all assays for early detection, DNA methylation based tests need to fulfill stringent criteria of sensitivity and specificity in the presence of possible confounding factors such as age, diet and hormones. The most effective way to detect the aberrant methylation is to analyze fluids that have been in physical contact with the site of the respective cancer. A large number of novel sources has been successfully tested including nipple aspirate fluid, breast-fine needle washing, bronchial brush samples, buccal cells, needle biopsies, pancreatic juice, peritoneal fluid, prostate fluid or ejaculate, brochialveolar lavages, saliva, exfoliative cells from bladder or cervix, urine, peritoneal fluid, stool or vaginal tampons and new approaches are still devised.

Exfoliative material from buccal cells obtained by oral rinses can be used for early detection and monitoring of disease recurrence in patients with oral cancer (Lopez et al., 2003) as can DNA extracted from saliva in head and neck cancer (Rosas et al., 2001). Ductal lavages through microcatheters can be used for the collection of epithelial cells from breast tissue. Methylation-specific PCR of three genes showed superior sensitivity for early detection compared to conventional screening as methylation was readily detected in two patients that were at high risk to develop breast cancers (Gail score) but had non-suspicious mammograms and these patients were subsequently diagnosed with breast cancer (Evron et al., 2001). The recovery of nipple aspirate fluid is a completely noninvasive process and the aspirate contains breast epithelial cells making it suitable for early stage breast cancer detection (Krassenstein et al., 2004). Peritoneal fluid can be used for the assessment of the methylation status and risk prediction of ovarian cancer patients (Ibanez de Caceres et al., 2004; Müller et al., 2004a).

Measurement of serum prostate-specific antigen (PSA) is a routine test for the early detection screening of prostate cancer. Although one of the best conventional tumor markers, detection of increased PSA levels does not have sufficient sensitivity and specificity to ensure a reliable diagnosis making secondary examinations mandatory and missing part of the cancer cases (Gray, 2005). CpG hypermethylation of the promoter region of the glutathione-*S*-transferase 1 (*GSTP1*) is a well established biomarker for hormone dependent cancers, especially prostate cancer. It is detectable in 90 % of prostate cancers but not in normal prostate cells or benign hyperplasia. It is already detectable in precursor lesions and has been detected in urine and ejaculate of men with prostate cancer and carries great potential as a clinical diagnostic test (Cairns et al., 2001; Goessl et al., 2000; Nakayama et al., 2004).

DAPK, *BCL2* and *TERT* were found to be methylated in superficial bladder tumors and hypermethylation of the respective promoters was exclusively detected in the urine sediments of bladder cancer patients while promoters remained unmethylated in cancer free people (Friedrich et al., 2004). The analysis of aberrant DNA methylation in a set of nine genes (*APC*, *ARF*, *CDH1*, *GSTP1*, *MGMT*, *CDKN2A*, *RARβ2*, *RASSF1A*, *TIMP3*) recently permit the detection of bladder cancer in urine sediment of 175 patients with a sensitivity of 82 % and a specificity of 96 % (Hoque et al., 2006). Urine can also be used as a noninvasive source of DNA for the detection of kidney cancer (Battagli et al., 2003; Hoque et al., 2004). Pancreatic juice harbors most of the epigenetic aberrations specific to the corresponding pancreatic adenocarcinoma (Fukushima et al., 2003). Methylation analysis of pancreatic juice for the early detection of pancreatic tumors or predisposing conditions provides therefore an alternative to imaging technologies and might detect lesions invisible to imaging.

Quantitative DNA methylation analysis of a set of 5 genes differentiated accurately between pancreatic cancers, patients with chronic pancreatitis and normal controls (Matsubayashi et al., 2006). Methylation profiles of individuals at high risk to develop cancers resembled strikingly that of patients with chronic pancreatitis. This quantitative methylation based staging might – when validated and confirmed in larger cohorts - facilitate decisions if and which further interventions will be necessary.

P16 methylation was readily detected in DNA obtained from exfoliative cells from sputum or bilateral bronchial lavage or from brushings taken at bronchoscopy from lung cancer patients suffering from various subtypes of lung cancer or heavy smokers (Ahrendt et al., 1999; Belinsky et al., 1998; Belinsky et al., 2002; Kersting et al., 2000). Aberrant methylation of the promoter regions of the *p16* and *MGMT* genes was detected in the sputum of former uranium mine workers, in one case three years prior to the clinical diagnosis of squamous cell lung carcinoma (Palmisano et al., 2000). Analysis of multiple genes detected methylation in at least one gene in 68 % bronchoalveolar lavage of lung cancer patients (Topaloglu et al., 2004). However, sputum analysis proved to be more sensitive than samples obtained through bronchial lavage or invasive bronchoscopy probably due to independence of the anatomic location where he sample is taken (Kersting et al., 2000). Sputum also proved superior in sensitivity when compared to plasma and concomitant methylation of several genes was significantly associated with increased lung cancer risk (Belinsky et al., 2005). Increased methylation frequencies in an improved panel of 14 genes analyzed in sputum correlated inversely with time to lung cancer diagnosis in a high risk cohort demonstrating the power of DNA methylation based screening for the identification of people at risk to develop lung cancer (Belinsky et al., 2006). Analysis can also be performed on formalin fixed paraffin embedded sputum specimens which possibly facilitates sample handling and storage (Olaussen et al., 2005).

Stool can be exploited as a noninvasive source of DNA for the detection of colorectal cancer. Human sequences have to be specifically amplified as the amount of human DNA in feces is largely exceeded by bacterial DNA. *SFRP2*, a gene frequently methylated in primary colorectal tumors, proved to be the target with the highest discriminative power between cancer patients and healthy controls and was detected with a sensitivity of 77-90 % in DNA extracted from stool (Müller et al., 2004b). In another study significant correlation between stool and colonic mucosa were found for two CpGs in the *ESR1* promoter (Belshaw et al., 2004), but the applicability of this specific gene for cancer diagnostics is questioned by the fact that this gene gets frequently methylated in normal ageing colonic mucosa (Issa et al., 1994).

2.2. Free Circulating DNA in Plasma/Serum

The amount of freely circulating DNA in plasma ranges from 10-30 ng/ml in healthy control subjects to up to 500 ng/ml in some cancer patients with the highest levels of circulating DNA found in patients with metastasized cancers. Tumors release a substantial amount of genomic DNA into the systemic circulation, probably through cellular necrosis and apoptosis although the exact source and mechanisms behind this phenomenon remain elusive (Jahr et al., 2001). The amount of DNA in plasma does not correlate with the type of tumor and a high variability between different types of tumors and/or individuals bearing the same tumors is observed. This freely circulating DNA contains the same genetic and epigenetic alterations that are specific to the primary tumor (Silva et al., 1999) and besides methylation analysis the quantity of plasma nucleic acids, the presence of mutations, viral sequences or microsatellite instability as well as measurement of RNAs have been used for early detection of cancer and/or monitoring of treatment or disease recurrence (Fleischhacker and Schmidt, 2007). As the analyzed gene-specific methylation patterns are in most cases absent in control patients, methylation analysis of DNA recovered from plasma and serum can be used as a biomarker for molecular diagnosis and prognosis in various types of malignancies. Aberrant promoter hypermethylation was first demonstrated in the serum of non-small cell lung cancer patients by the analysis of *p16*, *DAPK*, *GSTP1* and *MGMT* (Esteller et al., 1999). In a recent study methylation of at least one promoter was found in 51 % of sera from stage I lung cancers while serum protein based markers portended the presence of cancer in only 11 % of cases (Fujiwara et al., 2005). A large number of studies mainly focusing on one or a few genes have subsequently confirmed that methylation patterns of genes epigenetically deregulated in primary tumors can be found in the serum and/or plasma (Fleischhacker and Schmidt, 2007). Examples include methylation of *p16* in liver (Wong et al., 1999) and colorectal cancer patients (Zou et al., 2002), *GSTP1* in patients with prostate cancer (Goessl et al., 2000), *APC*, *RASSF1A* and *DAPK* in breast cancer (Dulaimi et al., 2004a) or *XIST* methylation in testicular cancer patients (Kawakami et al., 2004). If methylation is also found in disease-free persons but significantly increased in patients, accurate quantitative measurement of the methylation level is required to distinguish between the normal and the pathological situation. The ideal serum/plasma based screening test for early detection would permit identification of multiple common cancers in a single diagnostic assay by simultaneously interrogating DNA methylation levels at several loci some of which indicating solely the presence of a cancer and some which will be specific for certain type or even subtype of cancer. Identification and selection of these CpG sites with high explanatory power can be achieved using different DNA methylation profiling technologies such as RLGS or DMH.

3. Epigenetic Signatures for Tumor Classification

Characterizing and classifying CpG islands on the basis of the epigenetic status in cancer has practical applications. Distinct methylation profiles for various cancers have emerged in the last years. CpG islands become methylated to varying degrees in different types of tumors. While for example head and neck, breast or testicular tumors show overall low levels of methylation some other tumor types such as colon tumors, acute myeloid leukemias or gliomas are characterized by high levels of methylation, although some heterogeneity is observed in almost all tumor types (Costello et al., 2000). Methylation events in promoter or exonic CpG islands can be used as well as MINTs for the classification of different subtypes of human cancers. Methylation patterns can be shared by different types of tumors as well as being tumor type-specific and methylation profiling can therefore identify distinct subtypes of human cancers (Costello et al., 2000). Analysis of 12 genes in 600 samples representing 15 major tumor types confirmed the existence of tumor-specific patterns and demonstrated that small panels of markers can distinguish and identify the different tumor types (Esteller et al., 2001a). Hypermethylation at CpG islands leaves heritable marks on the genome probably due to growth selection pressure and differential susceptibility to methylation of CpG islands. Methylation patterns are therefore specific for a distinct tumor stage and progression of tumorigenesis can be potentially reconstructed from these epigenetic patterns. It should be noted that so far no single gene has been identified that is always methylated in a certain type of cancer thus panel of markers are required for accurate decisions. Therefore highly parallel analyses using microarray based DNA methylation technologies have been used to classify and stage various types of tumors by their methylation signature using in most cases surgically resected tissue samples.

The CpG island methylator phenotype (CIMP) defines a tumor subgroup which has characteristic clinicopathological features and locations probably arising from a distinct underlying biological mechanism (Issa, 2004). In CIMP multiple CpG islands become concordantly hypermethylated and the targets comprise promoter as well as non-promoter CpG islands. CIMP has been demonstrated in colorectal, gastric, liver, pancreatic, esophageal and ovarian cancers, acute lymphocytic and myeloid leukemia, glioblastomas and neuroblastomas (Abe et al., 2005; Issa, 2004; Toyota et al., 1999), but has so far been excluded for other cancers like breast cancer (Bae et al., 2004). For colorectal cancer, the "classic" CIMP panel includes *p16, MLH1, MINT1, MINT2, and MINT31* (Issa, 2004). These genes are specifically hypermethylated in cancer tissue and not in ageing normal colonic mucosa as many other genes that have originally been associated with CIMP. The CIMP phenotype is genetically characterized by mutations in *KRAS* or *BRAF* and paucity of *p53* mutations and in multiple malignancies associated with poor prognosis. Although the existence of a CIMP phenotype is a disputed question due to a certain degree of heterogeneity in the methylation patterns and the absence of clear bimodal distribution for the presence and pattern of methylation as observed for genetic mutations (Anacleto et al., 2005; Yamashita et al., 2003), a recent large-scale study analyzing ~200 CpG islands in ~300 tumors identified a refined CIMP panel (*CACNA1G*, *IGF2*, *NEUROG1*, *RUNX3* and *SOCS1*) that largely outperformed the classic panel in all aspects (Weisenberger et al., 2006). The new panel further substantiated the associations with the mentioned genetic and phenotypic factors and approached a bimodular distribution providing further evidence for CIMP as a distinct subtype of colorectal cancers. Colorectal cancer is by far the best studied pathology displaying a CIMP with clear phenotypic characteristics. Many of the other cancers display concordant methylation of a subset of genes and have therefore been defined as CIMP, but the methylation patterns can not be associated with specific clinical parameters (except poor prognosis) or genetic alterations.

Genome-wide profiling of DNA methylation patters is well suited to distinguish between specific tumor types. While by the analysis of 14 promoters similar methylation profiles were obtained for different subtypes of leukemia and lymphomas, these differed from multiple myelomas and monoclonal gammopathies (Takahashi et al., 2004a). A more refined analysis of 232 CpGs in 56 genes permitted however the clear differentiation between acute lymphoblastic leukemia (ALL) and acute myeloid leukemia (AML) while lymphocytes from healthy controls built a third cluster (Adorjan et al., 2002). Methylation profiling of 24 genes also established distinct profiles for gallbladder carcinoma and chronic cholecystis specimens (Takahashi et al., 2004b).

Already single gene analysis or analysis of a few genes might be sufficient in some cases to differentiate between different subtypes of a certain type of carcinomas. Papillary, clear cell renal cell carcinoma and oncocytomas were differentiated at the molecular levels by quantitative analysis of DNA methylation levels at the *RASSF1A* promoter (Gonzalgo et al., 2004). While oncocytomas were never methylated, papillary and clear cell tumors were methylated in 90 % of tumors. However, the methylation index was very different for the two subtypes with a mean index of 79 for the papillary and around 13 for the clear cell carcinoma demonstrating the power of quantitative methylation analysis but also the need for quantitative methods to analyze DNA methylation in clinical applications. Similarly epigenetic inactivation of *VHL* predominantly occurs in clear cell renal carcinoma, but not in the other subtypes (Battagli et al., 2003; Dulaimi et al., 2004b). Non-Hodgkin's lymphoma display a large genetic and epigenetic heterogeneity probably relating to the different stages of normal B-cell development. Methylation of the androgen receptor however is restricted to the subtype of germinal origin, mainly follicular lymphomas while most other loci showed a similar methylation pattern on the microarray (Yang et al., 2003). Two subtypes of non-Hodgkin's lymphoma (follicular and mantle cell lymphoma) could be differentiated based on their methylation profile (Shi et al., 2003).

Array-based methylation profiling demonstrated that methylation density can be linked to tumor progression as poorly differentiated breast tumors exhibited more frequently hypermethylated CpG islands than the moderate- or well-differentiated specimens (Yan et al., 2000). In a follow-up study distinct methylation profiles were identified clustering primary breast cancer tumors in function of their hormone rector status (Yan et al., 2001). Classifications based on the methylation status of several genes are able to predict the metastatic potential of a cancer as well as the response and sensitivity to a specific cancer drug or therapy. Methylation profiles were used to define tumor subgroups of ovarian cancers that correlated with time to disease recurrence after chemotherapy in patients and a selection of loci might be useful for the prediction of chemotherapy outcome (Wei et al., 2002) and progression free survival (Wei et al., 2006).

The most promising approach for classification as for early diagnosis following the initial phase of marker discovery using microarray based methods is to establish small to medium size panels of markers that have sufficiently high specificity to classify the tumors but can be routinely analyzed with simple diagnostic techniques such as real-time PCR available in clinical laboratories.

4. Methylation Patterns for the Detection of Tumor Recurrence and Patient Prognosis

Efficacy of therapeutic approaches is often limited when carcinoma are detected at an advanced or already metastatic stage. Biomarkers that evaluate the malignant potential of a tumor and are able to identify patients at high risk for rapid tumor progression are therefore of tremendous clinical value. Prognostic markers that predict decreased survival time might help to identify tumor subtypes where additional adjuvant therapy or a combination of therapies might be required and avoid thereby the loss of valuable time due to ineffective treatment. Disappearance or reappearance of DNA methylation patterns at specific promoters can also be used as an indication of relapse. The tumor suppressor gene *p16* is one of the best studied epigenetically inactivated tumor suppressor genes controlling the G1 to S phase transition in normal cells. Mutations occur but are rare in most tumor types. *P16* methylation has been associated with the stage of invasiveness and metastasis in different tumor types (Maesawa et al., 1996). Similarly, *DAPK* methylation was more frequent in patients with advanced stage head and neck cancers and also in patients with lymph node metastases (Sanchez-Cespedes et al., 2000).

Lung cancer is the major cancer responsible for death in the western world and survival rates have not much changed over the last two decades. Quantitative methylation analysis demonstrated the correlation of methylation of the *APC* promoter with reduced survival in different types of lung cancer (Brabender et al., 2001; Usadel et al., 2002). Similar results were obtained for *MLH1* (Hsu et al., 2005), *DAPK* methylation (Tang et al., 2000), *MGMT* (Brabender et al., 2003) and *p16* (Kim et al., 2001) in non-small cell lung cancer patients. In a larger recent study survival was reduced in patients with non-small cell lung cancer displaying *p16* methylation while methylated *E-cadherin* (*CDH1*) and *TIMP3* prolonged survival (Gu et al., 2006). In patients with malignant mesothelioma methylation of *APC, DAPK, RARβ, RASSF1A* in serum was a predictor for poor prognosis, while *p16* in contrast to the above described studies tended to be associated with prolonged survival (Fischer et al., 2006). This discrepancy might be due to a different tumor type studied but also demonstrates the need for large scale studies in respect to both number of genes and number of patients analyzed to identify reliable prognostic biomarkers.

In esophageal squamous cell carcinoma promoter methylation of *PGP9.5* is independently predictive for reduced survival and presence of metastases (Mandelker et al., 2005) as is methylation of *APC* which correlates with advanced stage and reduced survival of esophageal adenocarcinomas (Kawakami et al., 2000). Patients with esophageal adenocarcinoma that displayed methylation at four or more genes of a panel of seven genes (*APC*, *DAPK*, *E-cadherin*, *ESR1*, *MGMT*, *p16* and *TIMP3*) had significantly higher risk for early tumor recurrence and poor survival with hazard ratios of 2.5 and 2.7, respectively (Brock et al., 2003). However, none of the single genes including *APC* was significantly associated with survival.

Superficial bladder cancer recurs in about 50 % of cases after transurethral resection. Hypermethylation of more than two genes (out of seven analyzed) in this tumor type was significantly associated with the development of recurrence in the bladder of the patients. The methylation status of the death associated protein kinase (*DAPK*) promoter was found to be the best single gene predictor for recurrence and proposed as molecular marker for follow-up patients after surgery (Tada et al., 2002). In a similar way, low methylation at the antiapoptic protein-activating factor (*APAF-1*) and absence of methylation of *DAPK* were independently positively associated with recurrence-free and overall survival in renal cell carcinoma (Christoph et al., 2006). A further example is the predictive power of the preoperative, concurrent methylation of *p15* and *p16* for tumor recurrence and/or lung metastasis in patients with hepatocellular carcinoma that were followed up to 14 months after surgery (Wong et al., 2000b). The aggressiveness of the tumor can be correlated with the hypermethylation of specific genes as has for example been shown for the methylation of the γ-catenin promoter in renal carcinoma (Breault et al., 2005), *DAPK* methylation in stage I non-small cell lung cancer (Tang et al., 2000) or *E-cadherin* in prostate (Li et al., 2001) or gastric cancer (House et al., 2003). Methylation of the *ASC* (*TMS1*) gene in normal adjacent tissue of prostate cancer patients was also significantly associated with risk of recurrence (Collard et al., 2006). Hypermethylation proved to be an independent and superior predictive marker compared to the tumor-node-metastasis staging system for reduced survival and disease progression in the analyzed tumors. Methylation of 6 genes (*RARβ*, *RASSF1A*, *GSTP1*, *CDH13*, *APC*, and *CDH1*) correlated with higher tumor stage as defined by the Gleason score and increased preoperative PSA levels in prostate cancer patients both parameters indicating poor prognosis (Maruyama et al., 2002). *APC* and *GSTP1* methylation indicated an increased risk of relapse and thereby poor prognosis in patients that underwent radical prostatectomy as did methylation of *PTGS2* (Bastian et al., 2005a; Rosenbaum et al., 2005; Yegnasubramanian et al., 2004). *RASSF1A* methylation was also significantly associated with high-risk progression and poor survival of neuroblastomas (Yang et al., 2004).

In tamoxifen-treated breast cancers low methylation status of the estrogen receptor α (*ESR1*) correlated with reduced survival while increased methylation at the *ARH1* promoter was predictive for prolonged survival exclusively in non-tamoxifen treated breast cancers (Widschwendter et al., 2004c). *CYP1B1* methylation correlated with survival albeit less significantly in both classes of breast cancers. The presence of methylation was found to be a negative indicator in tamoxifen treated breast cancers but a protective effect was found in cancers not treated with tamoxifen.

Genome-wide methylation profiles obtained by differential methylation hybridization were used to define tumor subgroups of ovarian cancers that correlate with the time to disease recurrence after chemotherapy in patients and a selection of loci might be useful for prediction of chemotherapy outcome (Wei et al., 2002). In a recent study, 112 methylated loci were identified in ovarian cancer patients that enabled the prediction of progression free survival with an accuracy of 95 % (Wei et al., 2006).

Analysis of DNA methylation in circulating DNA molecules in serum and plasma as described above is one of the most promising tools for risk assessment of cancer patients. Besides early diagnosis, it has been proposed as a prognostic means to detect subclinical disease recurrence in disease-free patients. Analysis of circulating DNA in serum/plasma of cervical cancer patient showed an improved survival when *MYOD1* was unmethylated compared to patient with methylated alleles (Widschwendter et al., 2004b). Same was true for patients that had unmethylated promoters of the cell adhesion and metastases genes *CDH1* (*E-cadherin*) and *CDH13*, while methylation of both genes was associated with an enhanced relapse rate (Widschwendter et al., 2004a). In pretherapeutic serum of breast cancer patients methylation of *ESR1*, *APC*, *HSD174B4*, *HIC1* and *RASSF1A* was detected and methylation of *APC* and/or *RASSF1A* was a significant predictor for poor outcome in patients not receiving adjuvant therapy after surgery (Müller et al., 2003). Similarly, the presence of *p16* promoter methylation in plasma in combination with or without carcinoembryonic antigen levels correlated with the presence of nodal metastasis in invasive ductal breast cancer patients (Hu et al., 2003). Methylation profiling of 15 genes in peritoneal fluid collected at surgery predicted survival in ovarian cancer patients whereat patients with more genes methylated lived statistically longer than those with few genes methylated (Müller et al., 2004a). Chemotherapy itself can select for subpopulations of tumor cells that eventually will lead to the recurrence of tumors resistant to chemotherapy (acquired drug resistance, Teodoridis et al., 2004). Genome-wide profiling identified changes in the DNA methylation patterns in drug resistant ovarian cancer cell lines created by exposure of drug sensitive precursor cell lines to increasing concentrations of cisplatin (Wei et al., 2003). Acquisition of methylation at the *MLH1* promoter measured in the plasma of recurrent ovarian cancer patients treated with a combination of carboplatin and taxoids predicted independently poor overall survival while the prechemotherapeutic status did not correlate with survival (Gifford et al., 2004). The same trend had previously been observed in advanced breast cancer treated with a multidrug cocktail as primary chemotherapy where survival was inversely correlated with *MLH1* expression after chemotherapy, but not prechemotherapeutic expression status (Mackay et al., 2000).

P14 hypermethylation in the plasma of bladder cancer patients was found to be a potential prognostic marker for relapse, presence of multicentric foci, tumor size and muscle invasion (Dominguez et al., 2002). As in many other tumor types, *APC* and *E-cadherin* methylation identified a subgroup of gastric cancer patients with poor survival (Leung et al., 2005). Methylation of the *GSTP1* promoter detected in the preoperative serum of men that had previously undergone radical prostatectomy for the treatment of prostate cancer was the most significant predictor with a relative risk of 4.4 for early prostate specific antigen recurrence which is itself a marker for potential metastatic lesions and significantly associated with reduced survival (Bastian et al., 2005b). This noninvasive test - if adequately validated - would provide a tool for identification of patients with a high risk for relapse and that might benefit from additional adjuvant therapies.

As in the primary tumors (Esteller et al., 2001b), detection of *KRAS* mutations and *p16* hypermethylation in the plasma of colorectal cancer patients is a negative prognostic factor (Lecomte et al., 2002) and is more prevalent in patients with advanced Duke's stage (Zou et al., 2002) and methylation of either *HPP1* and/or *HLTF* predict poor survival as independent prognostic factors conferring a relative risk of death of 3.4 (Wallner et al., 2006).

P16 methylation in the serum of patients with conditions that predispose them to develop liver tumors such as cirrhosis was found a superior biomarker than the level of α-fetoprotein commonly used for disease monitoring but that might detect the presence of carcinoma only at late stage due to interindividual variation (Wong et al., 2000a). Concurrent methylation of *p15* and *p16* in the serum of hepatocellular carcinoma correlated with tumor recurrence or lung metastasis (Wong et al., 2000b)

Methylation of *p15* is a frequent event in hematological malignancies and has been correlated with chromosomal abnormalities, subtype, early relapse and reduced survival in acute myeloid and acute lymphoid leukemia. Methylation patterns in plasma reflect well the methylation status in bone marrow or peripheral blood cells (Christiansen et al., 2003; Wong et al., 2000c). In a recently completed phase I/II clinical trial of the combined administration of the DNA methylation inhibitor 5-deaza-2'deoxycytidine and the histone deacetylase inhibitor valproic acid in leukemia patients, low levels of *p15* promoter methylation correlated with a significantly higher response to therapy (Garcia-Manero et al., 2006).

5. Predicting and Monitoring Response and Effectiveness to Treatment

One objective of cancer research is to improve the treatment of cancer. Not every patient will respond to an administered therapy in the same way. In current practice cytotoxic chemotherapy is routinely administered as adjuvant therapy targeting potential disseminated tumor cells to most patients due to the lack of suitable biomarkers to predict treatment response. However, only few are at risk of recurrence thus patients are exposed to the toxic side effects of the treatment with a potentially low chance of benefit. A challenge to modern medicine is therefore to identify group of patients that will recover without adjuvant therapy after surgical removal of the tumor and for which the risk-benefit ratio might be in favor of withholding chemotherapy and on the other hand patients that have a high chance of clinical benefit from an adjuvant therapy. The identification of predictive markers for specific anticancer regimens such as endocrine treatments, chemotherapies or targeted drugs will be therefore of great value. A better selection of patients would increase time to progression and improve overall survival as individuals will have high chances of benefit from the first line treatment. The possibility to predict the efficacy of a specific treatment would not only be a benefit for the individual patient but to the entire health care system due to a significant decrease in expenses. DNA methylation might help to predict the clinical (sub)phenotypes in individual patients and may be the biomarker to individualize cancer treatment.

Aberrant hypermethylation at the estrogen receptor promoter improved the prediction of the lack of response to the antiestrogen tamoxifen compared to the protein expression status of the estrogen receptor in breast cancer patients (Widschwendter et al., 2004c). Further in a panel of ten genes, hypermethylation found to be predictive for tumor remission and favorable outcome of tamoxifen based adjuvant therapy in hormone receptor-positive patients with recurrent breast cancer (Martens et al., 2005). Hypermethylation of the phosphoserine aminotransferase (*PSAT1*) displayed the strongest single gene correlation with clinical benefit of adjuvant therapy while *S100A2* hypermethylation was an indicator for progressive disease probably due to tamoxifen resistance. Recently, methylation at the estrogen receptor β (*ERβ*) promoter and not at the estrogen receptor α (*ESR1*) was shown to be a positive predictor for tamoxifen response in recurrent breast cancers (Chang et al., 2005). Hypermethylation of the *PTEN* promoter was found to be a marker for tumor size and histological grade, but most importantly for ERBB2 overexpression (Garcia et al., 2004). ERBB2 (HER2) is a target for monoclonal antibody therapy using for example the FDA approved drug trastuzumab in combination with other pharmacological agents for the treatment of women with node-positive, ERBB2 overexpressing breast cancer.

It also becomes increasingly clear that epigenetic changes can be the molecular basis of inherent drug resistance or resistance acquired in response to chemotherapy. For example, hypermethylation induced inactivation of the antiapoptotic protein activating factor (*APAF-1*) renders metastatic melanoma resistant to chemotherapy (Soengas et al., 2001). Promoter methylation of the DNA repair gene *MLH1* impairs proper response to platinum based drugs such as cisplatin, doxorubicin or temozolomide resulting in increased resistance to chemotherapy (Plumb et al., 2000; Strathdee et al., 1999). Contrariwise hypermethylation of *FANCF* increases the sensitivity towards cisplatin in ovarian cancer cell lines, but tumors seem to select for unmethylated promoters to become ultimately resistant to platinum based chemotherapy (Taniguchi et al., 2003). This effect also raises important questions on the use of demethylating agents as part of an anticancer strategy (see below).

Hypermethylation of the O-6-methylguanine DNA methyltransferase (*MGMT*) promoter is one of the best described examples of epigenetic modifications increasing survival and predicting response to genotoxic alkylating agents such as carmustine, temozolomide or 1,3-bis(2-chloroethyl)-1-nitrosourea (BCNU) in glioblastoma patients (Esteller et al., 2000; Hegi et al., 2004; Paz et al., 2004). Promoter hypermethylation proved to be a more accurate predictor compared to the biochemical or immunohistochemical determination of enzyme activity. *MGMT* promoter hypermethylation is also a positive independent prognostic factor for overall and progression-free survival in patients with diffuse large B-cell lymphoma treated with a multidrug regimen including the alkylating agent cyclophosphamide (Esteller et al., 2002). However, initial response did not segregate with *MGMT* promoter methylation status. In colorectal cancer, the frequency of *MGMT* hypermethylation inversely correlated with tumor stage (Duke's staging) and likelihood of recurrence was significantly reduced in patients with methylated *MGMT* promoter after chemotherapy with fluoropyrimidines (Nagasaka et al., 2003). Methylation of at least one gene out of three (*GSTP1*, *MGMT* and *BRCA1*) improved the response to a platinum (and sometimes additional taxoid) based chemotherapy in advanced ovarian cancer (Teodoridis et al., 2005). However, the same gene-specific hypermethylation conferring improved prognosis or response to therapy in one cancer can be a predictor for poor diagnosis in other cancers as for example it is the case for *MGMT* in biliary tract cancer (Koga et al., 2005) or head and neck squamous cell carcinoma (Zuo et al., 2004). In male germ cell tumors *RASSF1A* methylation correlated with resistance to cisplatin, while *MGMT* methylation conferred sensitivity (Koul et al., 2004) and acquired resistance to the anticancer drug fotemustine in melanoma cells was accompanied by reactivation of the *MGMT* gene which might be due to changes in DNA methylation patterns (Christmann et al., 2001). Another example for increased sensitivity for anticancer drugs by epigenetic mechanisms is the hypermethylation of the checkpoint with fork head-associated and ring finger (*CFHR*) or the polo like kinase *SNK/PLK2* promoter conferring in its methylated state sensitivity to microtubule inhibitors such as docetaxel or paclitaxel (Satoh et al., 2003; Syed et al., 2006).

While most efforts concentrated on the identification of aberrant promoter hypermethylation and its correlation with clinical parameters, hypomethylation of genes that are methylated in the normal tissue can also be used to predict response to therapy. Quantitative analysis methods are a prerequisite to determine the degree of hypomethylation as the decrease is rarely complete. The CpG island situated in intron 1 of the *MJC* gene regulates the expression of the corresponding gene and is methylated in chemotherapeutic drug resistant ovarian cancer cell lines, but also tissue-specifically in ovarian surface epithelial cells (Strathdee et al., 2004). Partial demethylation of *MJC* was significantly correlated with higher rates of response to a cisplatin-based chemotherapy and gave better prognosis for stage III/IV epithelial ovarian cancer patients (Strathdee et al., 2005).

As for early detection, biologic material obtained through noninvasive procedures can be used for prognosis and prediction of treatment response. Hypermethylation in pretreatment serum of the *14-3-3σ* promoter was an independent positive predictor for survival and time to progression in stage III/IV lung cancer patients treated with a platinum and gemcitabine based chemotherapy especially in responders to therapy (Ramirez et al., 2005). The absence of an overall correlation between methylation status and response could be due to the involvement of multiple genes in drug response or the time point of sample drawing (pretherapeutic serum).

Analysis of DNA methylation in body fluids also permits monitoring of the response to treatment, which is of course not possible if biopsies would be required for each point of measurement. Persistent or *de novo* hypermethylation of the *RASSF1A* promoter in pre- and posttherapeutic (1 year) serum of breast cancer patients was shown to be an independent negative prognostic marker for the efficacy of adjuvant therapy with tamoxifen significantly increasing the relative risk for relapse and death, while decreasing methylation levels correlated with a response to treatment (Fiegl et al., 2005). The initial methylation status of the *RASSF1A* promoter in the pretherapeutic serum was not associated with treatment efficacy. Methylation of *RASSF1A* in the serum of stage IV melanoma patients also differentiated between responders and non-responders to biochemotherapy, where the latter group was more frequently methylated (Mori et al., 2005).

Similarly DNA methylation at the *APC* promoter was undetectable in lung cancer patients following curative surgery and independent of the preoperative methylation status of the promoter. (Re-)appearance of DNA methylation at the *APC* promoter detected in serum and plasma in the clinical follow-up correlated with disease recurrence (Usadel et al., 2002). In one patient efficacy of treatment could be monitored by *APC* methylation levels that followed the course of repeated remissions and recurrence. Similarly the methylation levels at *p16* measured in serum of hepatocellular cancer patients were greatly reduced or disappeared following surgery (Wong et al., 2003).

Gene-specific and genome-wide methylation profiles will help to reach decisions for personalized treatment and evaluate efficacy, benefit and risk of potential therapies faster and more reliable than today and thereby gain valuable time in the treatment of the different types cancers.

6. Epigenetic Therapy

As DNA methylation is a non-mutational and therefore - at least in principle - reversible modification, it has been shown that it can be used as point of departure for antineoplastic treatment by chemically or antisense oligonucleotide induced demethylation. As the genes are genetically intact, their normal function in apoptosis, cell cycle regulation and cell-cell adhesion can be restored. Inhibitors of DNA methyltransferases can be divided into nucleoside and non-nucleoside analogues. The FDA approved 5-azacytidine (azacitidine) or 5-aza-2'-deoxycytidine (decitabine) and their analogues such as zebularine (figure 2) belong to the first class and are incorporated into *de novo* synthesized DNA and/or RNA strands during replication and trap the maintenance DNA methyltransferase 1 (DNMT1) through the formation of a covalent bond between the catalytic site of the enzyme and the pyrimidine ring. This leads to a progressive and heritable hypomethylation during cell division due to the depletion of free enzyme from the cell (Juttermann et al., 1994). An additional mechanism of action of 5-aza-2'deoxycytidine that is independent of replication is the selective degradation of DNMT1 by the proteosomal pathway (Ghoshal et al., 2005). Decitabine needs to be phosphorylated to be active and is then preferentially incorporated into DNA. Phosphorylated 5-azacytidine is mainly found in *de novo* synthesized RNA but can also be reduced to the deoxy-diphophate and then incorporated into DNA. Both nucleoside analogues (especially the ribonucleotide) are unstable in aqueous solutions under physiological conditions which limits the bioavailability of the active compound (Yoo and Jones, 2006). Treatment of tumor cell lines with decitabine reactivates epigenetically inactivated genes such as *p16* in a heritable manner through demethylation, slows down tumor growth (Bender et al., 1998) and induces terminal cellular differentiation (Attadia, 1993). However, demethylation can be transient and in the absence of the drug the malignant cell population might reappear.

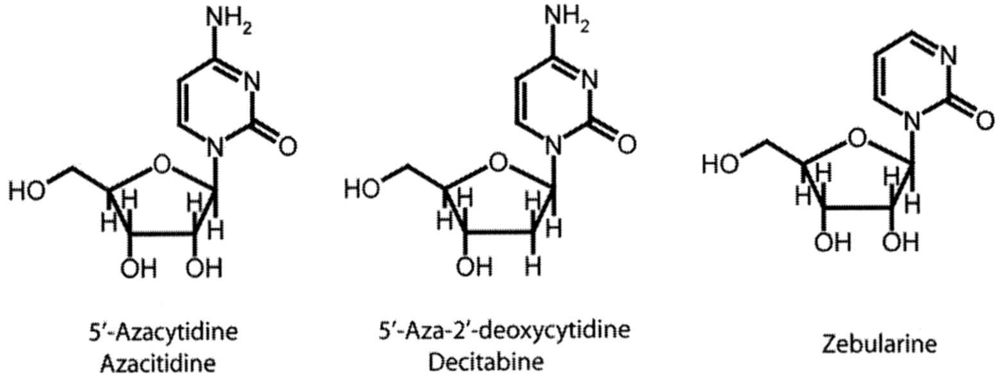

Figure 2. Chemical structure of the methylation inhibitory nucleoside analogues 5-azacytidine (azacitidine), 5-aza-2'-deoxycytidine (decitabine) and zebularine that are used as demethylating agents for the treatment of cancers alone or in combination with conventional cytotoxic drugs.

Only a few years after their synthesis in the 1960's (Piskala and Sorm, 1964; Pliml and Sorm, 1964), 5-azacytidine and decitabine were tested in clinical trials where they showed antitumor activity in patients with leukemia and other hematological malignancies while they proved to be ineffective for the treatment of various solid tumors (Goffin and Eisenhauer, 2002). However, the drugs given at high doses were not well tolerated and therapy showed no significant improvement compared with other therapies. The potential of 5-azacytidine was demonstrated in the treatment of myelodysplastic syndromes (MDS), a disease where the bone marrow is not able to produce a sufficient number of mature blood cells where it displayed higher response rates and significantly improved the quality of life of patients than the standard therapy consisting of supportive care and/or hematopoietic stem cell transplantation (Kornblith et al., 2002). In recent trials low doses of decitabine were repeatedly administered for the treatment of hematological malignancies such as leukemia and MDS to induce stable re-expression of previously epigenetically silenced tumor suppressor genes (Issa et al., 2004; Issa et al., 2005; Kantarjian et al., 2006). Thereby control of cellular differentiation and proliferation could be re-established and the malignant leukemic clone suppressed. These trials clearly demonstrated the superiority of repeated low doses over the previously used high-dose regimens. The US Food and Drug Administration (FDA) approved 5'-azacytidine (Vidaza^{TM}, Pharmion) as first agent for the treatment of several subtypes of MDS in 2004 (Kaminskas et al., 2005) and very recently also the injection of decitabine (Dacogen^{TM}, MGI Pharma) for the treatment of all French-American-British subtypes of MDS (Gore et al., 2006). However, it should be noted that the demethylation is transient and in the absence of the drug re-methylation occurs (Issa et al., 2005), which would possibly also allow a malignant cell population to reappear. Small molecules such as valproic acid or suberoylanilide hydroxamic acid (SAHA) inhibit the deacetylation of lysines in the tails of histones and represent a complementary strategy in epigenetic therapy (Carey and La Thangue, 2006). DNA methylation and repressive histone tail modifications are closely interrelated and both epigenetic modifications maintain the silent state of tumor suppressor genes in cancer. Cell line experiments have demonstrated synergistic effects of DNA methyltransferase inhibitors and histone deacetylase inhibitors for the induction of the re-expression of epigenetically silenced genes and induction of apoptosis (Cameron et al., 1999a; Zhu et al., 2001). It is hoped that the combined administration of both classes of drugs will further increase response rates and promising results from clinical phase I and II trials have recently been reported (Garcia-Manero et al., 2006).

Evidence for the efficacy of azacitidine and decitabine in solid tumors is much more scarce (Goffin and Eisenhauer, 2002). Azacitidine infusion led to measurable inhibition of DNA methylation in one patient with hepatocellular carcinoma out of 25 patients with solid tumors treated (Rudek et al., 2005). Infusion of decitabine was generally well tolerated in patient with incurable or metastatic solid tumors (mainly melanoma/breast) and demethylation was observed (Aparicio et al., 2003). However, DNA methylation changes were not consistent and also increased methylation occurred. High dose treatment with decitabine of patients with inoperable non-small cell lung cancer did not show significant antitumor activity in a phase I trial (Schwartsmann et al., 2000) and results of a combined therapy of decitabine administered with cisplatin in recurrent and metastasic squamous cell carcinoma of the cervix were inconclusive in a phase II trial (Pohlmann et al., 2002).

Epigenetic silencing of genes and pathways may lead to differences in response and sensitivity of tumors to chemotherapy. Another promising therapeutic approach is therefore to utilize demethylating compounds such as 5-aza-2'deoxycytidine as bio- or chemosensitizer of recalcitrant tumors that are either systemically or locally applied in combination with cytotoxic agents and/or biological response modifiers. Demethylation induced re-expression of the estrogen receptor (ER) could make patients with ER-negative breast tumors responsive to endocrine management as demonstrated in cell lines through the response to growth inhibitory effects of estrogen in formerly ER negative cells (Ferguson et al., 1995). This effect can be increased through combined treatment of cells with histone deacetylase and methyltransferase inhibitors (Sharma et al., 2005) and raises the prospect of a combination therapy of a demethylating agent with an estrogen receptor antagonist.

Gene silencing through promoter hypermethylation plays also a prominent role in the failure of anticancer therapy due to defective apoptosis programs and/or the development of drug resistance. Inherent and acquired drug resistance to chemo- or radiotherapy constitutes a major problem for the treatment of cancer. Demethylation of the caspase 8 promoter, a key determinant for the induction of cellular apoptosis, by decitabine has been shown to restore sensitivity to cytotoxic drugs and induces apoptosis in various tumor types (Fulda et al., 2001). Cells can also be sensitized to apoptosis through a combined administration of low doses of decitabine and interferon γ that synergistically induce caspase 8 re-expression and activate the tumor specific death receptor TRAIL (Fulda and Debatin, 2006). Demethylation of *RASSF1A* or the apoptosis-associated IFN response gene *XAF1* in renal cell carcinoma or melanoma cell cultures had similar effects and rendered cells sensitive to interferon induced apoptosis (Reu et al., 2006a; Reu et al., 2006b). Similarly, in gastric cancer cell line experiments demethylation of the *14-3-3σ* promoter with decitabine resulted in an increased sensitivity to the DNA damaging agent doxorubicin and cell cycle arrest in G2 (Suzuki et al., 2000).

A drawback of decitabine and azacitidine is their inherent instability in aqueous solution especially at acidic pH which prohibits oral administration of the drugs. Zebularine (1-(β-D-ribofuranosyl)-1,2-dihydropyrimidin-2-one, figure 2, Driscoll et al., 1991) is a stable nucleoside analogue that does not suffer from this drawback, is less toxic and can be administered orally to mice (Cheng et al., 2003). It has first been used as an additive to decitabine to prevent its enzymatic degradation by cytidine deaminase for which zebularine is a potent competitive inhibitor (Laliberte et al., 1992). Combined therapy with low doses of zebularine also sustained demethylated promoters in their unmethylated state in cancer cell lines (Cheng et al., 2004a) and reactivated preferentially cancer related antigens in cancer cells (Cheng et al., 2004b). Zebularine induced similar global DNA methylation changes when compared to decitabine, but differences were observed at the individual gene level. Both drugs similarly chemosensitized drug resistant ovarian cancer cell lines to the treatment with cisplatin (Balch et al., 2005). Zebularine also increases sensitivity of cancer cells to radiotherapy in cell culture and mouse xenograft models (Dote et al., 2005). However, complex metabolic pathways required for the incorporation of the analogue into DNA (Ben-Kasus et al., 2005) lead to limited bioavailability and require high doses of zebularine to achieve similar inhibition effects as decitabine, which makes it unsuitable for human clinical trials.

A frequently considered concern in the discussion of the application of global demethylating drugs in clinics is the potential inappropriate activation of genes and transposable elements as for the moment it is not possible to direct the demethylating activity to specific sequences. Activation of the second allele of imprinted genes or of the inactive X chromosome in women could have deleterious consequences, but these sequences seem to be more resistant to activation than hypermethylated promoters or repeat elements. An additional concern is the potential cancerogenic effect of such drugs as very low levels of DNMT1 in mouse models (hypomorphic allele) led to chromosomal instabilities and the occurrence of lymphomas (Eden et al., 2003). Although this instability has not been observed in humans, demethylation has been associated with autoimmune disorders (Richardson, 2003) and might promote carcinogenesis through gene activation. For example, treating breast cancer cells with hypomethylating agents increases their potential to metastasize which might be caused by the global hypomethylation leading to expression of genes involved in metastasis and invasion such as *S100A4* in pancreatic cancer or *uPa* in breast cancer (Szyf et al., 2004). Aberrant DNA methylation has a positive effect on treatment efficacy by increasing sensitivity to chemotherapy as above described for *MGMT*. Another already presented example is the selection of unmethylated *FANCF* promoters during platinum based chemotherapy after an initial hypermethylation step early during ovarian carcinogenesis where demethylation ultimately confers resistance to chemotherapy (Taniguchi et al., 2003). Decitabine induced hypomethylation and re-expression of genes involved in drug metabolism such as *MDR1* leads to acquired drug resistance to chemotherapeutic drugs (Kantharidis et al., 1997). All these observations question the use of demethylating agents in clinical practice.

A number of other nucleoside analogues are currently tested for demethylating and antitumor activity (Goffin and Eisenhauer, 2002; Yoo and Jones, 2006). Several non-nucleoside analogues have also been shown to inhibit DNA methylation and have been tested to avoid the cytotoxic side effects of azacitidine and decitabine. The vasodilator hydralazine used for the treatment of hypertensive disorders and procainamide administered for the treatment of cardiac arrhythmies have been positively tested for their potential as demethylating agents in cell lines and mouse models (Segura-Pacheco et al., 2003). A small phase I trial in which hydralazine was administered to cervical cancer patients at doses similar to those used for the treatment of hypertension showed at least partial demethylation and gene re-expression in a large proportion of informative samples (Zambrano et al., 2005). Hydralazine is now evaluated as a single agent therapy or in addition to chemotherapy and radiation in a phase II study of patients with hematological and solid malignancies (Arce et al., 2006). Procainamide specifically inhibits of DNA methyltransferase 1 (Lee et al., 2005) and reactivated silenced genes shown in nude mice model (Lin et al., 2001) but its antiarrhythmic properties might limit its use in patients. Dietary polyphenols such as (-)-epigallocatechin-3-gallate from green tea or genistein from soy beans show antitumor activities although their exact mechanism has not been elucidated. They have been associated with changed DNA methylation patterns, gene re-expression in cell culture and delay of tumor progression in mouse models (Fang et al., 2007; Mentor-Marcel et al., 2005). However, non-nucleoside demethylating compounds were much less effective in a comparative study using cell lines compared to the nucleosides analogues with decitabine being the most potent inhibitor of DNA methylation and inductor of gene expression (Chuang et al., 2005).

Chemically modified antisense oligonucleotide such as MG98 (MethylGene Inc.) hybridize to the mRNA of DNMT1, block translation and cause degradation of the double stranded RNA complex thereby reducing the amount of DNMT1 protein in a dose-dependent manner. They were shown to possess antitumor activity by demethylating and re-expressing tumor suppressor genes in cell lines (Fournel et al., 1999). The estrogen receptor was reactivated by treatment of cells with MG98 enabling in principle the described anti-estrogen therapies, however, re-expression was not due to demethylation of the CpG island (Yan et al., 2003). One out of two phase I clinical trials in patients with solid tumors showed a partial response for one patient with renal carcinoma (Davis et al., 2003; Stewart et al., 2003). However, in a recent phase II trial administering MG98 to patients with metastatic renal carcinoma no decrease of DNMT1 proteins levels or antitumor activity was observed and the trial was discontinued (Winquist et al., 2006).

Epigenetic therapy holds great promise to improve cancer treatment especially as chemo- or biosensitizer in combination with conventional cytotoxic drugs. Although most DNMT inhibitors show a beneficial effect in hematological malignancies, treatment of solid tumors is much less successful probably to limited bioavailability. The design of new drugs for example more stable nucleoside analogues and improved delivery to the tumor site could help to overcome this problem. One major improvement when using DNA methylation inhibitors in anticancer therapy would be the ability to target specific genes/sequences for reactivation. Although not yet possible, it has been shown that the fusion of zinc finger proteins and mutant bacterial methyltransferases permits the targeted methylation of predetermined sequences in a stable and heritable manner (Smith and Ford, 2006).

7. Conclusion

Epigenetic analysis is likely to have an increasingly important part to play in the diagnosis, prognostic assessment and treatment of malignant disease. In the last years a multitude of epigenetic markers has been discovered. However, with an estimation of about 10 % of all genes epigenetically deregulated in a specific type of cancer, we only have scratched the tip of the iceberg. Identification of new genes involved in carcinogenesis is greatly facilitated but the ever more powerful methods and technological platforms that are available and many commercial companies now provide tools that make DNA methylation analysis easily accessible for non-specialists. This also reflects the increasing popularity of DNA methylation analysis for diagnostics, staging and monitoring of cancer and other diseases. One of the most important tasks now is to validate the already known biomarkers in larger cohorts as most studies analyzed tens of samples instead of the hundreds or thousands required to definitely confirm the utility of a biomarker. Another task is to translate the knowledge on epigenetic changes and their correlation with pathophysiological parameters into clinical practice. Predictive epigenetic biomarkers will allow a personalized treatment of disease based on the individual methylation profile. While for hematological malignancies low doses of demethylating agents seem to be a promising treatment, epigenetically active drugs can be used to chemosensitize solid tumors for conventional cytotoxic drugs.

8. References

Abe, M., Ohira, M., Kaneda, A., Yagi, Y., Yamamoto, S., Kitano, Y., Takato, T., Nakagawara, A., and Ushijima, T. (2005). CpG island methylator phenotype is a strong determinant of poor prognosis in neuroblastomas. *Cancer Res. 65*, 828-834.

Adorjan, P., Distler, J., Lipscher, E., Model, F., Muller, J., Pelet, C., Braun, A., Florl, A.R., Gutig, D., Grabs, G., *et al.* (2002). Tumour class prediction and discovery by microarray-based DNA methylation analysis. *Nucleic Acids Res. 30*, e21.

Ahrendt, S.A., Chow, J.T., Xu, L.H., Yang, S.C., Eisenberger, C.F., Esteller, M., Herman, J.G., Wu, L., Decker, P.A., Jen, J., and Sidransky, D. (1999). Molecular detection of tumor cells in bronchoalveolar lavage fluid from patients with early stage lung cancer. *J. Natl. Cancer Inst. 91*, 332-339.

Akey, D.T., Akey, J.M., Zhang, K., and Jin, L. (2002). Assaying DNA methylation based on high-throughput melting curve approaches. *Genomics. 80*, 376-384.

Anacleto, C., Leopoldino, A.M., Rossi, B., Soares, F.A., Lopes, A., Rocha, J.C., Caballero, O., Camargo, A.A., Simpson, A.J., and Pena, S.D. (2005). Colorectal cancer "methylator phenotype": fact or artifact? *Neoplasia. 7*, 331-335.

Aparicio, A., Eads, C.A., Leong, L.A., Laird, P.W., Newman, E.M., Synold, T.W., Baker, S.D., Zhao, M., and Weber, J.S. (2003). Phase I trial of continuous infusion 5-aza-2'-deoxycytidine. *Cancer Chemother. Pharmacol. 51*, 231-239.

Arce, C., Segura-Pacheco, B., Perez-Cardenas, E., Taja-Chayeb, L., Candelaria, M., and Duennas-Gonzalez, A. (2006). Hydralazine target: from blood vessels to the epigenome. *J. Transl. Med. 4*, 10.

Attadia, V. (1993). Effects of 5-aza-2'-deoxycytidine on differentiation and oncogene expression in the human monoblastic leukemia cell line U-937. *Leukemia. 7 Suppl 1*, 9-16.

Bae, Y.K., Brown, A., Garrett, E., Bornman, D., Fackler, M.J., Sukumar, S., Herman, J.G., and Gabrielson, E. (2004). Hypermethylation in histologically distinct classes of breast cancer. *Clin. Cancer Res. 10*, 5998-6005.

Balch, C., Yan, P., Craft, T., Young, S., Skalnik, D.G., Huang, T.H., and Nephew, K.P. (2005). Antimitogenic and chemosensitizing effects of the methylation inhibitor zebularine in ovarian cancer. *Mol. Cancer Ther. 4*, 1505-1514.

Balmain, A., Gray, J., and Ponder, B. (2003). The genetics and genomics of cancer. *Nat. Genet. 33 Suppl*, 238-244.

Bastian, P.J., Ellinger, J., Wellmann, A., Wernert, N., Heukamp, L.C., Muller, S.C., and von Ruecker, A. (2005a). Diagnostic and prognostic information in prostate cancer with the help of a small set of hypermethylated gene loci. *Clin. Cancer Res. 11*, 4097-4106.

Bastian, P.J., Palapattu, G.S., Lin, X., Yegnasubramanian, S., Mangold, L.A., Trock, B., Eisenberger, M.A., Partin, A.W., and Nelson, W.G. (2005b). Preoperative serum DNA GSTP1 CpG island hypermethylation and the risk of early prostate-specific antigen recurrence following radical prostatectomy. *Clin. Cancer Res. 11*, 4037-4043.

Battagli, C., Uzzo, R.G., Dulaimi, E., Ibanez de Caceres, I., Krassenstein, R., Al-Saleem, T., Greenberg, R.E., and Cairns, P. (2003). Promoter hypermethylation of tumor suppressor genes in urine from kidney cancer patients. *Cancer Res. 63*, 8695-8699.

Belinsky, S.A., Klinge, D.M., Dekker, J.D., Smith, M.W., Bocklage, T.J., Gilliland, F.D., Crowell, R.E., Karp, D.D., Stidley, C.A., and Picchi, M.A. (2005). Gene promoter methylation in plasma and sputum increases with lung cancer risk. *Clin. Cancer Res. 11*, 6505-6511.

Belinsky, S.A., Liechty, K.C., Gentry, F.D., Wolf, H.J., Rogers, J., Vu, K., Haney, J., Kennedy, T.C., Hirsch, F.R., Miller, Y., *et al.* (2006). Promoter hypermethylation of multiple genes in sputum precedes lung cancer incidence in a high-risk cohort. *Cancer Res. 66*, 3338-3344.

Belinsky, S.A., Nikula, K.J., Palmisano, W.A., Michels, R., Saccomanno, G., Gabrielson, E., Baylin, S.B., and Herman, J.G. (1998). Aberrant methylation of p16(INK4a) is an early event in lung cancer and a potential biomarker for early diagnosis. *Proc. Natl. Acad. Sci. USA. 95*, 11891-11896.

Belinsky, S.A., Palmisano, W.A., Gilliland, F.D., Crooks, L.A., Divine, K.K., Winters, S.A., Grimes, M.J., Harms, H.J., Tellez, C.S., Smith, T.M., *et al.* (2002). Aberrant promoter methylation in bronchial epithelium and sputum from current and former smokers. *Cancer Res. 62*, 2370-2377.

Belshaw, N.J., Elliott, G.O., Williams, E.A., Bradburn, D.M., Mills, S.J., Mathers, J.C., and Johnson, I.T. (2004). Use of DNA from human stools to detect aberrant CpG island methylation of genes implicated in colorectal cancer. *Cancer Epidemiol. Biomarkers Prev. 13*, 1495-1501.

Ben-Kasus, T., Ben-Zvi, Z., Marquez, V.E., Kelley, J.A., and Agbaria, R. (2005). Metabolic activation of zebularine, a novel DNA methylation inhibitor, in human bladder carcinoma cells. *Biochem. Pharmacol. 70*, 121-133.

Bender, C.M., Pao, M.M., and Jones, P.A. (1998). Inhibition of DNA methylation by 5-aza-2'-deoxycytidine suppresses the growth of human tumor cell lines. *Cancer Res. 58*, 95-101.

Bestor, T.H. (2000). The DNA methyltransferases of mammals. *Hum. Mol. Genet. 9*, 2395-2402.

Bian, Y.S., Osterheld, M.C., Fontolliet, C., Bosman, F.T., and Benhattar, J. (2002). p16 inactivation by methylation of the CDKN2A promoter occurs early during neoplastic progression in Barrett's esophagus. *Gastroenterology. 122*, 1113-1121.

Bibikova, M., Lin, Z., Zhou, L., Chudin, E., Garcia, E.W., Wu, B., Doucet, D., Thomas, N.J., Wang, Y., Vollmer, E., *et al.* (2006). High-throughput DNA methylation profiling using universal bead arrays. *Genome Res. 16*, 383-393.

Bird, A. (2002). DNA methylation patterns and epigenetic memory. *Genes Dev. 16*, 6-21.

Bird, A.P., and Southern, E.M. (1978). Use of restriction enzymes to study eukaryotic DNA methylation: I. The methylation pattern in ribosomal DNA from Xenopus laevis. *J. Mol. Biol. 118*, 27-47.

Brabender, J., Usadel, H., Danenberg, K.D., Metzger, R., Schneider, P.M., Lord, R.V., Wickramasinghe, K., Lum, C.E., Park, J., Salonga, D., *et al.* (2001). Adenomatous polyposis coli gene promoter hypermethylation in non-small cell lung cancer is associated with survival. *Oncogene. 20*, 3528-3532.

Brabender, J., Usadel, H., Metzger, R., Schneider, P.M., Park, J., Salonga, D., Tsao-Wei, D.D., Groshen, S., Lord, R.V., Takebe, N., *et al.* (2003). Quantitative O(6)-methylguanine DNA methyltransferase methylation analysis in curatively resected non-small cell lung cancer: associations with clinical outcome. *Clin. Cancer Res. 9*, 223-227.

Breault, J.E., Shiina, H., Igawa, M., Ribeiro-Filho, L.A., Deguchi, M., Enokida, H., Urakami, S., Terashima, M., Nakagawa, M., Kane, C.J., *et al.* (2005). Methylation of the gamma-catenin gene is associated with poor prognosis of renal cell carcinoma. *Clin. Cancer Res. 11*, 557-564.

Brena, R.M., Auer, H., Kornacker, K., Hackanson, B., Raval, A., Byrd, J.C., and Plass, C. (2006a). Accurate quantification of DNA methylation using combined bisulfite restriction analysis coupled with the Agilent 2100 Bioanalyzer platform. *Nucleic Acids Res. 34*, e17.

Brena, R.M., Huang, T.H., and Plass, C. (2006b). Quantitative assessment of DNA methylation: potential applications for disease diagnosis, classification, and prognosis in clinical settings. *J. Mol. Med. 84*, 365-377.

Brock, M.V., Gou, M., Akiyama, Y., Muller, A., Wu, T.T., Montgomery, E., Deasel, M., Germonpre, P., Rubinson, L., Heitmiller, R.F., *et al.* (2003). Prognostic importance of promoter hypermethylation of multiple genes in esophageal adenocarcinoma. *Clin. Cancer Res. 9*, 2912-2919.

Burri, N., and Chaubert, P. (1999). Complex methylation patterns analyzed by single-strand conformation polymorphism. *Biotechniques. 26*, 232-234.

Cairns, P., Esteller, M., Herman, J.G., Schoenberg, M., Jeronimo, C., Sanchez-Cespedes, M., Chow, N.H., Grasso, M., Wu, L., Westra, W.B., and Sidransky, D. (2001). Molecular detection of prostate cancer in urine by GSTP1 hypermethylation. *Clin. Cancer Res. 7*, 2727-2730.

Cameron, E.E., Bachman, K.E., Myohanen, S., Herman, J.G., and Baylin, S.B. (1999a). Synergy of demethylation and histone deacetylase inhibition in the re-expression of genes silenced in cancer. *Nat. Genet. 21*, 103-107.

Cameron, E.E., Baylin, S.B., and Herman, J.G. (1999b). p15(INK4B) CpG island methylation in primary acute leukemia is heterogeneous and suggests density as a critical factor for transcriptional silencing. *Blood. 94*, 2445-2451.

Carey, N., and La Thangue, N.B. (2006). Histone deacetylase inhibitors: gathering pace. *Curr. Opin. Pharmacol. 6*, 369-375.

Chang, H.G., Kim, S.J., Chung, K.W., Noh, D.Y., Kwon, Y., Lee, E.S., and Kang, H.S. (2005). Tamoxifen-resistant breast cancers show less frequent methylation of the estrogen receptor beta but not the estrogen receptor alpha gene. *J. Mol. Med. 83*, 132-139.

Chang, S.C., Tucker, T., Thorogood, N.P., and Brown, C.J. (2006). Mechanisms of X-chromosome inactivation. *Front. Biosci. 11*, 852-866.

Cheng, J.C., Matsen, C.B., Gonzales, F.A., Ye, W., Greer, S., Marquez, V.E., Jones, P.A., and Selker, E.U. (2003). Inhibition of DNA methylation and reactivation of silenced genes by zebularine. *J. Natl. Cancer Inst. 95*, 399-409.

Cheng, J.C., Weisenberger, D.J., Gonzales, F.A., Liang, G., Xu, G.L., Hu, Y.G., Marquez, V.E., and Jones, P.A. (2004a). Continuous zebularine treatment effectively sustains demethylation in human bladder cancer cells. *Mol. Cell Biol. 24*, 1270-1278.

Cheng, J.C., Yoo, C.B., Weisenberger, D.J., Chuang, J., Wozniak, C., Liang, G., Marquez, V.E., Greer, S., Orntoft, T.F., Thykjaer, T., and Jones, P.A. (2004b). Preferential response of cancer cells to zebularine. *Cancer Cell. 6*, 151-158.

Christiansen, D.H., Andersen, M.K., and Pedersen-Bjergaard, J. (2003). Methylation of p15INK4B is common, is associated with deletion of genes on chromosome arm 7q and predicts a poor prognosis in therapy-related myelodysplasia and acute myeloid leukemia. *Leukemia. 17*, 1813-1819.

Christmann, M., Pick, M., Lage, H., Schadendorf, D., and Kaina, B. (2001). Acquired resistance of melanoma cells to the antineoplastic agent fotemustine is caused by reactivation of the DNA repair gene MGMT. *Int. J. Cancer. 92*, 123-129.

Christoph, F., Weikert, S., Kempkensteffen, C., Krause, H., Schostak, M., Miller, K., and Schrader, M. (2006). Regularly methylated novel pro-apoptotic genes associated with recurrence in transitional cell carcinoma of the bladder. *Int. J. Cancer. 119*, 1396-1402.

Chuang, J.C., Yoo, C.B., Kwan, J.M., Li, T.W., Liang, G., Yang, A.S., and Jones, P.A. (2005). Comparison of biological effects of non-nucleoside DNA methylation inhibitors versus 5-aza-2'-deoxycytidine. *Mol. Cancer Ther. 4*, 1515-1520.

Clark, S.J., Harrison, J., Paul, C.L., and Frommer, M. (1994). High sensitivity mapping of methylated cytosines. *Nucleic Acids Res. 22*, 2990-2997.

Colella, S., Shen, L., Baggerly, K.A., Issa, J.P., and Krahe, R. (2003). Sensitive and quantitative universal Pyrosequencing methylation analysis of CpG sites. *Biotechniques. 35*, 146-150.

Collard, R.L., Harya, N.S., Monzon, F.A., Maier, C.E., and O'Keefe, D.S. (2006). Methylation of the ASC gene promoter is associated with aggressive prostate cancer. *Prostate. 66*, 687-695.

Costello, J.F., Fruhwald, M.C., Smiraglia, D.J., Rush, L.J., Robertson, G.P., Gao, X., Wright, F.A., Feramisco, J.D., Peltomaki, P., Lang, J.C., *et al.* (2000). Aberrant CpG-island methylation has non-random and tumour-type-specific patterns. *Nat. Genet. 24*, 132-138.

Cottrell, S.E., Distler, J., Goodman, N.S., Mooney, S.H., Kluth, A., Olek, A., Schwope, I., Tetzner, R., Ziebarth, H., and Berlin, K. (2004). A real-time PCR assay for DNA-methylation using methylation-specific blockers. *Nucleic Acids Res. 32*, e10.

Davis, A.J., Gelmon, K.A., Siu, L.L., Moore, M.J., Britten, C.D., Mistry, N., Klamut, H., D'Aloisio, S., MacLean, M., Wainman, N., *et al.* (2003). Phase I and pharmacologic study of the human DNA methyltransferase antisense oligodeoxynucleotide MG98 given as a 21-day continuous infusion every 4 weeks. *Invest. New Drugs. 21*, 85-97.

Deng, D., Deng, G., Smith, M.F., Zhou, J., Xin, H., Powell, S.M., and Lu, Y. (2002). Simultaneous detection of CpG methylation and single nucleotide polymorphism by denaturing high performance liquid chromatography. *Nucleic Acids Res. 30*, e13.

Dominguez, G., Carballido, J., Silva, J., Silva, J.M., Garcia, J.M., Menendez, J., Provencio, M., Espana, P., and Bonilla, F. (2002). p14ARF promoter hypermethylation in plasma DNA as an indicator of disease recurrence in bladder cancer patients. *Clin. Cancer Res. 8*, 980-985.

Dote, H., Cerna, D., Burgan, W.E., Carter, D.J., Cerra, M.A., Hollingshead, M.G., Camphausen, K., and Tofilon, P.J. (2005). Enhancement of in vitro and in vivo tumor cell radiosensitivity by the DNA methylation inhibitor zebularine. *Clin. Cancer Res. 11*, 4571-4579.

Driscoll, J.S., Marquez, V.E., Plowman, J., Liu, P.S., Kelley, J.A., and Barchi, J.J., Jr. (1991). Antitumor properties of 2(1H)-pyrimidinone riboside (zebularine) and its fluorinated analogues. *J. Med. Chem. 34*, 3280-3284.

Dulaimi, E., Hillinck, J., Ibanez de Caceres, I., Al-Saleem, T., and Cairns, P. (2004a). Tumor suppressor gene promoter hypermethylation in serum of breast cancer patients. *Clin. Cancer Res. 10*, 6189-6193.

Dulaimi, E., Ibanez de Caceres, I., Uzzo, R.G., Al-Saleem, T., Greenberg, R.E., Polascik, T.J., Babb, J.S., Grizzle, W.E., and Cairns, P. (2004b). Promoter hypermethylation profile of kidney cancer. *Clin. Cancer Res. 10*, 3972-3979.

Eads, C.A., Danenberg, K.D., Kawakami, K., Saltz, L.B., Blake, C., Shibata, D., Danenberg, P.V., and Laird, P.W. (2000a). MethyLight: a high-throughput assay to measure DNA methylation. *Nucleic Acids Res. 28*, e32.

Eads, C.A., Lord, R.V., Kurumboor, S.K., Wickramasinghe, K., Skinner, M.L., Long, T.I., Peters, J.H., DeMeester, T.R., Danenberg, K.D., Danenberg, P.V., *et al.* (2000b). Fields of aberrant CpG island hypermethylation in Barrett's esophagus and associated adenocarcinoma. *Cancer Res. 60*, 5021-5026.

Eads, C.A., Lord, R.V., Wickramasinghe, K., Long, T.I., Kurumboor, S.K., Bernstein, L., Peters, J.H., DeMeester, S.R., DeMeester, T.R., Skinner, K.A., and Laird, P.W. (2001). Epigenetic patterns in the progression of esophageal adenocarcinoma. *Cancer Res. 61*, 3410-3418.

Eden, A., Gaudet, F., Waghmare, A., and Jaenisch, R. (2003). Chromosomal instability and tumors promoted by DNA hypomethylation. *Science. 300*, 455.

Ehrich, M., Nelson, M.R., Stanssens, P., Zabeau, M., Liloglou, T., Xinarianos, G., Cantor, C.R., Field, J.K., and van den Boom, D. (2005). Quantitative high-throughput analysis of DNA methylation patterns by base-specific cleavage and mass spectrometry. *Proc. Natl. Acad. Sci. USA. 102*, 15785-15790.

Ehrlich, M. (2002). DNA methylation in cancer: too much, but also too little. *Oncogene. 21*, 5400-5413.

Ehrlich, M., Gama-Sosa, M.A., Huang, L.H., Midgett, R.M., Kuo, K.C., McCune, R.A., and Gehrke, C. (1982). Amount and distribution of 5-methylcytosine in human DNA from different types of tissues of cells. *Nucleic Acids Res. 10*, 2709-2721.

El-Maarri, O., Herbiniaux, U., Walter, J., and Oldenburg, J. (2002). A rapid, quantitative, non-radioactive bisulfite-SNuPE- IP RP HPLC assay for methylation analysis at specific CpG sites. *Nucleic Acids Res. 30*, e25.

Esteller, M., Corn, P.G., Baylin, S.B., and Herman, J.G. (2001a). A gene hypermethylation profile of human cancer. *Cancer Res. 61*, 3225-3229.

Esteller, M., Gaidano, G., Goodman, S.N., Zagonel, V., Capello, D., Botto, B., Rossi, D., Gloghini, A., Vitolo, U., Carbone, A., *et al.* (2002). Hypermethylation of the DNA repair gene O(6)-methylguanine DNA methyltransferase and survival of patients with diffuse large B-cell lymphoma. *J. Natl. Cancer Inst. 94*, 26-32.

Esteller, M., Garcia-Foncillas, J., Andion, E., Goodman, S.N., Hidalgo, O.F., Vanaclocha, V., Baylin, S.B., and Herman, J.G. (2000). Inactivation of the DNA-repair gene MGMT and the clinical response of gliomas to alkylating agents. *N. Engl. J. Med. 343*, 1350-1354.

Esteller, M., Gonzalez, S., Risques, R.A., Marcuello, E., Mangues, R., Germa, J.R., Herman, J.G., Capella, G., and Peinado, M.A. (2001b). K-ras and p16 aberrations confer poor prognosis in human colorectal cancer. *J. Clin. Oncol. 19*, 299-304.

Esteller, M., Sanchez-Cespedes, M., Rosell, R., Sidransky, D., Baylin, S.B., and Herman, J.G. (1999). Detection of aberrant promoter hypermethylation of tumor suppressor genes in serum DNA from non-small cell lung cancer patients. *Cancer Res. 59*, 67-70.

Evron, E., Dooley, W.C., Umbricht, C.B., Rosenthal, D., Sacchi, N., Gabrielson, E., Soito, A.B., Hung, D.T., Ljung, B., Davidson, N.E., and Sukumar, S. (2001). Detection of breast cancer cells in ductal lavage fluid by methylation-specific PCR. *Lancet. 357*, 1335-1336.

Fang, M., Chen, D., and Yang, C.S. (2007). Dietary polyphenols may affect DNA methylation. *J. Nutr. 137*, 223S-228S.

Feil, R. (2006). Environmental and nutritional effects on the epigenetic regulation of genes. *Mutat. Res. 600*, 46-57.

Feinberg, A.P., and Vogelstein, B. (1983). Hypomethylation distinguishes genes of some human cancers from their normal counterparts. *Nature. 301*, 89-92.

Feltus, F.A., Lee, E.K., Costello, J.F., Plass, C., and Vertino, P.M. (2006). DNA motifs associated with aberrant CpG island methylation. *Genomics. 87*, 572-579.

Ferguson, A.T., Lapidus, R.G., Baylin, S.B., and Davidson, N.E. (1995). Demethylation of the estrogen receptor gene in estrogen receptor-negative breast cancer cells can reactivate estrogen receptor gene expression. *Cancer Res. 55*, 2279-2283.

Fiegl, H., Millinger, S., Mueller-Holzner, E., Marth, C., Ensinger, C., Berger, A., Klocker, H., Goebel, G., and Widschwendter, M. (2005). Circulating tumor-specific DNA: a marker for monitoring efficacy of adjuvant therapy in cancer patients. *Cancer Res. 65*, 1141-1145.

Fischer, J.R., Ohnmacht, U., Rieger, N., Zemaitis, M., Stoffregen, C., Kostrzewa, M., Buchholz, E., Manegold, C., and Lahm, H. (2006). Promoter methylation of RASSF1A, RARbeta and DAPK predict poor prognosis of patients with malignant mesothelioma. *Lung Cancer. 54*, 109-116.

Fleischhacker, M., and Schmidt, B. (2007). Circulating nucleic acids (CNAs) and cancer--a survey. *Biochim. Biophys. Acta. 1775*, 181-232.

Fournel, M., Sapieha, P., Beaulieu, N., Besterman, J.M., and MacLeod, A.R. (1999). Down-regulation of human DNA-(cytosine-5) methyltransferase induces cell cycle regulators p16(ink4A) and p21(WAF/Cip1) by distinct mechanisms. *J. Biol. Chem. 274*, 24250-24256.

Fraga, M.F., and Esteller, M. (2002). DNA methylation: a profile of methods and applications. *Biotechniques. 33*, 632-649.

Friedrich, M.G., Weisenberger, D.J., Cheng, J.C., Chandrasoma, S., Siegmund, K.D., Gonzalgo, M.L., Toma, M.I., Huland, H., Yoo, C., Tsai, Y.C., *et al.* (2004). Detection of methylated apoptosis-associated genes in urine sediments of bladder cancer patients. *Clin. Cancer Res. 10*, 7457-7465.

Frigola, J., Song, J., Stirzaker, C., Hinshelwood, R.A., Peinado, M.A., and Clark, S.J. (2006). Epigenetic remodeling in colorectal cancer results in coordinate gene suppression across an entire chromosome band. *Nat. Genet. 38*, 540-549.

Frommer, M., McDonald, L.E., Millar, D.S., Collis, C.M., Watt, F., Grigg, G.W., Molloy, P.L., and Paul, C.L. (1992). A genomic sequencing protocol that yields a positive display of 5-methylcytosine residues in individual DNA strands. *Proc. Natl. Acad. Sci. USA.. 89*, 1827-1831.

Fujiwara, K., Fujimoto, N., Tabata, M., Nishii, K., Matsuo, K., Hotta, K., Kozuki, T., Aoe, M., Kiura, K., Ueoka, H., and Tanimoto, M. (2005). Identification of epigenetic aberrant promoter methylation in serum DNA is useful for early detection of lung cancer. *Clin. Cancer Res. 11*, 1219-1225.

Fukushima, N., Walter, K.M., Uek, T., Sato, N., Matsubayashi, H., Cameron, J.L., Hruban, R.H., Canto, M., Yeo, C.J., and Goggins, M. (2003). Diagnosing pancreatic cancer using methylation specific PCR analysis of pancreatic juice. *Cancer Biol. Ther. 2*, 78-83.

Fulda, S., and Debatin, K.M. (2006). 5-Aza-2'-deoxycytidine and IFN-gamma cooperate to sensitize for TRAIL-induced apoptosis by upregulating caspase-8. *Oncogene. 25*, 5125-5133.

Fulda, S., Kufer, M.U., Meyer, E., van Valen, F., Dockhorn-Dworniczak, B., and Debatin, K.M. (2001). Sensitization for death receptor- or drug-induced apoptosis by re-expression of caspase-8 through demethylation or gene transfer. *Oncogene. 20*, 5865-5877.

Garcia-Manero, G., Kantarjian, H.M., Sanchez-Gonzalez, B., Yang, H., Rosner, G., Verstovsek, S., Rytting, M., Wierda, W.G., Ravandi, F., Koller, C., *et al.* (2006). Phase 1/2 study of the combination of 5-aza-2'-deoxycytidine with valproic acid in patients with leukemia. *Blood. 108*, 3271-3279.

Garcia, J.M., Silva, J., Pena, C., Garcia, V., Rodriguez, R., Cruz, M.A., Cantos, B., Provencio, M., Espana, P., and Bonilla, F. (2004). Promoter methylation of the PTEN gene is a common molecular change in breast cancer. *Genes Chromosomes Cancer. 41*, 117-124.

Ghoshal, K., Datta, J., Majumder, S., Bai, S., Kutay, H., Motiwala, T., and Jacob, S.T. (2005). 5-Aza-deoxycytidine induces selective degradation of DNA methyltransferase 1 by a proteasomal pathway that requires the KEN box, bromo-adjacent homology domain, and nuclear localization signal. *Mol. Cell Biol. 25*, 4727-4741.

Gifford, G., Paul, J., Vasey, P.A., Kaye, S.B., and Brown, R. (2004). The acquisition of hMLH1 methylation in plasma DNA after chemotherapy predicts poor survival for ovarian cancer patients. *Clin. Cancer Res. 10*, 4420-4426.

Goelz, S.E., Vogelstein, B., Hamilton, S.R., and Feinberg, A.P. (1985). Hypomethylation of DNA from benign and malignant human colon neoplasms. *Science. 228*, 187-190.

Goessl, C., Krause, H., Muller, M., Heicappell, R., Schrader, M., Sachsinger, J., and Miller, K. (2000). Fluorescent methylation-specific polymerase chain reaction for DNA-based detection of prostate cancer in bodily fluids. *Cancer Res. 60*, 5941-5945.

Goffin, J., and Eisenhauer, E. (2002). DNA methyltransferase inhibitors-state of the art. *Ann. Oncol. 13*, 1699-1716.

Gonzalgo, M.L., and Jones, P.A. (1997). Rapid quantitation of methylation differences at specific sites using methylation-sensitive single nucleotide primer extension (Ms-SNuPE). *Nucleic Acids Res. 25*, 2529-2531.

Gonzalgo, M.L., Liang, G., Spruck, C.H., 3rd, Zingg, J.M., Rideout, W.M., 3rd, and Jones, P.A. (1997). Identification and characterization of differentially methylated regions of genomic DNA by methylation-sensitive arbitrarily primed PCR. *Cancer Res. 57*, 594-599.

Gonzalgo, M.L., Yegnasubramanian, S., Yan, G., Rogers, C.G., Nicol, T.L., Nelson, W.G., and Pavlovich, C.P. (2004). Molecular profiling and classification of sporadic renal cell carcinoma by quantitative methylation analysis. *Clin. Cancer Res. 10*, 7276-7283.

Gore, S.D., Jones, C., and Kirkpatrick, P. (2006). Decitabine. *Nat. Rev. Drug Discov. 5*, 891-892.

Gray, M.A. (2005). Clinical use of serum prostate-specific antigen: a review. *Clin. Lab. 51*, 127-133.

Gu, J., Berman, D., Lu, C., Wistuba, II, Roth, J.A., Frazier, M., Spitz, M.R., and Wu, X. (2006). Aberrant promoter methylation profile and association with survival in patients with non-small cell lung cancer. *Clin. Cancer Res. 12*, 7329-7338.

Guo, M., House, M.G., Hooker, C., Han, Y., Heath, E., Gabrielson, E., Yang, S.C., Baylin, S.B., Herman, J.G., and Brock, M.V. (2004). Promoter hypermethylation of resected bronchial margins: a field defect of changes? *Clin. Cancer Res. 10*, 5131-5136.

Hedenfalk, I., Duggan, D., Chen, Y., Radmacher, M., Bittner, M., Simon, R., Meltzer, P., Gusterson, B., Esteller, M., Kallioniemi, O.P., *et al.* (2001). Gene-expression profiles in hereditary breast cancer. *N. Engl. J. Med. 344*, 539-548.

Hegi, M.E., Diserens, A.C., Godard, S., Dietrich, P.Y., Regli, L., Ostermann, S., Otten, P., Van Melle, G., de Tribolet, N., and Stupp, R. (2004). Clinical trial substantiates the predictive value of O-6-methylguanine-DNA methyltransferase promoter methylation in glioblastoma patients treated with temozolomide. *Clin. Cancer Res. 10*, 1871-1874.

Herman, J.G., Graff, J.R., Myohanen, S., Nelkin, B.D., and Baylin, S.B. (1996). Methylation-specific PCR: a novel PCR assay for methylation status of CpG islands. *Proc. Natl. Acad. Sci. USA. 93*, 9821-9826.

Holliday, R., and Grigg, G.W. (1993). DNA methylation and mutation. *Mutat. Res. 285*, 61-67.

Holst, C.R., Nuovo, G.J., Esteller, M., Chew, K., Baylin, S.B., Herman, J.G., and Tlsty, T.D. (2003). Methylation of p16(INK4a) promoters occurs in vivo in histologically normal human mammary epithelia. *Cancer Res. 63*, 1596-1601.

Hoque, M.O., Begum, S., Topaloglu, O., Chatterjee, A., Rosenbaum, E., Van Criekinge, W., Westra, W.H., Schoenberg, M., Zahurak, M., Goodman, S.N., and Sidransky, D. (2006). Quantitation of promoter methylation of multiple genes in urine DNA and bladder cancer detection. *J. Natl. Cancer Inst. 98*, 996-1004.

Hoque, M.O., Begum, S., Topaloglu, O., Jeronimo, C., Mambo, E., Westra, W.H., Califano, J.A., and Sidransky, D. (2004). Quantitative detection of promoter hypermethylation of multiple genes in the tumor, urine, and serum DNA of patients with renal cancer. *Cancer Res. 64*, 5511-5517.

House, M.G., Guo, M., Efron, D.T., Lillemoe, K.D., Cameron, J.L., Syphard, J.E., Hooker, C.M., Abraham, S.C., Montgomery, E.A., Herman, J.G., and Brock, M.V. (2003). Tumor suppressor gene hypermethylation as a predictor of gastric stromal tumor behavior. *J. Gastrointest. Surg. 7*, 1004-1014; discussion 1014.

Hsu, H.S., Wen, C.K., Tang, Y.A., Lin, R.K., Li, W.Y., Hsu, W.H., and Wang, Y.C. (2005). Promoter hypermethylation is the predominant mechanism in hMLH1 and hMSH2 deregulation and is a poor prognostic factor in nonsmoking lung cancer. *Clin. Cancer Res. 11*, 5410-5416.

Hu, X.C., Wong, I.H., and Chow, L.W. (2003). Tumor-derived aberrant methylation in plasma of invasive ductal breast cancer patients: clinical implications. *Oncol. Rep. 10*, 1811-1815.

Huang, T.H., Perry, M.R., and Laux, D.E. (1999). Methylation profiling of CpG islands in human breast cancer cells. *Hum. Mol. Genet. 8*, 459-470.

Ibanez de Caceres, I., Battagli, C., Esteller, M., Herman, J.G., Dulaimi, E., Edelson, M.I., Bergman, C., Ehya, H., Eisenberg, B.L., and Cairns, P. (2004). Tumor cell-specific BRCA1 and RASSF1A hypermethylation in serum, plasma, and peritoneal fluid from ovarian cancer patients. *Cancer Res. 64*, 6476-6481.

Issa, J.P. (2004). CpG island methylator phenotype in cancer. *Nat. Rev. Cancer. 4*, 988-993.

Issa, J.P., Ahuja, N., Toyota, M., Bronner, M.P., and Brentnall, T.A. (2001). Accelerated age-related CpG island methylation in ulcerative colitis. *Cancer Res. 61*, 3573-3577.

Issa, J.P., Garcia-Manero, G., Giles, F.J., Mannari, R., Thomas, D., Faderl, S., Bayar, E., Lyons, J., Rosenfeld, C.S., Cortes, J., and Kantarjian, H.M. (2004). Phase 1 study of low-dose prolonged exposure schedules of the hypomethylating agent 5-aza-2'-deoxycytidine (decitabine) in hematopoietic malignancies. *Blood. 103*, 1635-1640.

Issa, J.P., Gharibyan, V., Cortes, J., Jelinek, J., Morris, G., Verstovsek, S., Talpaz, M., Garcia-Manero, G., and Kantarjian, H.M. (2005). Phase II study of low-dose decitabine in patients with chronic myelogenous leukemia resistant to imatinib mesylate. *J. Clin. Oncol. 23*, 3948-3956.

Issa, J.P., Ottaviano, Y.L., Celano, P., Hamilton, S.R., Davidson, N.E., and Baylin, S.B. (1994). Methylation of the oestrogen receptor CpG island links ageing and neoplasia in human colon. *Nat. Genet. 7*, 536-540.

Jahr, S., Hentze, H., Englisch, S., Hardt, D., Fackelmayer, F.O., Hesch, R.D., and Knippers, R. (2001). DNA fragments in the blood plasma of cancer patients: quantitations and evidence for their origin from apoptotic and necrotic cells. *Cancer Res. 61*, 1659-1665.

Jones, P.A., and Baylin, S.B. (2002). The fundamental role of epigenetic events in cancer. *Nat. Rev. Genet. 3*, 415-428.

Juttermann, R., Li, E., and Jaenisch, R. (1994). Toxicity of 5-aza-2'-deoxycytidine to mammalian cells is mediated primarily by covalent trapping of DNA methyltransferase rather than DNA demethylation. *Proc. Natl. Acad. Sci. USA. 91*, 11797-11801.

Kaminskas, E., Farrell, A., Abraham, S., Baird, A., Hsieh, L.S., Lee, S.L., Leighton, J.K., Patel, H., Rahman, A., Sridhara, R., *et al.* (2005). Approval summary: azacitidine for treatment of myelodysplastic syndrome subtypes. *Clin. Cancer Res. 11*, 3604-3608.

Kaminsky, Z.A., Assadzadeh, A., Flanagan, J., and Petronis, A. (2005). Single nucleotide extension technology for quantitative site-specific evaluation of metC/C in GC-rich regions. *Nucleic Acids Res. 33*, e95.

Kanai, Y., Ushijima, S., Tsuda, H., Sakamoto, M., and Hirohashi, S. (2000). Aberrant DNA methylation precedes loss of heterozygosity on chromosome 16 in chronic hepatitis and liver cirrhosis. *Cancer Lett. 148*, 73-80.

Kantarjian, H., Issa, J.P., Rosenfeld, C.S., Bennett, J.M., Albitar, M., DiPersio, J., Klimek, V., Slack, J., de Castro, C., Ravandi, F., *et al.* (2006). Decitabine improves patient outcomes in myelodysplastic syndromes: results of a phase III randomized study. *Cancer. 106*, 1794-1803.

Kantharidis, P., El-Osta, A., deSilva, M., Wall, D.M., Hu, X.F., Slater, A., Nadalin, G., Parkin, J.D., and Zalcberg, J.R. (1997). Altered methylation of the human MDR1 promoter is associated with acquired multidrug resistance. *Clin. Cancer Res. 3*, 2025-2032.

Kawakami, K., Brabender, J., Lord, R.V., Groshen, S., Greenwald, B.D., Krasna, M.J., Yin, J., Fleisher, A.S., Abraham, J.M., Beer, D.G., *et al.* (2000). Hypermethylated APC DNA in plasma and prognosis of patients with esophageal adenocarcinoma. *J. Natl. Cancer Inst. 92*, 1805-1811.

Kawakami, K., Ruszkiewicz, A., Bennett, G., Moore, J., Grieu, F., Watanabe, G., and Iacopetta, B. (2006). DNA hypermethylation in the normal colonic mucosa of patients with colorectal cancer. *Br. J. Cancer. 94*, 593-598.

Kawakami, T., Okamoto, K., Ogawa, O., and Okada, Y. (2004). XIST unmethylated DNA fragments in male-derived plasma as a tumour marker for testicular cancer. *Lancet. 363*, 40-42.

Kersting, M., Friedl, C., Kraus, A., Behn, M., Pankow, W., and Schuermann, M. (2000). Differential frequencies of p16(INK4a) promoter hypermethylation, p53 mutation, and K-ras mutation in exfoliative material mark the development of lung cancer in symptomatic chronic smokers. *J. Clin. Oncol. 18*, 3221-3229.

Kim, D.H., Nelson, H.H., Wiencke, J.K., Zheng, S., Christiani, D.C., Wain, J.C., Mark, E.J., and Kelsey, K.T. (2001). p16(INK4a) and histology-specific methylation of CpG islands by exposure to tobacco smoke in non-small cell lung cancer. *Cancer Res. 61*, 3419-3424.

Knudson, A.G. (2001). Two genetic hits (more or less) to cancer. Nat Rev Cancer *1*, 157-162.

Koga, Y., Kitajima, Y., Miyoshi, A., Sato, K., Kitahara, K., Soejima, H., and Miyazaki, K. (2005). Tumor progression through epigenetic gene silencing of O(6)-methylguanine-DNA methyltransferase in human biliary tract cancers. *Ann. Surg. Oncol. 12*, 354-363.

Kondo, Y., Kanai, Y., Sakamoto, M., Mizokami, M., Ueda, R., and Hirohashi, S. (2000). Genetic instability and aberrant DNA methylation in chronic hepatitis and cirrhosis--A comprehensive study of loss of heterozygosity and microsatellite instability at 39 loci and DNA hypermethylation on 8 CpG islands in microdissected specimens from patients with hepatocellular carcinoma. *Hepatology. 32*, 970-979.

Kornblith, A.B., Herndon, J.E., 2nd, Silverman, L.R., Demakos, E.P., Odchimar-Reissig, R., Holland, J.F., Powell, B.L., DeCastro, C., Ellerton, J., Larson, R.A., *et al.* (2002). Impact of azacytidine on the quality of life of patients with myelodysplastic syndrome treated in a randomized phase III trial: a Cancer and Leukemia Group B study. *J. Clin. Oncol. 20*, 2441-2452.

Koul, S., McKiernan, J.M., Narayan, G., Houldsworth, J., Bacik, J., Dobrzynski, D.L., Assaad, A.M., Mansukhani, M., Reuter, V.E., Bosl, G.J., *et al.* (2004). Role of promoter hypermethylation in Cisplatin treatment response of male germ cell tumors. *Mol. Cancer. 3*, 16.

Krassenstein, R., Sauter, E., Dulaimi, E., Battagli, C., Ehya, H., Klein-Szanto, A., and Cairns, P. (2004). Detection of breast cancer in nipple aspirate fluid by CpG island hypermethylation. *Clin. Cancer Res. 10*, 28-32.

Laird, P.W. (2003). Early detection: The power and the promise of DNA methylation markers. *Nat. Rev. Cancer. 3*, 253-266.

Laird, P.W. (2005). Cancer epigenetics. *Hum. Mol. Genet. 14 Spec No 1*, R65-76.

Laliberte, J., Marquez, V.E., and Momparler, R.L. (1992). Potent inhibitors for the deamination of cytosine arabinoside and 5-aza-2'-deoxycytidine by human cytidine deaminase. *Cancer Chemother. Pharmacol. 30*, 7-11.

Lecomte, T., Berger, A., Zinzindohoue, F., Micard, S., Landi, B., Blons, H., Beaune, P., Cugnenc, P.H., and Laurent-Puig, P. (2002). Detection of free-circulating tumor-associated DNA in plasma of colorectal cancer patients and its association with prognosis. *Int. J. Cancer. 100*, 542-548.

Lee, B.H., Yegnasubramanian, S., Lin, X., and Nelson, W.G. (2005). Procainamide is a specific inhibitor of DNA methyltransferase 1. *J. Biol. Chem. 280*, 40749-40756.

Lehmann, U., Berg-Ribbe, I., Wingen, L.U., Brakensiek, K., Becker, T., Klempnauer, J., Schlegelberger, B., Kreipe, H., and Flemming, P. (2005). Distinct methylation patterns of benign and malignant liver tumors revealed by quantitative methylation profiling. *Clin. Cancer Res. 11*, 3654-3660.

Leung, W.K., To, K.F., Chu, E.S., Chan, M.W., Bai, A.H., Ng, E.K., Chan, F.K., and Sung, J.J. (2005). Potential diagnostic and prognostic values of detecting promoter hypermethylation in the serum of patients with gastric cancer. *Br. J. Cancer. 92*, 2190-2194.

Lewis, C.M., Cler, L.R., Bu, D.W., Zochbauer-Muller, S., Milchgrub, S., Naftalis, E.Z., Leitch, A.M., Minna, J.D., and Euhus, D.M. (2005). Promoter hypermethylation in benign breast epithelium in relation to predicted breast cancer risk. *Clin. Cancer Res. 11*, 166-172.

Li, L.C., Zhao, H., Nakajima, K., Oh, B.R., Ribeiro Filho, L.A., Carroll, P., and Dahiya, R. (2001). Methylation of the E-cadherin gene promoter correlates with progression of prostate cancer. *J. Urol. 166*, 705-709.

Lin, X., Asgari, K., Putzi, M.J., Gage, W.R., Yu, X., Cornblatt, B.S., Kumar, A., Piantadosi, S., DeWeese, T.L., De Marzo, A.M., and Nelson, W.G. (2001). Reversal of GSTP1 CpG island hypermethylation and reactivation of pi-class glutathione S-transferase (GSTP1) expression in human prostate cancer cells by treatment with procainamide. *Cancer Res. 61*, 8611-8616.

Lopez, M., Aguirre, J.M., Cuevas, N., Anzola, M., Videgain, J., Aguirregaviria, J., and Martinez de Pancorbo, M. (2003). Gene promoter hypermethylation in oral rinses of leukoplakia patients--a diagnostic and/or prognostic tool? *Eur. J. Cancer. 39*, 2306-2309.

Mackay, H.J., Cameron, D., Rahilly, M., Mackean, M.J., Paul, J., Kaye, S.B., and Brown, R. (2000). Reduced MLH1 expression in breast tumors after primary chemotherapy predicts disease-free survival. *J. Clin. Oncol. 18*, 87-93.

Maekita, T., Nakazawa, K., Mihara, M., Nakajima, T., Yanaoka, K., Iguchi, M., Arii, K., Kaneda, A., Tsukamoto, T., Tatematsu, M., *et al.* (2006). High levels of aberrant DNA methylation in Helicobacter pylori-infected gastric mucosae and its possible association with gastric cancer risk. *Clin. Cancer Res. 12*, 989-995.

Maesawa, C., Tamura, G., Nishizuka, S., Ogasawara, S., Ishida, K., Terashima, M., Sakata, K., Sato, N., Saito, K., and Satodate, R. (1996). Inactivation of the CDKN2 gene by homozygous deletion and de novo methylation is associated with advanced stage esophageal squamous cell carcinoma. *Cancer Res. 56*, 3875-3878.

Mandelker, D.L., Yamashita, K., Tokumaru, Y., Mimori, K., Howard, D.L., Tanaka, Y., Carvalho, A.L., Jiang, W.W., Park, H.L., Kim, M.S., *et al.* (2005). PGP9.5 promoter methylation is an independent prognostic factor for esophageal squamous cell carcinoma. *Cancer Res. 65*, 4963-4968.

Martens, J.W., Nimmrich, I., Koenig, T., Look, M.P., Harbeck, N., Model, F., Kluth, A., Bolt-de Vries, J., Sieuwerts, A.M., Portengen, H., *et al.* (2005). Association of DNA methylation of phosphoserine aminotransferase with response to endocrine therapy in patients with recurrent breast cancer. *Cancer Res. 65*, 4101-4117.

Maruyama, R., Toyooka, S., Toyooka, K.O., Virmani, A.K., Zochbauer-Muller, S., Farinas, A.J., Minna, J.D., McConnell, J., Frenkel, E.P., and Gazdar, A.F. (2002). Aberrant promoter methylation profile of prostate cancers and its relationship to clinicopathological features. *Clin. Cancer Res. 8*, 514-519.

Matsubayashi, H., Canto, M., Sato, N., Klein, A., Abe, T., Yamashita, K., Yeo, C.J., Kalloo, A., Hruban, R., and Goggins, M. (2006). DNA methylation alterations in the pancreatic juice of patients with suspected pancreatic disease. *Cancer Res. 66*, 1208-1217.

Mentor-Marcel, R., Lamartiniere, C.A., Eltoum, I.A., Greenberg, N.M., and Elgavish, A. (2005). Dietary genistein improves survival and reduces expression of osteopontin in the prostate of transgenic mice with prostatic adenocarcinoma (TRAMP). *J. Nutr. 135*, 989-995.

Mori, T., O'Day, S.J., Umetani, N., Martinez, S.R., Kitago, M., Koyanagi, K., Kuo, C., Takeshima, T.L., Milford, R., Wang, H.J., *et al.* (2005). Predictive utility of circulating methylated DNA in serum of melanoma patients receiving biochemotherapy. *J. Clin. Oncol. 23*, 9351-9358.

Müller, H.M., Millinger, S., Fiegl, H., Goebel, G., Ivarsson, L., Widschwendter, A., Muller-Holzner, E., Marth, C., and Widschwendter, M. (2004a). Analysis of methylated genes in peritoneal fluids of ovarian cancer patients: a new prognostic tool. *Clin. Chem. 50*, 2171-2173.

Müller, H.M., Oberwalder, M., Fiegl, H., Morandell, M., Goebel, G., Zitt, M., Muhlthaler, M., Ofner, D., Margreiter, R., and Widschwendter, M. (2004b). Methylation changes in faecal DNA: a marker for colorectal cancer screening? *Lancet. 363*, 1283-1285.

Müller, H.M., Widschwendter, A., Fiegl, H., Ivarsson, L., Goebel, G., Perkmann, E., Marth, C., and Widschwendter, M. (2003). DNA methylation in serum of breast cancer patients: an independent prognostic marker. *Cancer Res. 63*, 7641-7645.

Nagasaka, T., Sharp, G.B., Notohara, K., Kambara, T., Sasamoto, H., Isozaki, H., MacPhee, D.G., Jass, J.R., Tanaka, N., and Matsubara, N. (2003). Hypermethylation of O6-methylguanine-DNA methyltransferase promoter may predict nonrecurrence after chemotherapy in colorectal cancer cases. *Clin. Cancer Res. 9*, 5306-5312.

Nakayama, M., Gonzalgo, M.L., Yegnasubramanian, S., Lin, X., De Marzo, A.M., and Nelson, W.G. (2004). GSTP1 CpG island hypermethylation as a molecular biomarker for prostate cancer. *J. Cell Biochem. 91*, 540-552.

Nguyen, C., Liang, G., Nguyen, T.T., Tsao-Wei, D., Groshen, S., Lubbert, M., Zhou, J.H., Benedict, W.F., and Jones, P.A. (2001). Susceptibility of nonpromoter CpG islands to de novo methylation in normal and neoplastic cells. *J. Natl. Cancer Inst. 93*, 1465-1472.

Novak, P., Jensen, T., Oshiro, M.M., Wozniak, R.J., Nouzova, M., Watts, G.S., Klimecki, W.T., Kim, C., and Futscher, B.W. (2006). Epigenetic inactivation of the HOXA gene cluster in breast cancer. *Cancer Res. 66*, 10664-10670.

Olaussen, K.A., Soria, J.C., Park, Y.W., Kim, H.J., Kim, S.H., Ro, J.Y., Andre, F., and Jang, S.J. (2005). Assessing abnormal gene promoter methylation in paraffin-embedded sputum from patients with NSCLC. *Eur. J. Cancer. 41*, 2112-2119.

Palmisano, W.A., Divine, K.K., Saccomanno, G., Gilliland, F.D., Baylin, S.B., Herman, J.G., and Belinsky, S.A. (2000). Predicting lung cancer by detecting aberrant promoter methylation in sputum. *Cancer Res. 60*, 5954-5958.

Pao, M.M., Tsutsumi, M., Liang, G., Uzvolgyi, E., Gonzales, F.A., and Jones, P.A. (2001). The endothelin receptor B (EDNRB) promoter displays heterogeneous, site specific methylation patterns in normal and tumor cells. *Hum. Mol. Genet. 10*, 903-910.

Paz, M.F., Yaya-Tur, R., Rojas-Marcos, I., Reynes, G., Pollan, M., Aguirre-Cruz, L., Garcia-Lopez, J.L., Piquer, J., Safont, M.J., Balana, C., *et al.* (2004). CpG island hypermethylation of the DNA repair enzyme methyltransferase predicts response to temozolomide in primary gliomas. *Clin. Cancer Res. 10*, 4933-4938.

Piskala, A., and Sorm, F. (1964). Nucleic acids components and their analogues. LI. Synthesis of 1-glycosyl derivates of 5-azauracil and 5-azacytosine. *Coll. Czeck. Chem. Commun. 29*, 2060-2076.

Pliml, J., and Sorm, F. (1964). Synthesis of 2'-deoxy-D-ribofuranosyl-5-azacytosine. *Coll. Czeck. Chem. Commun. 29*, 2576-2577.

Plumb, J.A., Strathdee, G., Sludden, J., Kaye, S.B., and Brown, R. (2000). Reversal of drug resistance in human tumor xenografts by 2'-deoxy-5-azacytidine-induced demethylation of the hMLH1 gene promoter. *Cancer Res. 60*, 6039-6044.

Pohlmann, P., DiLeone, L.P., Cancella, A.I., Caldas, A.P., Dal Lago, L., Campos, O., Jr., Monego, E., Rivoire, W., and Schwartsmann, G. (2002). Phase II trial of cisplatin plus decitabine, a new DNA hypomethylating agent, in patients with advanced squamous cell carcinoma of the cervix. *Am. J. Clin. Oncol. 25*, 496-501.

Ramirez, J.L., Rosell, R., Taron, M., Sanchez-Ronco, M., Alberola, V., de Las Penas, R., Sanchez, J.M., Moran, T., Camps, C., Massuti, B., *et al.* (2005). 14-3-3sigma methylation in pretreatment serum circulating DNA of cisplatin-plus-gemcitabine-treated advanced non-small-cell lung cancer patients predicts survival: The Spanish Lung Cancer Group. *J. Clin. Oncol. 23*, 9105-9112.

Reik, W., Dean, W., and Walter, J. (2001). Epigenetic reprogramming in mammalian development. *Science. 293*, 1089-1093.

Reik, W., and Walter, J. (2001). Genomic imprinting: parental influence on the genome. *Nat. Rev. Genet. 2*, 21-32.

Reu, F.J., Bae, S.I., Cherkassky, L., Leaman, D.W., Lindner, D., Beaulieu, N., MacLeod, A.R., and Borden, E.C. (2006a). Overcoming resistance to interferon-induced apoptosis of renal carcinoma and melanoma cells by DNA demethylation. *J. Clin. Oncol. 24*, 3771-3779.

Reu, F.J., Leaman, D.W., Maitra, R.R., Bae, S.I., Cherkassky, L., Fox, M.W., Rempinski, D.R., Beaulieu, N., MacLeod, A.R., and Borden, E.C. (2006b). Expression of RASSF1A, an epigenetically silenced tumor suppressor, overcomes resistance to apoptosis induction by interferons. *Cancer Res. 66*, 2785-2793.

Richardson, B. (2003). DNA methylation and autoimmune disease. *Clin. Immunol. 109*, 72-79.

Robertson, K.D. (2005). DNA methylation and human disease. *Nat. Rev. Genet. 6*, 597-610.

Rosas, S.L., Koch, W., da Costa Carvalho, M.G., Wu, L., Califano, J., Westra, W., Jen, J., and Sidransky, D. (2001). Promoter hypermethylation patterns of p16, O6-methylguanine-DNA-methyltransferase, and death-associated protein kinase in tumors and saliva of head and neck cancer patients. *Cancer Res. 61*, 939-942.

Rosenbaum, E., Hoque, M.O., Cohen, Y., Zahurak, M., Eisenberger, M.A., Epstein, J.I., Partin, A.W., and Sidransky, D. (2005). Promoter hypermethylation as an independent prognostic factor for relapse in patients with prostate cancer following radical prostatectomy. *Clin. Cancer Res. 11*, 8321-8325.

Rudek, M.A., Zhao, M., He, P., Hartke, C., Gilbert, J., Gore, S.D., Carducci, M.A., and Baker, S.D. (2005). Pharmacokinetics of 5-azacitidine administered with phenylbutyrate in patients with refractory solid tumors or hematologic malignancies. *J. Clin. Oncol. 23*, 3906-3911.

Rush, L.J., and Plass, C. (2002). Restriction landmark genomic scanning for DNA methylation in cancer: past, present, and future applications. *Anal. Biochem. 307*, 191-201.

Sanchez-Cespedes, M., Esteller, M., Wu, L., Nawroz-Danish, H., Yoo, G.H., Koch, W.M., Jen, J., Herman, J.G., and Sidransky, D. (2000). Gene promoter hypermethylation in tumors and serum of head and neck cancer patients. *Cancer Res. 60*, 892-895.

Satoh, A., Toyota, M., Itoh, F., Sasaki, Y., Suzuki, H., Ogi, K., Kikuchi, T., Mita, H., Yamashita, T., Kojima, T., *et al.* (2003). Epigenetic inactivation of CHFR and sensitivity to microtubule inhibitors in gastric cancer. *Cancer Res. 63*, 8606-8613.

Schumacher, A., and Petronis, A. (2006). Epigenetics of complex diseases: from general theory to laboratory experiments. *Curr. Top Microbiol. Immunol. 310*, 81-115.

Schwartsmann, G., Schunemann, H., Gorini, C.N., Filho, A.F., Garbino, C., Sabini, G., Muse, I., DiLeone, L., and Mans, D.R. (2000). A phase I trial of cisplatin plus decitabine, a new DNA-hypomethylating agent, in patients with advanced solid tumors and a follow-up early phase II evaluation in patients with inoperable non-small cell lung cancer. *Invest. New Drugs. 18*, 83-91.

Segura-Pacheco, B., Trejo-Becerril, C., Perez-Cardenas, E., Taja-Chayeb, L., Mariscal, I., Chavez, A., Acuna, C., Salazar, A.M., Lizano, M., and Duenas-Gonzalez, A. (2003). Reactivation of tumor suppressor genes by the cardiovascular drugs hydralazine and procainamide and their potential use in cancer therapy. *Clin. Cancer Res. 9*, 1596-1603.

Sharma, D., Blum, J., Yang, X., Beaulieu, N., Macleod, A.R., and Davidson, N.E. (2005). Release of methyl CpG binding proteins and histone deacetylase 1 from the Estrogen receptor alpha (ER) promoter upon reactivation in ER-negative human breast cancer cells. *Mol. Endocrinol. 19*, 1740-1751.

Shen, L., Kondo, Y., Rosner, G.L., Xiao, L., Hernandez, N.S., Vilaythong, J., Houlihan, P.S., Krouse, R.S., Prasad, A.R., Einspahr, J.G., *et al.* (2005). MGMT promoter methylation and field defect in sporadic colorectal cancer. *J. Natl. Cancer Inst. 97*, 1330-1338.

Shi, H., Maier, S., Nimmrich, I., Yan, P.S., Caldwell, C.W., Olek, A., and Huang, T.H. (2003). Oligonucleotide-based microarray for DNA methylation analysis: principles and applications. *J. Cell Biochem. 88*, 138-143.

Silva, J.M., Dominguez, G., Garcia, J.M., Gonzalez, R., Villanueva, M.J., Navarro, F., Provencio, M., San Martin, S., Espana, P., and Bonilla, F. (1999). Presence of tumor DNA in plasma of breast cancer patients: clinicopathological correlations. *Cancer Res. 59*, 3251-3256.

Simon, R. (2005). Roadmap for developing and validating therapeutically relevant genomic classifiers. *J. Clin. Oncol. 23*, 7332-7341.

Singer-Sam, J., LeBon, J.M., Tanguay, R.L., and Riggs, A.D. (1990). A quantitative HpaII-PCR assay to measure methylation of DNA from a small number of cells. *Nucleic Acids Res. 18*, 687.

Smith, A.E., and Ford, K.G. (2006). Specific targeting of cytosine methylation to DNA sequences in vivo. *Nucleic Acids Res. 35*, 740-754.

Soengas, M.S., Capodieci, P., Polsky, D., Mora, J., Esteller, M., Opitz-Araya, X., McCombie, R., Herman, J.G., Gerald, W.L., Lazebnik, Y.A., *et al.* (2001). Inactivation of the apoptosis effector Apaf-1 in malignant melanoma. *Nature. 409*, 207-211.

Stewart, D.J., Donehower, R.C., Eisenhauer, E.A., Wainman, N., Shah, A.K., Bonfils, C., MacLeod, A.R., Besterman, J.M., and Reid, G.K. (2003). A phase I pharmacokinetic and pharmacodynamic study of the DNA methyltransferase 1 inhibitor MG98 administered twice weekly. *Ann. Oncol. 14*, 766-774.

Stransky, N., Vallot, C., Reyal, F., Bernard-Pierrot, I., de Medina, S.G., Segraves, R., de Rycke, Y., Elvin, P., Cassidy, A., Spraggon, C., *et al.* (2006). Regional copy number-independent deregulation of transcription in cancer. *Nat. Genet. 38*, 1386-1396.

Strathdee, G., Davies, B.R., Vass, J.K., Siddiqui, N., and Brown, R. (2004). Cell type-specific methylation of an intronic CpG island controls expression of the MCJ gene. *Carcinogenesis 25*, 693-701.

Strathdee, G., MacKean, M.J., Illand, M., and Brown, R. (1999). A role for methylation of the hMLH1 promoter in loss of hMLH1 expression and drug resistance in ovarian cancer. *Oncogene 18*, 2335-2341.

Strathdee, G., Vass, J.K., Oien, K.A., Siddiqui, N., Curto-Garcia, J., and Brown, R. (2005). Demethylation of the MCJ gene in stage III/IV epithelial ovarian cancer and response to chemotherapy. *Gynecol. Oncol. 97*, 898-903.

Suzuki, H., Itoh, F., Toyota, M., Kikuchi, T., Kakiuchi, H., and Imai, K. (2000). Inactivation of the 14-3-3 sigma gene is associated with 5' CpG island hypermethylation in human cancers. *Cancer Res. 60*, 4353-4357.

Syed, N., Smith, P., Sullivan, A., Spender, L.C., Dyer, M., Karran, L., O'Nions, J., Allday, M., Hoffmann, I., Crawford, D., *et al.* (2006). Transcriptional silencing of Polo-like kinase 2 (SNK/PLK2) is a frequent event in B-cell malignancies. *Blood. 107*, 250-256.

Szyf, M., Pakneshan, P., and Rabbani, S.A. (2004). DNA methylation and breast cancer. *Biochem. Pharmacol. 68*, 1187-1197.

Tada, Y., Wada, M., Taguchi, K., Mochida, Y., Kinugawa, N., Tsuneyoshi, M., Naito, S., and Kuwano, M. (2002). The association of death-associated protein kinase hypermethylation with early recurrence in superficial bladder cancers. *Cancer Res. 62*, 4048-4053.

Takahashi, T., Shivapurkar, N., Reddy, J., Shigematsu, H., Miyajima, K., Suzuki, M., Toyooka, S., Zochbauer-Muller, S., Drach, J., Parikh, G., *et al.* (2004a). DNA methylation profiles of lymphoid and hematopoietic malignancies. *Clin. Cancer Res. 10*, 2928-2935.

Takahashi, T., Shivapurkar, N., Riquelme, E., Shigematsu, H., Reddy, J., Suzuki, M., Miyajima, K., Zhou, X., Bekele, B.N., Gazdar, A.F., and Wistuba, II (2004b). Aberrant promoter hypermethylation of multiple genes in gallbladder carcinoma and chronic cholecystitis. *Clin. Cancer Res. 10*, 6126-6133.

Tang, X., Khuri, F.R., Lee, J.J., Kemp, B.L., Liu, D., Hong, W.K., and Mao, L. (2000). Hypermethylation of the death-associated protein (DAP) kinase promoter and aggressiveness in stage I non-small-cell lung cancer. *J. Natl. Cancer Inst. 92*, 1511-1516.

Taniguchi, T., Tischkowitz, M., Ameziane, N., Hodgson, S.V., Mathew, C.G., Joenje, H., Mok, S.C., and D'Andrea, A.D. (2003). Disruption of the Fanconi anemia-BRCA pathway in cisplatin-sensitive ovarian tumors. *Nat. Med. 9*, 568-574.

Teodoridis, J.M., Hall, J., Marsh, S., Kannall, H.D., Smyth, C., Curto, J., Siddiqui, N., Gabra, H., McLeod, H.L., Strathdee, G., and Brown, R. (2005). CpG island methylation of DNA damage response genes in advanced ovarian cancer. *Cancer Res. 65*, 8961-8967.

Teodoridis, J.M., Strathdee, G., and Brown, R. (2004). Epigenetic silencing mediated by CpG island methylation: potential as a therapeutic target and as a biomarker. *Drug Resist. Updat. 7*, 267-278.

Thomassin, H., Kress, C., and Grange, T. (2004). MethylQuant: a sensitive method for quantifying methylation of specific cytosines within the genome. *Nucleic Acids Res. 32*, e168.

Topaloglu, O., Hoque, M.O., Tokumaru, Y., Lee, J., Ratovitski, E., Sidransky, D., and Moon, C.S. (2004). Detection of promoter hypermethylation of multiple genes in the tumor and bronchoalveolar lavage of patients with lung cancer. *Clin. Cancer Res. 10*, 2284-2288.

Tost, J., Dunker, J., and Gut, I.G. (2003a). Analysis and quantification of multiple methylation variable positions in CpG islands by Pyrosequencing. *Biotechniques. 35*, 152-156.

Tost, J., El Abdalaoui, H., and Gut, I.G. (2006). Serial pyrosequencing for quantitative DNA methylation analysis. *Biotechniques. 40*, 721-726.

Tost, J., Schatz, P., Schuster, M., Berlin, K., and Gut, I.G. (2003b). Analysis and accurate quantification of CpG methylation by MALDI mass spectrometry. *Nucleic Acids Res. 31*, e50.

Toyota, M., Ho, C., Ahuja, N., Jair, K.W., Li, Q., Ohe-Toyota, M., Baylin, S.B., and Issa, J.P. (1999). Identification of differentially methylated sequences in colorectal cancer by methylated CpG island amplification. *Cancer Res. 59*, 2307-2312.

Uhlmann, K., Brinckmann, A., Toliat, M.R., Ritter, H., and Nürnberg, P. (2002). Evaluation of a potential epigenetic biomarker by quantitative methyl-single nucleotide polymorphism analysis. *Electrophoresis. 23*, 4072-4079.

Umbricht, C.B., Evron, E., Gabrielson, E., Ferguson, A., Marks, J., and Sukumar, S. (2001). Hypermethylation of 14-3-3 sigma (stratifin) is an early event in breast cancer. *Oncogene. 20*, 3348-3353.

Usadel, H., Brabender, J., Danenberg, K.D., Jeronimo, C., Harden, S., Engles, J., Danenberg, P.V., Yang, S., and Sidransky, D. (2002). Quantitative adenomatous polyposis coli promoter methylation analysis in tumor tissue, serum, and plasma DNA of patients with lung cancer. *Cancer Res. 62*, 371-375.

Ushijima, T., Watanabe, N., Okochi, E., Kaneda, A., Sugimura, T., and Miyamoto, K. (2003). Fidelity of the methylation pattern and its variation in the genome. *Genome Res. 13*, 868-874.

Waki, T., Tamura, G., Tsuchiya, T., Sato, K., Nishizuka, S., and Motoyama, T. (2002). Promoter methylation status of E-cadherin, hMLH1, and p16 genes in nonneoplastic gastric epithelia. *Am. J. Pathol. 161*, 399-403.

Wallner, M., Herbst, A., Behrens, A., Crispin, A., Stieber, P., Goke, B., Lamerz, R., and Kolligs, F.T. (2006). Methylation of serum DNA is an independent prognostic marker in colorectal cancer. *Clin. Cancer Res. 12*, 7347-7352.

Weber, M., Davies, J.J., Wittig, D., Oakeley, E.J., Haase, M., Lam, W.L., and Schubeler, D. (2005). Chromosome-wide and promoter-specific analyses identify sites of differential DNA methylation in normal and transformed human cells. *Nat. Genet. 37*, 853-862.

Wei, S.H., Balch, C., Paik, H.H., Kim, Y.S., Baldwin, R.L., Liyanarachchi, S., Li, L., Wang, Z., Wan, J.C., Davuluri, R.V., *et al.* (2006). Prognostic DNA methylation biomarkers in ovarian cancer. *Clin. Cancer Res. 12*, 2788-2794.

Wei, S.H., Brown, R., and Huang, T.H. (2003). Aberrant DNA methylation in ovarian cancer: is there an epigenetic predisposition to drug response? *Ann. N Y Acad. Sci. 983*, 243-250.

Wei, S.H., Chen, C.M., Strathdee, G., Harnsomburana, J., Shyu, C.R., Rahmatpanah, F., Shi, H., Ng, S.W., Yan, P.S., Nephew, K.P., *et al.* (2002). Methylation microarray analysis of late-stage ovarian carcinomas distinguishes progression-free survival in patients and identifies candidate epigenetic markers. *Clin. Cancer Res. 8*, 2246-2252.

Weisenberger, D.J., Siegmund, K.D., Campan, M., Young, J., Long, T.I., Faasse, M.A., Kang, G.H., Widschwendter, M., Weener, D., Buchanan, D., *et al.* (2006). CpG island methylator phenotype underlies sporadic microsatellite instability and is tightly associated with BRAF mutation in colorectal cancer. *Nat. Genet. 38*, 787-793.

Widschwendter, A., Ivarsson, L., Blassnig, A., Muller, H.M., Fiegl, H., Wiedemair, A., Muller-Holzner, E., Goebel, G., Marth, C., and Widschwendter, M. (2004a). CDH1 and CDH13 methylation in serum is an independent prognostic marker in cervical cancer patients. *Int. J. Cancer. 109*, 163-166.

Widschwendter, A., Muller, H.M., Fiegl, H., Ivarsson, L., Wiedemair, A., Muller-Holzner, E., Goebel, G., Marth, C., and Widschwendter, M. (2004b). DNA methylation in serum and tumors of cervical cancer patients. *Clin. Cancer Res. 10*, 565-571.

Widschwendter, M., Siegmund, K.D., Muller, H.M., Fiegl, H., Marth, C., Muller-Holzner, E., Jones, P.A., and Laird, P.W. (2004c). Association of breast cancer DNA methylation profiles with hormone receptor status and response to tamoxifen. *Cancer Res. 64*, 3807-3813.

Winquist, E., Knox, J., Ayoub, J.P., Wood, L., Wainman, N., Reid, G.K., Pearce, L., Shah, A., and Eisenhauer, E. (2006). Phase II trial of DNA methyltransferase 1 inhibition with the antisense oligonucleotide MG98 in patients with metastatic renal carcinoma: a

National Cancer Institute of Canada Clinical Trials Group investigational new drug study. *Invest. New Drugs. 24*, 159-167.

Wong, I.H., Lo, Y.M., Lai, P.B., and Johnson, P.J. (2000a). Relationship of p16 methylation status and serum alpha-fetoprotein concentration in hepatocellular carcinoma patients. *Clin. Chem. 46*, 1420-1422.

Wong, I.H., Lo, Y.M., Yeo, W., Lau, W.Y., and Johnson, P.J. (2000b). Frequent p15 promoter methylation in tumor and peripheral blood from hepatocellular carcinoma patients. *Clin. Cancer Res. 6*, 3516-3521.

Wong, I.H., Lo, Y.M., Zhang, J., Liew, C.T., Ng, M.H., Wong, N., Lai, P.B., Lau, W.Y., Hjelm, N.M., and Johnson, P.J. (1999). Detection of aberrant p16 methylation in the plasma and serum of liver cancer patients. *Cancer Res. 59*, 71-73.

Wong, I.H., Ng, M.H., Huang, D.P., and Lee, J.C. (2000c). Aberrant p15 promoter methylation in adult and childhood acute leukemias of nearly all morphologic subtypes: potential prognostic implications. *Blood. 95*, 1942-1949.

Wong, I.H., Zhang, J., Lai, P.B., Lau, W.Y., and Lo, Y.M. (2003). Quantitative analysis of tumor-derived methylated p16INK4a sequences in plasma, serum, and blood cells of hepatocellular carcinoma patients. *Clin. Cancer Res. 9*, 1047-1052.

Worm, J., Aggerholm, A., and Guldberg, P. (2001). In-tube DNA methylation profiling by fluorescence melting curve analysis. *Clin. Chem. 47*, 1183-1189.

Xiong, Z., and Laird, P.W. (1997). COBRA: a sensitive and quantitative DNA methylation assay. *Nucleic Acids Res. 25*, 2532-2534.

Yamashita, K., Dai, T., Dai, Y., Yamamoto, F., and Perucho, M. (2003). Genetics supersedes epigenetics in colon cancer phenotype. *Cancer Cell. 4*, 121-131.

Yan, L., Nass, S.J., Smith, D., Nelson, W.G., Herman, J.G., and Davidson, N.E. (2003). Specific inhibition of DNMT1 by antisense oligonucleotides induces re-expression of estrogen receptor-alpha (ER) in ER-negative human breast cancer cell lines. *Cancer Biol. Ther. 2*, 552-556.

Yan, P.S., Chen, C.M., Shi, H., Rahmatpanah, F., Wei, S.H., Caldwell, C.W., and Huang, T.H. (2001). Dissecting complex epigenetic alterations in breast cancer using CpG island microarrays. *Cancer Res. 61*, 8375-8380.

Yan, P.S., Perry, M.R., Laux, D.E., Asare, A.L., Caldwell, C.W., and Huang, T.H. (2000). CpG island arrays: an application toward deciphering epigenetic signatures of breast cancer. *Clin. Cancer Res. 6*, 1432-1438.

Yan, P.S., Venkataramu, C., Ibrahim, A., Liu, J.C., Shen, R.Z., Diaz, N.M., Centeno, B., Weber, F., Leu, Y.W., Shapiro, C.L., *et al.* (2006). Mapping geographic zones of cancer risk with epigenetic biomarkers in normal breast tissue. *Clin. Cancer Res. 12*, 6626-6636.

Yang, H., Chen, C.M., Yan, P., Huang, T.H., Shi, H., Burger, M., Nimmrich, I., Maier, S., Berlin, K., and Caldwell, C.W. (2003). The androgen receptor gene is preferentially hypermethylated in follicular non-Hodgkin's lymphomas. *Clin. Cancer Res. 9*, 4034-4042.

Yang, Q., Zage, P., Kagan, D., Tian, Y., Seshadri, R., Salwen, H.R., Liu, S., Chlenski, A., and Cohn, S.L. (2004). Association of epigenetic inactivation of RASSF1A with poor outcome in human neuroblastoma. *Clin. Cancer Res. 10*, 8493-8500.

Yegnasubramanian, S., Kowalski, J., Gonzalgo, M.L., Zahurak, M., Piantadosi, S., Walsh, P.C., Bova, G.S., De Marzo, A.M., Isaacs, W.B., and Nelson, W.G. (2004). Hypermethylation of CpG islands in primary and metastatic human prostate cancer. *Cancer Res. 64*, 1975-1986.

Yoo, C.B., and Jones, P.A. (2006). Epigenetic therapy of cancer: past, present and future. *Nat. Rev. Drug Discov. 5*, 37-50.

Yuasa, Y. (2002). DNA methylation in cancer and ageing. *Mech. Ageing Dev. 123*, 1649-1654.

Zambrano, P., Segura-Pacheco, B., Perez-Cardenas, E., Cetina, L., Revilla-Vazquez, A., Taja-Chayeb, L., Chavez-Blanco, A., Angeles, E., Cabrera, G., Sandoval, K., *et al.* (2005). A phase I study of hydralazine to demethylate and reactivate the expression of tumor suppressor genes. *BMC Cancer. 5*, 44.

Zeschnigk, M., Bohringer, S., Price, E.A., Onadim, Z., Masshofer, L., and Lohmann, D.R. (2004). A novel real-time PCR assay for quantitative analysis of methylated alleles (QAMA): analysis of the retinoblastoma locus. *Nucleic Acids Res. 32*, e125.

Zhang, X., Yazaki, J., Sundaresan, A., Cokus, S., Chan, S.W., Chen, H., Henderson, I.R., Shinn, P., Pellegrini, M., Jacobsen, S.E., and Ecker, J.R. (2006). Genome-wide high-resolution mapping and functional analysis of DNA methylation in arabidopsis. *Cell. 126*, 1189-1201.

Zhu, W.G., Lakshmanan, R.R., Beal, M.D., and Otterson, G.A. (2001). DNA methyltransferase inhibition enhances apoptosis induced by histone deacetylase inhibitors. *Cancer Res. 61*, 1327-1333.

Zou, H.Z., Yu, B.M., Wang, Z.W., Sun, J.Y., Cang, H., Gao, F., Li, D.H., Zhao, R., Feng, G.G., and Yi, J. (2002). Detection of aberrant p16 methylation in the serum of colorectal cancer patients. *Clin. Cancer Res. 8*, 188-191.

Zuo, C., Ai, L., Ratliff, P., Suen, J.Y., Hanna, E., Brent, T.P., and Fan, C.Y. (2004). O6-methylguanine-DNA methyltransferase gene: epigenetic silencing and prognostic value in head and neck squamous cell carcinoma. *Cancer Epidemiol. Biomarkers Prev. 13*, 967-975.

In: DNA Methylation Research Trends
Editor: Taku B. Kobayashi, pp. 135-182
ISBN: 978-1-60021-721-0

Chapter VI

DNA Methylation Changes in Pathogenesis and Therapy of Hematopoietic Malignancies

Edgar Jost, Rainhardt Osieka and Oliver Galm*
Medizinische Klinik IV, Universitaetsklinikum Aachen, RWTH Aachen
52074 Aachen, Germany

Abstract

For a long time, cancer has been regarded as a genetically driven disease involving mutations of tumor suppressor genes and oncogenes as well as chromosomal aberrations. New concepts have recognized the fundamental role of the epigenome in the pathogenesis of human malignancies. Epigenetic changes refer to a heritable change in the pattern of gene expression without alterations in the primary nucleotide sequence. DNA methylation of cytosines within CpG dinucleotides and post-translational acetylation and methylation of histones are the most common epigenetic modifications with an impact on chromatin structure and transcriptional activity. Hypermethylation of CpG islands in the promoter region of genes is a well characterized epigenetic modification associated with transcriptional silencing of cancer-related genes and plays a crucial role in carcinogenesis. In solid tumors and hematopoietic malignancies, a large number of genes involved in different pathways have been described to be affected by CpG island methylation in association with transcriptional inactivation. Altered DNA methylation patterns may be clinically useful as biomarkers for cancer detection, assessment of prognosis and prediction of response to therapy. In contrast to genetic alterations, epigenetic changes in cancer are potentially reversible by pharmacological inhibition of DNA methylation and histone deacetylation. Thus, the increasing knowledge about the altered epigenome in malignant cells offers new strategies in targeted anti-cancer therapy. Clinical trials using epigenetically targeted therapies have

* Corresponding author: Oliver Galm, M.D. Medizinische Klinik IV; Universitaetsklinikum Aachen; RWTH Aachen; Pauwelsstrasse 30; 52074 Aachen; Germany; Tel: +49-241-8089806; Fax: +49-241-8082449; Email: oliver.galm@post.rwth-aachen.de

yielded promising results in acute and chronic myelogenous leukemia as well as in myelodysplastic syndromes. The exploration of our growing knowledge about epigenetic aberrations will contribute to develop novel strategies in diagnosis and treatment of hematopoietic malignancies for the future.

List of Abbreviations

ABL	Abelson
AML	Acute myelogenous leukemia
ALL	Acute lymphocytic leukemia
APL	Acute promyelocytic leukemia
ATRA	All-trans-retinoic acid
AZA	5-Azacytidine
Bcl	B-cell CLL/lymphoma 2
BCR	Breakpoint cluster region
BMI1	polycomb group RING finger protein 4
BSC	Best supportive care
CB	Clinical benefit
CDK	Cyclin dependant kinase
CLL	Chronic lymphocytic leukemia
CML	Chronic myelogenous leukemia
CIMP	CpG island methylator phenotype
CMML	Chronic myelomonocytic leukemia
CR	Complete remission
CRBP1	Cellular retinol binding protein 1
DAC	5-Aza-2'-deoxycytidine
DAPK1	Death-associated protein kinase 1
DLBCL	Diffuse large B-cell lymphoma
DNMT	DNA methyltransferase
ER	Estrogen receptor
EZH	Enhancer of zeste homologue
FAS	TNF receptor superfamily member 6
GST	Gluthation S-transferase
HAT	Histone acetyltransferase
HDAC	Histone deacetylase
HER-2/neu	v-erb-b2 erythroblastic leukemia viral oncogene homologue 2
HIC1	Hypermethylated in cancer-1
HIF-1α	Hypoxia inducible factor-1α
hMLH1	Human MutL homologue-1
hMSH2	Human MutS homologue-2
hMSH6	Human MutS homologue-6
ICAM	Intercellular adhesion molecule
JAK	Janus kinase

LOH	Loss of heterozygosity
MDS	Myelodysplastic syndrome
MDM2	Mouse double minute 2, human homologue
miRNA	Micro RNA
MM	Multiple myeloma
MSI	Microsatellite instability
MSP	Methylation specific polymerase chain reaction
MTD	Maximal tolerated dose
MYC	v-myc myelocytomatosis viral oncogene homolog
N	Number of patients
NHL	Non-Hodgkin's lymphoma
NR	Not reported
O^6MGMT	O^6 methylguanine DNA methyltransferase
OR	Overall response
PB	Phenylbutyrate
PcG	Polycomb group protein
PD	Progressive disease
PR	Partial remission
pRb	Retinoblastoma gene
PRC	Polycomb repressive complex
RAR	Retinoic acid receptor
RAS	Rat sarcoma viral oncogene, human homologue
SD	Stable disease
SFRP	Secreted frizzled-related protein
siRNA	Small inhibitory RNA
SOCS	Suppressor of cytokine signaling
TIMP	Tissue inhibitor of matrix metalloproteinases
TNF	Tumor necrosis factor
TRAIL	TNF related apoptosis inducing ligand
TSA	Trichostatin A
VEGF	Vascular endothelial growth factor
VHL	Von Hippel-Lindau
VPA	Valproic acid
Wnt	Wingless-type MMTV integration site
WRN	Werner syndrome gene
ZAP70	Zeta associated protein 70

Introduction

Cancer has been recognized as a genetic disease, and a large number of molecular disturbances have been identified in malignant cells. Tumorigenesis is regarded as a multistep process involving multiple genetic alterations that contribute to the progressive cellular transformation [1, 2]. Genetic aberrations are irreversible and involve changes in the DNA sequence such as mutations, deletions, amplifications and translocations. These may result in activation of oncogenes, inactivation of tumor suppressor genes, or the formation of novel chimeric genes with a modified function contributing to the malignant phenotype of the affected cells. Well-characterized examples for genetic events in carcinogenesis are point mutations of the epithelial growth factor receptor and the RAS gene family [3, 4], deletion of the apoptosis control gene p53 [5], and the balanced reciprocal translocation t(9;22) with the fusion gene BCR/ABL [6, 7]. The identification of genetic alterations as the molecular basis of cancer has not only revealed mechanistic insights in the pathogenesis of malignant diseases, but has also provided opportunities for targeted anti-cancer therapies [8, 9]. Examples for targeted approaches are the antibody trastuzumab in breast cancer with overexpression and amplification of HER-2/neu [10] and the tyrosine kinase inhibitor imatinib in BCR/ABL-positive chronic myelogenous leukemia (CML) [11, 12]. Furthermore, the differentiation block of the malignant cell clone in acute promyelocytic leukemia (APL) with the translocation t(15;17) and the fusion gene PML/RARα may be overcome by the oral application of all-trans retinoic acid (ATRA) [13].

There is now increasing evidence that - in addition to genetic aberrations - epigenetic events also play a critical role in the pathogenesis of human cancer [14-17]. Epigenetics refers to heritable stable changes in gene expression controlled by mechanisms other than alterations in the primary nucleotide sequence. Fundamental epigenetic transcriptional regulatory mechanisms involve DNA methylation [18], covalent methylation and acetylation of histone proteins [19, 20], RNA interference [21, 22] and polycomb group proteins (PcG) [23].

Epigenetic alterations in tumor cells represent attractive novel therapeutic targets that are different from those of conventional anti-cancer drugs. In solid tumors and hematopoietic malignancies, a great number of genes involved in different pathways have been described to be affected by CpG island methylation in association with transcriptional silencing [14-16]. In hematopoietic malignancies, the analysis of DNA methylation changes can serve as a biomarker for the detection and classification of the malignant disease [24], as a prognostic marker [25, 26] and for the follow-up of minimal residual disease [27]. In a near future, epigenetic disturbances may also be used for the prediction of response and eventually as a guide for targeted therapy [28]. Clinical trials using a strategy based on the modification of epigenetic changes with demethylating agents and histone deacetylase (HDAC) inhibitors may help to generate new approaches in diagnosis, risk stratification, prevention and treatment of acute and chronic leukemias, malignant lymphomas, as well as myelodysplastic and myeloproliferative disorders.

Epigenetic Changes in Hematopoietic Malignancies

The Physiological Role of DNA Methylation

DNA methylation in mammals is largely limited to cytosines that are part of the symmetrical dinucleotide CpG. 5-methylcytosine results from the addition of a methyl group to cytosine at the C5 position of the pyrimidine ring [29, 30]. This enzymatic reaction is catalyzed through DNA methyltransferases (DNMTs) and uses S-adenosyl-methionine as a methyl group donor. The distribution of CpG dinucleotides in the human genome is not uniform, probably owing to the high mutagenic potential of 5-methylcytosine. If the spontaneous deamination of 5-methylcytosine to thymidine is not recognized by DNA repair enzymes, a permanent cytosine-to-thymidine change remains. Over the course of time, this process has contributed to a progressive depletion of CpG dinucleotides in the human genome. However, there are clusters of CpG sites that have not been affected by this depletion process. These DNA regions are termed CpG islands and are often located near the promoter regions of approximately half of the genes in the genome [31]. In the bulk of the genome, most of the CpG dinucleotides outside CpG islands are predominantly methylated. In contrast, the CpG sites within CpG islands, especially those associated with promoter regions, are normally unmethylated. These methylation patterns in normal cells are thought to contribute to maintaining the large amount of noncoding DNA in a transcriptionally inert state [16, 30, 32]. The lack of methylation in promoter-associated CpG islands allows the expression of the gene, if the appropriate transcription factors are present, and the chromatin structure allows access to them [18]. The exception to this unmethylated state of CpG islands in normal cells involves genes encompassed within regions of X-chromosome inactivation in females and the silenced alleles of imprinted genes [33-35].

Histone Acetylation and Methylation

The nucleosome is the fundamental repeating unit of chromatin and consists of approximately 146 base pairs of DNA wrapped around an octamer of histone core proteins. These histone proteins are termed H2A, H2B, H3 and H4 [36]. Post-translational modifications of histones are essential for controlling and maintaining the chromatin structure and represent a major category of epigenetic transcriptional control. The histone tails on the nucleosome surface may undergo post-translational modifications that form a code specifying patterns of gene expression by helping to determine the transcriptional activity of a particular gene [20, 37, 38]. In this regard, acetylation of key histone amino acid residues has been extensively studied and is maintained by a balance between the activities of histone acetyltransferases (HATs) and HDACs. In general, acetylated histones associate preferentially with transcriptionally active chromatin, while histone deacetylation is characteristic for chromatin condensation and transcriptional repression [19].

Another key histone modification involves methylation of selected amino acid sites, and this step is controlled by various histone methyltransferases. This modification appears to have multiple effects on chromatin function and can be a marker for both transcriptionally active and inactive chromatin regions. Two important methylation events involving lysines 9 and 27 on the N-terminus of histone H3 (H3-K9 and H3-K27) are mostly associated with transcriptional silencing and repression, while methylation of lysine 4 of histone H3 (H3-K4) is often associated with an open chromatin structure surrounding promoter regions of active genes [39]. It could be shown for several genes that methylation of H3-K9 is enriched at promoter regions affected by CpG island hypermethylation, while this histone site is acetylated and unmethylated, when the same gene is being actively transcribed and the CpG island is unmethylated [14, 16, 40]. The relationship between acetylation and methylation of this histone site is a key regulatory event, since the two events appear to be inhibitory to one another [19]. In turn, acetylation of H3-K9 appears to facilitate another key transcriptional activation mark, methylation of H3-K4, and this latter modification is enriched at genes with unmethylated versus methylated CpG islands [14, 41]. The different patterns of those covalent modifications of histone tails that function to regulate gene transcription are referred to as the "histone code". It is presently less well understood which events exactly govern the setting of this histone code.

RNA Interference

RNA mediated silencing is an evolutionarily conserved mechanism initially described in C. elegans and termed RNA interference [42]. In animals, sequence-specific gene silencing can be induced by double-stranded micro RNA fragments (miRNA) by two mechanisms. miRNA fragments may act by targeting the complementary sequence in the 3' untranslated region of messengerRNA (mRNA) and induce a translational repression [43]. A further type of gene silencing induced by miRNAs is a trigger function for DNA and histones packaging into condensed and transcriptionally silent heterochromatin [44]. The resulting transcriptional silencing may be mediated by DNA methylation but also by mechanisms independent of DNA modifications [45]. The genome of vertebrates contains about 1000 miRNAs.

In human acute myeloid leukemia (AML), the AML/ETO fusion transcript is found in 10-15 % of patients and caused by a translocation between chromosomes 8 and 21. The exact role of this fusion gene has been elucidated by an AML1/ETO-specific small inhibitory RNA (siRNA), a second type of RNA interference induced by small double stranded ribonucleotide fragments [46]. The siRNA was introduced by electroporation in cell lines with the t(8;21) translocation. A suppression of the fusion protein could be achieved, and an increase in susceptibility to hematopoietic growth factors with an induction of differentiation was observed. Transfected cells displayed a change in cell surface marker profile and an upregulation of the CCAAT/enhancer binding protein alpha (C/EBPalpha) as a sign of differentiation. Thus, siRNAs are a useful tool to study the role of specific cancer-related genes. This knowledge can also be used to identify relevant targets for molecular therapy. Preliminary in vitro studies suggest a great potential of siRNAs for the treatment of cancer as well as virus-induced diseases. A corresponding result could been obtained by the incubation of BCR/ABL-positive CML cell lines with siRNA [47]. Treatment with a fusion transcript-specific siRNA induced an increased sensitivity to the tyrosine kinase inhibitor imatinib and irradiation as well as apoptosis of the transfected cells.

Polycomb Silencers

PcGs are epigenetic gene silencers and have also been implicated in neoplastic development. PcGs are classified in two groups according to their association in multimeric complexes and are termed polycomb repressive complexes (PRC) [23]. PRC2 is an initiation complex for gene repression, and its enhancers of zeste homolog EZH1 and EZH2 components possess a histone methyltransferase activity with trimethylation of lysine 9 and lysine 27 of histone H3 [48-50]. Additionally, methyltransferase activity of this complex has been observed for lysine 26 of the linker histone H1 with a corresponding influence on higher chromatin structure [51]. Additionally, EZH2 contributes to the recruitment of DNMTs with consequent promoter hypermethylation and silencing of cancer-related genes [52]. The PRC1 complex is more diverse and contributes to the maintenance of gene repression. The sequence-specific binding of this complex is obtained by zinc finger domains, and there is recognition of methylated histones. Ubiquitin ligase activity and methyl lysine binding of the PRC1 complex are due to the polycomb group RING finger protein 4 (BMI1) and the chromobox homolog proteins (CBX2, -4, -8) [53-55]. A regulatory role for RNA interference and the nucleosome clustering role of PcG have been described [56]. Gene repression of the PRC1 complex is explained in part by the ubiquitin E3 ligase activity on lysine 119 of histone H2 [53]. Figure 1 summarizes major steps and consequences of PcG activity.

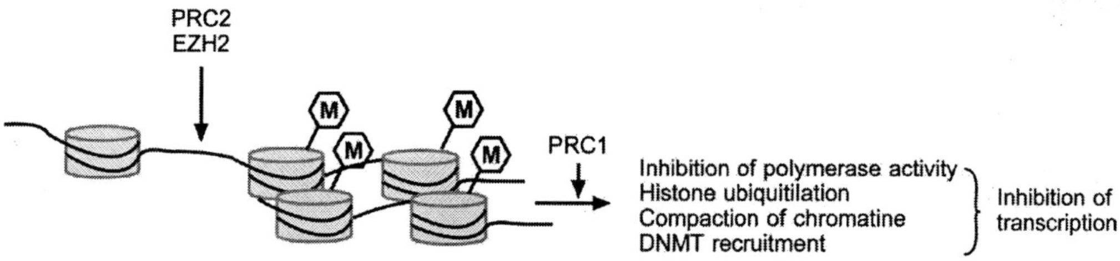

Figure 1. Major steps in the activity of PcGs are the induction of transcriptional silencing by the PRC2 complex with the incorporated EZH proteins and the maintenance role in inhibition of transcription of the PRC1 complex. Consequences of this PcG activity are the inhibition of RNA polymerase activity and the recruitment of DNMTs.

Multiple proteins of the PRC1 and PRC2 complexes have been implicated in the development of human cancer. A high expression of BMI1, part of the PRC1 complex, contributes to the neoplastic transformation of T- and B-lymphocytes to the corresponding non-Hodgkin's lymphomas (NHL) by interaction with the MYC protein [57]. MYC acts as a transcription factor for multiple genes involved in neoplastic cellular transformation. In a mouse model of leukemia, the crucial effect of BMI1 on the proliferation could be demonstrated. Increased expression levels of BMI1 were also observed in breast cancer, B-cell NHL, medulloblastoma, neuroblastoma and non-small cell lung cancer [57-59]. EZH2, part of the PRC2 complex, is upregulated in human B-cell NHL, Hodgkin's disease, melanoma, prostate, bladder, liver and breast cancer and has been reported to be associated with a poorer prognosis and/or advanced disease [57, 60-64].

An additional observation is a change in PRC stoichiometry in human tumors with an additional repressive activity, in part explained by the recruitment of the sirtuin histone deacetylase to the complex [65].

Recent findings support the hypothesis that PcGs are involved in the maintenance of tumor stem cells through defective specification of cells toward a stem cell state [66]. Even though discussed controversially at the moment, PcGs may not only be involved in the preservation of a neoplastic stem cell phenotype in many human tumors but also in embryonic and adult stem cells. The maintenance of stem cell fate and the PcG-mediated silencing of tumor suppressor genes by DNMT recruitment or other mechanisms seem to be a major aspect of this protein group in tumorigenesis.

DNA Methylation Changes in Cancer

Abnormal DNA methylation patterns in cancer have been recognized for more than 20 years [67]. The genome of tumor cells is characterized by global hypomethylation which predominantly affects introns and the body of genes as well as repetitive DNA sequences [68]. The lack of methylation may contribute to a malignant phenotype by interfering with the functional stability of chromosomes, reactivation of transposable elements, and disruption of normal gene imprinting patterns [69]. Genome-wide demethylation may further favor mitotic recombination with the consecutive loss of heterozygosity (LOH) and promotion of chromosomal rearrangements. Additionally, hypomethylation in centromeric DNA regions may contribute to aneuploidy in malignant cells [70]. Thus, deranged methylation patterns and genomic instability in tumor cells are functionally linked.

Additionally, virtually all cancer types show hypermethylation of promoter-associated CpG islands. This aberrant methylation of promoter regions is associated with transcriptional inactivation and represents an important mechanism of gene silencing in the pathogenesis of neoplasia [14-16, 18, 30, 71]. The epigenetic silencing of tumor suppressor genes is at least as frequent as genetic disturbances in human cancer [18].

As proposed in Knudson's two-hit hypothesis [72], the disruption of the function of a tumor suppressor gene requires a complete loss of function of both gene copies. In this context, abnormal promoter hypermethylation may exert the same effects as a coding region mutation or deletion in one copy of the gene and can thus be regarded as an additional pathway to contribute to the loss of one or both alleles [16]. Figure 2 shows different mechanisms of inactivation of tumor suppressor genes by genetic and epigenetic alterations.

DNMTs are required for establishing and maintaining cellular DNA methylation patterns [73, 74]. These enzymes are not only involved in catalyzing the methylation process, but they also interact with HDACs and have the potential to recruit them to sites of gene promoters [75, 76]. Deacetylation of histones is an important factor for both chromatin status and transcriptional activity [77]. The binding of methylcytosine-binding proteins, transcriptional corepressors and chromatin remodeling proteins to sites of DNA methylation represents another way to recruit HDACs to hypermethylated DNA regions. All these components of the transcriptional machinery may interact and finally result in a repressive chromatin conformation [35, 78, 79]. Figure 3 shows a summary of the differences in DNA methylation patterns between normal and cancer cells.

DNA methylation and histone modifications are fundamental components in a complex epigenetic network and have a major impact on transcriptional activity and chromatin structure [14, 16, 41]. Both epigenetic processes may be pharmacologically targeted for the translational goal of reversing these changes in cancer cells for therapeutic purposes. In the collaboration between DNA methylation and histone deacetylation resulting in transcriptional inactivation, DNA methylation appears to be dominant and may serve to keep the affected genes in a permanently silent state [14, 16, 80]. In experimental systems, hypermethylated DNA regions must be at least partially demethylated before gene transcription is restored and other chromatin and histone marks for transcriptional repression are reversed [81].

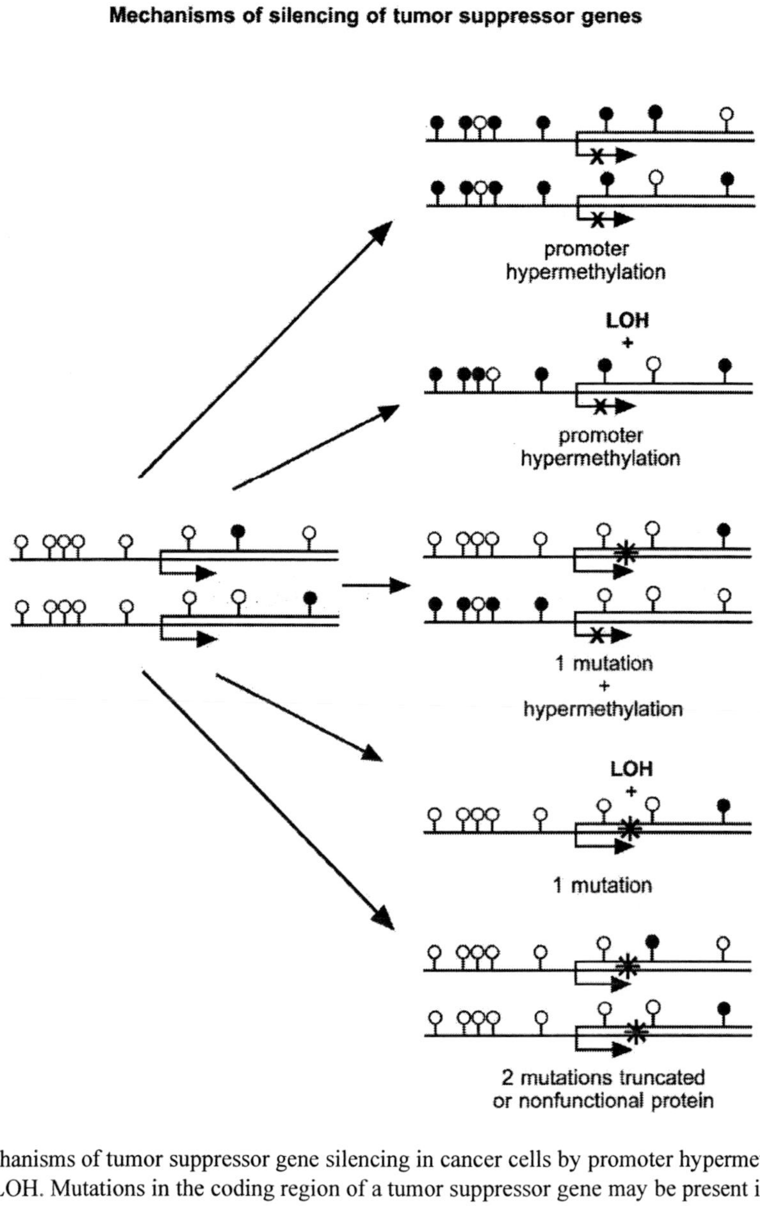

Figure 2. Mechanisms of tumor suppressor gene silencing in cancer cells by promoter hypermethylation, mutations or LOH. Mutations in the coding region of a tumor suppressor gene may be present in a germ cell in inherited cancers or as a somatic event in sporadic cancers. According to Knudson's two-hit hypothesis, a second mutation is required to result in a complete loss of gene function. The second copy of the gene may also be lost by deletion of a part of the chromosome, known as LOH. The affected gene can also be silenced by promoter hypermethylation. This epigenetic mechanism may occur in combination with LOH or other genetic events. Open circles represent unmethylated CpG sites, closed circles represent methylated CpG sites. The arrow indicates the transcriptional start site, and the arrow with X symbolizes transcriptional silencing. * = mutation.

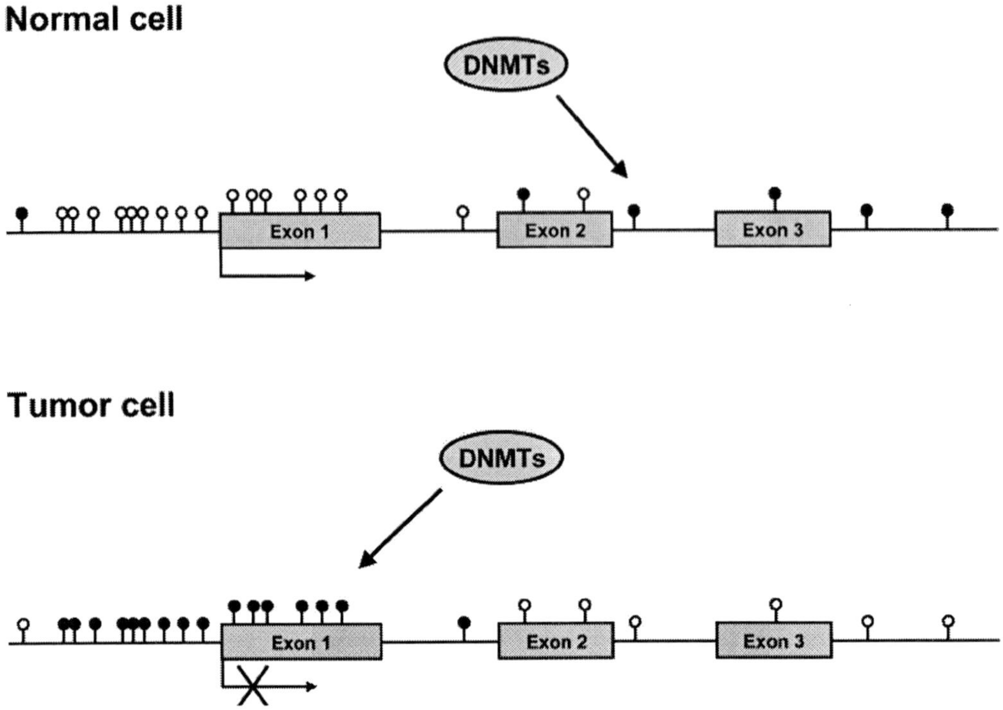

Figure 3. DNA methylation patterns in normal cells and cancer cells. A characteristic CpG island near a gene promoter region is constituted by a CpG-rich region that in normal cells is mostly unmethylated (open circles). The areas downstream of the promoter within the body of the gene show a low frequency of CpG dinucleotides, and those downstream CpG sites are often methylated (closed circles). This DNA methylation pattern allows the gene to be actively transcribed (arrow = transcriptional start site). In a cancer cell, the promoter-associated CpG island is often hypermethylated and transcriptionally silenced (arrow with X), whereas the downstream CpG sites are hypomethylated. Each lollipop indicates a CpG dinucleotide; open circles represent unmethylated cytosines, closed circles represent methylated cytosines. DNMTs = DNA methyltransferases.

Pathways Affected by Epigenetic Changes in Hematopoietic Malignancies

The most important factors contributing to the cancer phenotype of a tissue are limitless replicative potential, sustained angiogenesis, evasion from apoptosis, self sufficiency in growth signals, insensitivity to anti-growth signals as well as tissue invasion and metastatic potential. A large number of genes affecting these fundamental cellular pathways have been shown to undergo aberrant CpG island methylation in association with transcriptional silencing in human cancer [16, 18, 82]. Although some genes are affected in multiple tumor types (e.g. the cell cycle regulator p16), methylation profiling studies have shown that the spectrum of hypermethylated genes in malignant cells tends to be specific for the tissue of origin of the tumor. Additionally, there may be characteristic methylation patterns in various types of human cancer [16, 18, 82]. The methylation profile in hematopoietic malignancies differs from solid tumors regarding the affected genes, but again the full spectrum of cellular pathways may be deranged [83-85]. It can be hypothesized that epigenetic silencing of many genes leads to a survival advantage to the affected cells, since fundamental cancer-related pathways may be affected by epigenetic silencing of tumor suppressor genes [86-101].

In NHL, the genes most frequently affected by epigenetic silencing are death associated protein kinase-1 (DAPK1), p57, p16, O^6methylguanosinemethyltransferase (O^6MGMT), gluthatione S-transferase (GSTπ), retinoic acid receptor beta 2 (RARβ2), and the cellular retinol binding protein 1 (CRBP1) [25, 83, 93, 102, 103]. Multiple myeloma (MM) is characterized by aberrant methylation of p16, suppressor of cytokine signaling 1 (SOCS-1), E-cadherin, p73 and DAPK1 [104-110]. In CML, p15 and the Abelson gene (ABL) are commonly affected by hypermethylation [111, 112]. The methylation profile in acute lymphoblastic leukemia (ALL) comprises predominantly p15, E-cadherin, p73, DAPK1 and O^6MGMT [113, 114]. In AML, the most frequently involved genes are SOCS-1, p15, E-cadherin, p73, RARβ2, DAPK1, hypermethylated in cancer 1 (HIC1) and CRBP1 [32, 115-118].

Cell Cycle Control

Tumor cells are characterized by uncontrolled proliferation. Modifications in molecules that regulate the cell cycle have been implicated in the formation and progression of cancer [119]. Mutations in genes involved in cell cycle progression such as p53, cyclins, the cyclin-dependent kinase (CDK) inhibitors p15, p16, p21 and the retinoblastoma protein (pRb) are observed in virtually all human cancers [120]. Mutations, deletions and epigenetic alterations have been described as causes for the loss of function of various negative regulators of the cell cycle [121, 122]. Downregulation of CDK inhibitors is often obtained by epigenetic events such as promoter hypermethylation. CpG island hypermethylation-associated silencing has been described for pRb, p15, p16 and p73 [118, 123, 124].

Two main pathways control the progression of the cell cycle from G1 to S phase. The first involves pRb, p16 and cyclin D1 (pRb pathway) [123] and the second comprises p14, MDM2 and p53 (p53 pathway) [125]. p53 is also a major factor in regulation of apoptosis and will be focused upon below. Relevant factors involved in cell cycle control are summarized in figure 4 [126-128].

Inactivation of the pRb pathway, responsible for the progression from G1 to S phase of the cell cycle, has been described in many human cancers [129]. Inherited allelic loss of pRb is associated with a high tumor formation risk [130]. Mutations, deletions and promoter hypermethylation may be responsible for a loss of function of pRb [131]. pRb binds to the E2F transcription factor, and when E2F is released from pRb, it induces the transcription of factors necessary for the progression of the cell cycle such as cyclin E1, cyclin A2, dehydrofolate reductase and thymidine kinase. pRb also plays a role in the physiological regulation of hematopoiesis. p16 functions as a tumor suppressor by binding to CDK4 as well as cyclin D1. As an inhibitor of the pRb phosphorylation, p16 contributes to an increased stability of the pRb/E2F complex [132, 133]. Inactivation of p16 as well as increased activity of cyclin D1 result in a higher phosphorylation rate of pRb and contribute to an evasion from cell cycle arrest. Loss of function owing to small chromosomal deletions of p16 has been described in gliomas, mesothelioma and familial melanoma [121, 134]. Even more frequently, p16 is silenced by promoter hypermethylation in common solid tumors of lung, breast, bladder, colon and esophagus as well as in MM and NHL.

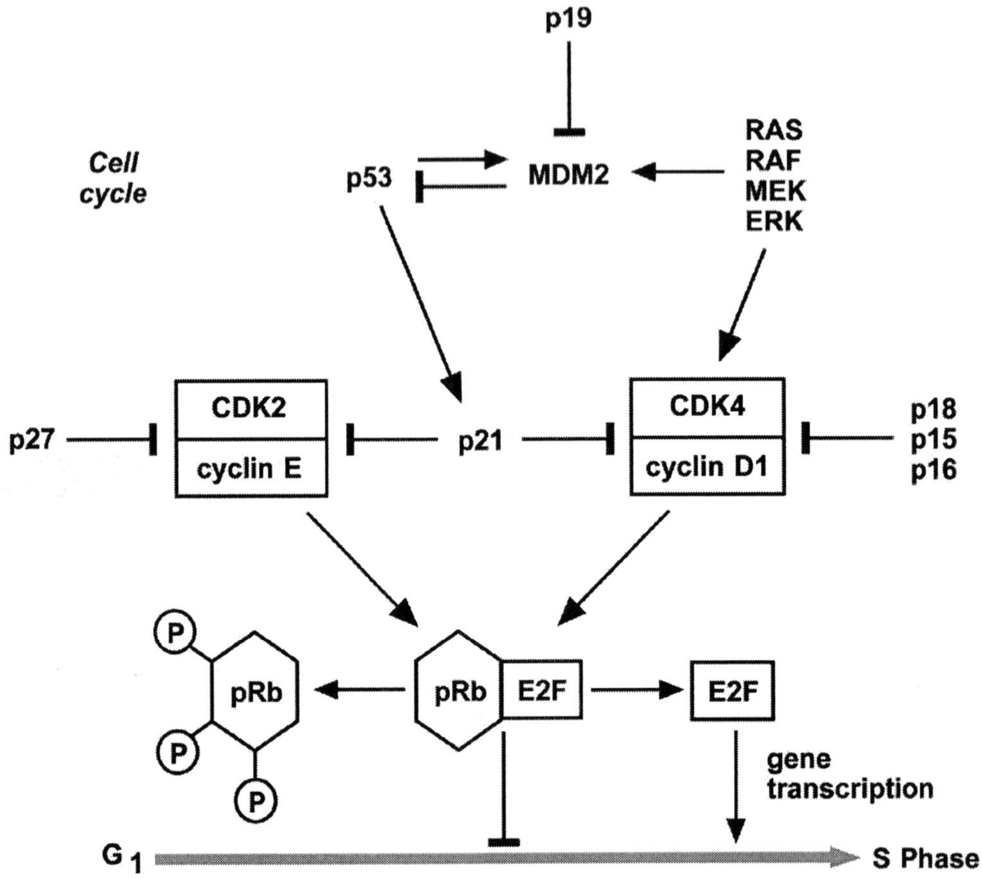

Figure 4. The transition of the G1 to the S phase of the cell cycle is controlled by the pRb and E2F factors. The complex of these two proteins is repressive for cell cycle progression. By the influence of the CDK2/cyclin E complex or the CDK4/cyclin D1 complex, E2F is deliberated and exercises the role of a transcription factor. Multiple proteins such as p15, p16, p18, p21, p27 and p53 control the activity of these CDK/cyclin complexes and have a tumor suppressor activity. These negative regulators may be silenced by promoter hypermethylation.

p15 as an additional regulator of the cell cycle is a CDK inhibitor and shares this activity with p16, p18 and p19 [135, 136]. The inhibiting function is most intense against the CDK/cyclin complexes 4 and 6. Contrary to p16, p15 is frequently silenced by promoter hypermethylation in hematopoietic malignancies such as AML, ALL and myelodysplastic syndromes (MDS) [118]. For p15, a correlation between promoter hypermethylation and p15 expression could be shown in AML and MDS, and for the methylated patients, rapid disease progression and impaired prognosis were reported [137]. Studies for correlation of p15 promoter methylation, expression levels and response to demethylating agents are under investigation. Gore et al. have observed quantitative correlations between promoter hypermethylation of p15 by bisulfite sequencing and expression of p15 by real-time polymerase chain reaction. They could also demonstrate an increase in global histone acetylation induced by the treatment with the demethylating agent 5-azacytidine (AZA) [28].

p57 is a negative regulator of CDK/cyclin complexes with a tumor suppressor activity. Silencing of p57 has been observed in low grade lymphoma, chronic lymphocytic leukemia (CLL) [102] and rarely in cases of acute leukemia [138]. Downregulation of p57 by genomic imprinting is responsible for the Beckwith-Wiedemann syndrome [139], a hereditary syndrome with overgrowth, macrocrania and poor coordination.

Inactivation of p21 and p14 by promoter hypermethylation have been reported to occur only at a low frequency in hematopoietic malignancies such as AML [140, 141].

Regulation of Apoptosis

Apoptosis is a genetically controlled cell death process and plays an essential role in the balance between cell proliferation and cell death [142]. Apoptosis is considered to be an important factor in the evolution of cancer cells, and defects in the apoptotic pathway are frequent in human tumors [143]. Tumor cells can avoid apoptosis by overexpression of anti-apoptotic proteins or presence of oncogenic factors with a pro-survival activity. The loss of function of proteins with a tumor-suppressor activity involved in the apoptotic pathway is the most frequent aberration in programmed cell death in cancer cells. When apoptosis is initiated through the intrinsic pathway, the initial signal originates in the mitochondria [144]. The result of the intrinsic pathway is a permeabilization of the outer mitochondrial membrane. The extrinsic pathway is initiated by extracellular signals and is receptor-mediated [145]. This process ends with the activation of caspases, proteolytic enzymes responsible for a catabolic cell metabolism and DNA fragmentation typical for apoptotic cells [146].

The central signal for the intrinsic pathway is activation and stabilization of p53 [126]. An increased p53 activity may be initiated by chemotherapy- or radiotherapy-induced DNA damage. p53 is a major player in cell cycle control, DNA damage repair and regulation of apoptosis. For the proapoptotic function, p53 enhances the transcription and expression of the cytokine receptors of the extrinsic pathway (Fas, TRAIL receptor) or proapoptotic proteins of the intrinsic pathway (Bax or Noxa) [147-150]. Furthermore, p53 induces the expression of MDM2, a negative regulator of p53 and p73. The physiological counterparts of p53 are the proteins of the bcl2 family. This group of proteins comprises bcl2, bcl-x1, Mcl-1 and Bcl-W with anti-apoptotic properties and Bax and Bak as proapoptotic proteins [151, 152].

The extrinsic pathway is induced by cytokines from the tumor necrosis factor (TNF) family. These cytokines activate caspases-3, -6 and -7 with a direct cell death effect or the activation of caspase-8 via the activation of Bax and Bak proteins. Intrinsic and extrinsic pathways are illustrated in figure 5.

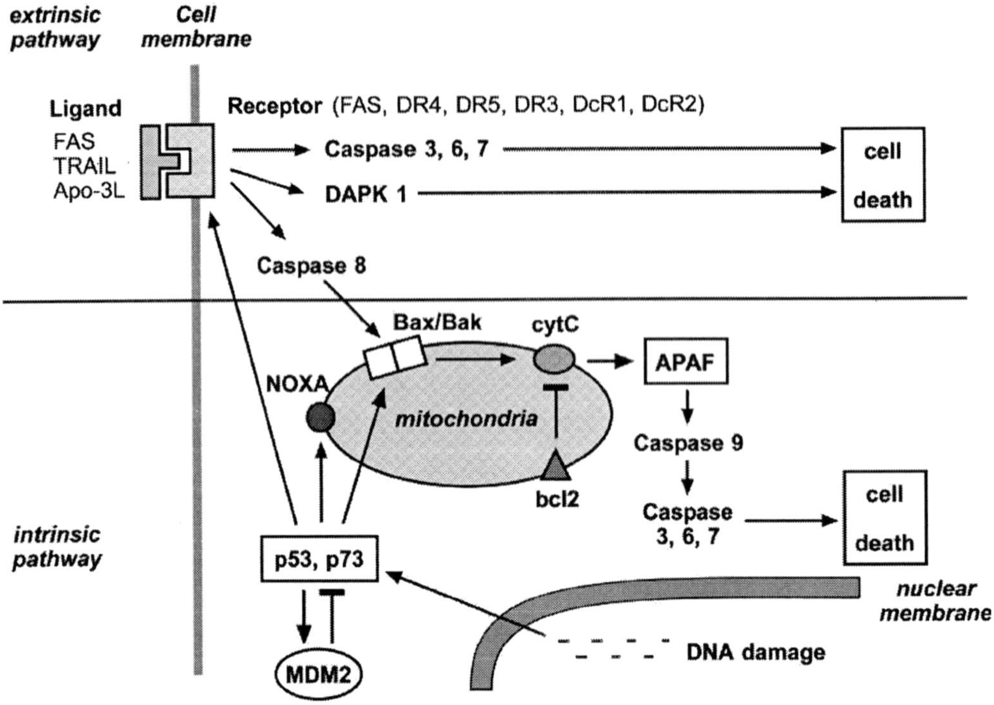

Figure 5. Apoptosis can follow two major pathways: the intrinsic pathway, initiated by DNA damage or the extrinsic pathway, initiated by external signals. The extrinsic pathway causes cell death by activation of mitochondrial signals, caspases and DAPK1. The activation of the intrinsic pathway leads to mitochondrial changes and to an enhanced sensitivity of the cell to external apoptosis-inducing signals. p53 is a key protein with activity in external and internal pathway induction. Factors that may be influenced by DNA methylation changes are p53, p73, DAPK1 and APAF.

Studies in cancer revealed that downregulation of proteins ranging from the induction to the execution phase of apoptosis may be mediated by DNA promoter hypermethylation [153].

Inactivation of the proapoptotic factor p53 is mostly caused by deletions or mutations. Although the p53 gene has a CpG island in the promoter region, hypermethylation of this region has only been described in a rather small proportion of ALL and not in other malignancies [154, 155].

p14 is a stabilizing factor for p53, and promoter hypermethylation of this gene has been observed in nearly all solid tumors, but among hematopoietic malignancies only in ALL [156]. p73 is a factor that activates the transcription of p21, MDM2, cyclin G and Bax. Downregulation of p73 by promoter hypermethylation could be observed in a large proportion of acute leukemias and malignant lymphomas [157, 158].

DAPK1 is a cofactor involved in the induction of programmed cell death by TNFalpha, interferon gamma and Fas [159]. Cells without DAPK1 activity are protected from Fas ligand or TNFalpha induced cell death. DAPK1 has a serine/threonine kinase activity and a death domain. Hypermethylation of the promoter region of DAPK1 has been described in high percentages of B- and T-cell lymphomas as well as in acute leukemias [93].

DNA Damage Repair

Cancer cells display high mitotic activity with an augmented risk for disturbances in DNA sequence. An increased error rate in the genome may alter the transcriptional activity of the genetic code and impair the division of a cancer cell [160]. Additionally, treatment of cancer with chemotherapeutic agents and radiotherapy causes DNA damage with a possible induction of cell death by apoptosis. For these two reasons, a cancer cell needs a high activity of DNA damage repair enzymes to maintain a functionally active genome despite a high mitotic index with an increased level of transcriptional activity. Inherited defects in DNA damage repair enzymes are well-known risks factors for familial cancer syndromes such as p53 in Li-Fraumeni syndrome [161] or the human MutL homologue-1 (hMLH1) or the human MutS homologue-2 and -6 (hMSH2 and hMSH6) in hereditary non-polyposis colon cancer (HNPCC) [162]. Following DNA damage, p53 will be activated and induce apoptosis or, with p21 as a cofactor, induce cell cycle arrest in G1 phase enabling the repair mechanisms to correct the errors [129, 163].

O^6MGMT is an alkyltransferase with the ability to remove methyl groups from guanine. This mechanism is particularly important after exposure of a tumor to alkylating agents such as temozolomide and cyclophosphamide which transfer methyl groups to the N^7, O^3 and O^6 positions of guanine. In glioblastomas, tumors with a downregulation of O^6MGMT activity by hypermethylation of its promoter region had a better prognosis [164]. The increased DNA damage rate achieved by temozolomide was associated with a higher death rate of the cancer cells and a better response to therapy [165]. Hypermethylation of O^6MGMT has been demonstrated in 24 % of MM and 10 % of monoclonal gammopathy of undetermined significance [166]. CpG island hypermethylation of O^6MGMT was also observed in about half of diffuse large B-cell lymphomas (DLBCL) [25]. Promoter methylation was associated with a better survival and recognized as an independent prognostic factor in DLBCL patients treated with cyclophosphamide-containing regimens.

DNA mismatch repair proteins recognize errors occurring during DNA strand duplication in the S Phase of mitosis. Important proteins involved are hMLH1, hMSH2 and hMSH6. Defects in these genes are also responsible for microsatellite instability (MSI) often observed in colon cancer [162]. Among hematopoietic malignancies, CpG hypermethylation has been described for hMLH1 in less then 10 % cases of MM [166] and AML, even when MSI or loss of expression of hMLH1 by immunohistochemistry was detected at a high frequency of 20-50 % [167, 168]. MSI in association with CpG island methylation of the hMLH1 gene has also been found in Richter's syndrome, the transformation of CLL into DLBCL [169].

Angiogenesis, Cell Adhesion and Tissue Invasion

Tumor vasculature provides an important basis for cancer cell growth and metastasis [1]. Angiogenesis is caused by an increased expression of pro-angiogenic factors by the cancer cells and a downregulation of angiogenesis inhibitors [170]. Hypoxic conditions allow the activation of the transcription factor hypoxia-inducible factor-1α (HIF-1α) which induces the expression of the vascular endothelial growth factor (VEGF) and other angiogenic proteins

[171]. The inactivation of the tumor suppressor genes p53 and the von Hippel-Lindau (VHL) gene may also contribute to the downregulation of angiostatic factors and upregulation of pro-angiogenic proteins [172]. In addition to endothelial cells, stromal components such as pericytes, the basement membrane and the extracellular matrix play also a role in cancer-associated neo-angiogenesis [173]. Endothelial cells produce matrix metalloproteinases locally degrading extracellular components. This allows endothelial cell migration and proliferation in the direction of the developing cancer. Owing to an imbalance between pro-angiogenic and angiostatic factors, the structure of the new tumor vessels is disturbed compared to normal vessels [174, 175]. Tumor vessels do not present a strong endothelial cell monolayer, and leakage with increased interstitial fluid pressure is observed [176]. Pericytes do not form a tight association with endothelial cells, and the basement membrane is often defective [173]. As a result, the tumor neovasculature is often highly disorganized, permeable, irregular and tortuous. A summary of factors and interactions involved in tumor neoangiogenesis is given in figure 6.

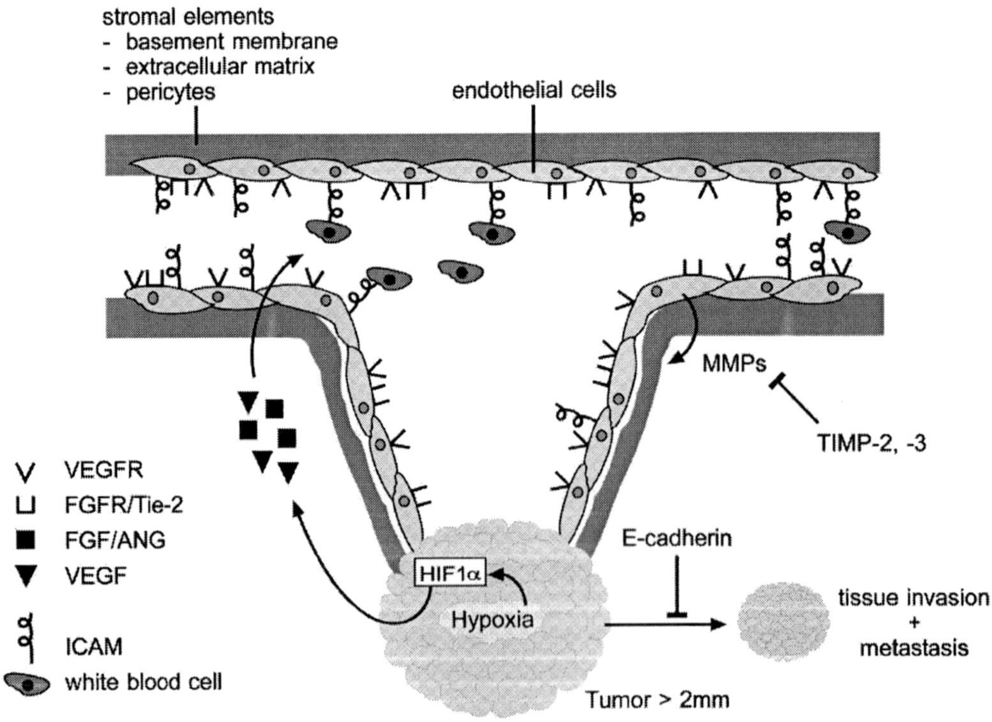

Figure 6. Neoangiogenesis is induced by hypoxia in growing tumors. HIF-1α is responsible for the production of enzymes protecting the cell from a low oxygen concentration and activates the production of proangiogenic proteins such as VEGF, FGF and angiopoietins. These factors induce changes in endothelial cells, pericytes, basement membrane and extracellular matrix. This results finally in neovessels, providing the tumor with oxygen and nutrients. Neoangiogenesis contributes also to the spreading of the tumor resulting from abnormal vessel structure. Additionally, an anergy against the immune response by a low density of ICAM molecules on the endothelial membrane is related to tumor-induced neovessels. DNA methylation can directly influence factors involved in the reorganization of the extracellular matrix such as TIMP-2 and TIMP-3 and proteins important for tissue invasion and metastasis such as E-cadherin. However, treatment with demethylating agents may also influence endothelial cell anergy.

The VEGF family comprises VEGF-A, -B, -C, -D and the placenta growth factor. They are responsible for endothelial cell survival, proliferation and migration as well as vascular permeability [175, 177]. VEGF mainly binds to the vascular endothelial growth factor receptors VEGFR-1 and -2. Other pro-angiogenic factors comprise angiopoietins, fibroblast growth factor, platelet-derived growth factor, transforming growth factor beta, interleukin 8, epidermal growth factor, hepatocyte growth factor and pleiotrophin. In particular, the angiopoietins Ang-1 and Ang-2 are frequently produced by tumor cells and induce angiogenesis by binding to the receptor tyrosine kinase Tie-2. Ang-1 is responsible for endothelial cell migration, adhesion and recruitment of pericytes and smooth muscle cells, whereas Ang-2 acts as a destabilizer for vessels [178-180].

Physiological angiogenesis inhibitors comprise interferons, platelet factor-4, tumstatin, angiostatin, endostatin, thrombospondin-1 and the tissue inhibitors of matrix metalloproteinases TIMP-2 and TIMP-3 [181-183].

Angiogenesis in growing tumors is involved in the inhibition of an anti-tumor immune response. An inflammatory reaction requires the recruitment and migration of leukocytes from circulating blood to the site of inflammation. Therefore, endothelial cells express adhesion molecules and selectins on their surface [184]. In inflammatory tissue, the presentation of the adhesion molecules is enhanced by TNF, interleukin-1 and interferons. In tumor tissues, the adhesion molecules are downregulated, preventing an appropriate anti-tumor immune response [184]. This so-called "endothelial cell anergy" contributes to the escape of tumor cells from immune response.

Endothelial cells in a tumor are not part of the malignant clone but they are influenced by the cancer tissue. Despite the non-malignant phenotype of the cells, in vitro data have provided evidence for an influence of the DNMT inhibitor 5-aza-2'-deoxycytidine (DAC) on angiogenesis [185]. DNMT inhibitors comprise DAC, AZA and zebularine and act as demethylating agents. In a mouse model, angiogenesis induced by a growing malignant tumor was markedly decreased by DAC and zebularine. DAC as well as the HDAC inhibitor trichostatin A (TSA) induced the transcription of growth inhibiting genes such as JUNB. These genes are not suppressed in tumor-conditioned endothelial cells by promoter hypermethylation. Regulation by histone methylation or acetylation as well as pathway-associated reexpression are possible mechanisms for this observation. Reexpression of the intercellular adhesion molecule-1 (ICAM-1) which is important for the anergy of the tumor tissue regarding an immune response could also be observed after treatment with DAC or TSA. Again, the suppression of gene expression in tumor-conditioned endothelial cells was due to methylation changes in histones and not to promoter hypermethylation.

Promoter region hypermethylation of TIMP-2 and TIMP-3 have been found in multiple solid and hematopoietic tumors [181, 186, 187]. TIMP-2 CpG islandmethylation was observed in more than 30 % of tissue samples from indolent and aggressive B-cell NHL [101]. TIMP-3 was hypermethylated in primary central nervous system lymphoma and in several solid tumors such as renal cell carcinoma [188].

Impaired cell adhesion is a further important aspect in the malignant phenotype of cancer cells with a contribution to tissue invasion and metastasis. Enhanced cell-cell and cell-stroma adhesion may result in destruction of the histological structure of a tissue. E-cadherin and the associated catenins, which connect cadherins to actin filaments, play a major role in cell adhesiveness in epithelial but also in other tissues. In hematopoiesis, expression of E-cadherin is important for erythroblast and normoblast maturation [189]. E-cadherin may be downregulated by different mechanisms such as mutations with loss of function of the E-cadherin gene or increased phosphorylation of beta-catenin by the c-erbB-2 protein or dysadherin, a cell membrane glycoprotein. However, DNA hypermethylation of E-cadherin is also frequently observed in human tumors [190, 191], especially in precancerous conditions [109].

Epigenetic downregulation of E-cadherin has been observed in AML, ALL, primary testicular lymphoma as well as in MM [118, 192]. Simultaneous promoter hypermethylation of E-cadherin, p16 and DAPK1 was associated with plasma cell leukemia, suggesting that an accumulation of epigenetic events may contribute to a more aggressive phenotype in MM [110].

Growth Factor Response and Signal Transduction

Cellular response to growth factor involves membrane bound or cytoplasmic receptors. These receptors often exert a kinase activity to initiate signal transduction, and the cascade ends with the binding of transcription factors in promoter regions of specific genes. The response to growth signals such as cytokines and hematopoietic growth factors is subjected to control mechanisms by counterregulatory proteins. Hematopoietic growth factors bind to membrane bound receptors, and signal transduction is often transmitted by the JAK/STAT (Janus kinase/signal transducers and activators of transcription) pathway [193]. In myeloproliferative syndromes (MPS), Janus kinase 2 (JAK2) is constitutively active owing to an autoactivating mutation that contributes to the hypersensitivity to hematopoietic growth factors [194, 195].

JAK proteins are under the control of SOCS proteins [196, 197] and SH2 domain containing tyrosine phosphatases such as SHP-1 [198]. Loss of function of SOCS-1 by biallelic mutation contributes to an increased JAK2 phosphorylation rate [199]. Downregulation of SOCS-1 in association with CpG island hypermethylation has been described in MM, MPS, MDS and AML [107, 117, 200, 201]. SHP-1, another well described negative regulator of JAK2, was tested positive for promoter hypermethylation in MM, AML, ALL, MDS and NHL [200, 202].

Recent studies have demonstrated a role for the Wnt signal transduction pathway in the self renewal of hematopoietic stem cells and normal lymphocyte development. The importance of the Wnt proteins could in particular be established in gastrointestinal cancer [203]. One important group of antagonists of Wnt are the secreted frizzled-related proteins (SFRP), and loss of function of SFRPs has been first described in colorectal cancer [204-208]. SFRP 1, 2, 4 and 5 are characterized by a CpG island in the promoter region and loss of expression of these proteins due to hypermethylation has also been observed in CLL and ALL [208].

ABL is a cytoplasmic tyrosine kinase and was the first gene for that a correlation between a chromosomal translocation and a specific disease, CML, could be elaborated [6]. Whereas one copy of the ABL gene in the cell is involved in the fusion gene BCR/ABL, the other copy of the gene remains functional during the chronic phase of CML. After a prolonged chronic phase of CML or when progression to blast crisis or acute leukemia of the disease occurs, promoter hypermethylation of the proximal promoter of the ABL gene is observed [112, 209]. This process may be reversed by a treatment with interferon-alpha.

The retinoic acid receptor RARbeta2 is a transcription factor involved in growth, differentiation and death of hematopoietic cells. CpG promoter hypermethylation with low expression levels of RARbeta2 has been demonstrated in MPS and AML [84, 210]. CRBP1, a further component in the transcription control exerted by retinoic acid, may also be silenced by promoter hypermethylation in malignant lymphomas and acute leukemias [99].

Additional Genes with DNA Methylation and Relevance in Signature

Glutathione is a molecule involved in detoxification, and glutathione S-transferases (GST) are a family of enzymes with an important role in protecting cells from cytotoxic and carcinogenic agents. Hypermethylation of the isotype GSTP1 is associated with loss of expression of the enzyme and observed in many solid tumors [89, 115]. Mutations in the Werner syndrome gene (WRN) are found in the corresponding disease with premature aging and a high incidence of malignant tumors. The gene is thought to be a tumor suppressor gene with a role in DNA repair, and silencing of the expression is associated with an increased sensitivity to topoisomerase type I inhibitors as irinotecan and topotecan. The importance of the WRN gene is known in colon cancer. CpG island hypermethylation of the promoter region was found frequently in NHL and rarely in ALL and AML [211].

DNA methylation contributes to the naturally occurring downregulation of genes after the embryonic and fetal period in early live. These proteins may serve as a target for a tumor-specific immune response when reexpressed during adult live [212]. As a consequence, the administration of demethylating agents may contribute to the development of an anticancer immune reaction. Of interest are the cancer testis antigens and melanoma antigens (MAGE, GAGE, BAGE). Demethylating agents may thus be part of tumor vaccine treatment strategies [213, 214].

In 2001, the human genome project has published the sequence of the human genetic code [215, 216]. About 23.000 genes were identified, and many of them have still an unknown function. Using array assays, activation or downregulation of every of these genes can be studied in different cancer types and treatment situations [217]. Additionally, arrays specific for the detection of methylated genes and the human epigenome project currently under way will define a great number of additional genes with promoter hypermethylation [218]. The precise role of these genes in tumorigenesis will be specified as well as the influence of demethylating agents and other treatment options in the disease course.

DNA Methylation as Prognostic Factor and Biomarker

Since the introduction of methylation-specific polymerase chain reaction (MSP), detection of promoter methylation became possible by testing tissue samples from large patient cohorts [219]. This permitted the analysis of DNA methylation as a prognostic marker in the context of clinical trials with conventional chemotherapeutic drugs or demethylating agents [220]. In particular, the effect of demethylating agents on the methylation status of specific genes could be observed [28, 137].

MSP and other techniques such as combined bisulfite restriction analysis (COBRA), bisulfite sequencing or quantitative MSP may be used for the detection of minimal residual disease [221]. Until now, no conclusive results could be obtained in hematopoietic malignancies.

Furthermore, detection of promoter methylation has been suggested to indicate risk assessment or susceptibility for colon cancer or as a screening method for occult cancer [222, 223]. Analysis of promoter methylation has also been applied in sentinel lymph node biopsies for the detection of low number of tumor cells for further operative or adjuvant treatment decisions [224].

In 2000, Esteller et al. published the first large study to prove the importance of methylation analysis for the prognosis of glioblastoma [164]. Hypermethylation of the promoter region of the O^6MGMT gene was associated with a better prognosis in glioblastomas treated with the alkylating agent carmustine. A silenced O^6MGMT gene presumably allowed the cytotoxic effect of carmustine to be realized in tumor cells by a less efficient DNA repair. Similar effects were observed in patients with DLBCL treated with a cyclophosphamide-containing regimen [25]. The presence of O^6MGMT methylation was associated with a statistically significant increase in overall survival and progression-free survival. O^6MGMT promoter hypermethylation was a strong and independent prognostic factor. These results could be confirmed in other trials, and promoter hypermethylation of O^6MGMT gene could be related to low expression level of the corresponding protein. No correlation between O^6MGMT promoter hypermethylation and mutation in the p53 gene could be established [225]. Whereas p15 had no prognostic impact on survival, DNA hypermethylation of the p16 promoter was found to be a risk factor for poor prognosis in intermediate and high risk DLBCL [103].

Richter's syndrome is the transformation of CLL into an aggressive DLBCL. A high incidence of MSI related to the promoter hypermethylation of hMLH1 has been observed in Richter's syndrome [169]. Recently, the expression of zeta associated protein (ZAP70) has been described as an important prognostic factor in CLL [24]. The expression of ZAP70 is assessed routinely by flow cytometry. Reduced expression of ZAP70 on the cell surface of lymphocytes from CLL patients could be related to the methylation status of the intron 1-exon 2 boundary region of the corresponding gene. Methylation analysis can be used as an alternative to flow cytometry to predict the prognosis in CLL patients. Additionally, aberrant hypermethylation of the fragile histidine triad (FHIT) gene could be found as an independent strong adverse prognostic factor in MM [226].

Multiple studies have been conducted to assess the role of CpG island methylation as a prognostic factor in myeloid malignancies [220]. In MDS, a study analyzed promoter hypermethylation of the p15, E-cadherin, HIC1 and estrogen receptor (ER) gene in CD34-positive hematopoietic precursor cells [135, 227]. Statistical analysis revealed that promoter methylation of one or more genes was an independent negative prognostic factor with a median survival of 17 months versus 63 months in the low methylation group. CpG island methylation of the p15 promoter region correlated with a high leukemic transformation rate in early MDS, and hypermethylation in all four genes was more frequent in advanced compared to early-stage disease. Patients with CML present a BCR/ABL fusion protein in the cell cytoplasm in balance with the normal ABL protein, and this equilibrium is probably important for the maintenance of the prolonged chronic phase of the disease [111]. Loss of expression of the normal ABL copy owing to promoter hypermethylation of the gene is often observed during leukemic transformation of the disease [112]. However, studies to prove the prognostic impact of ABL promoter hypermethylation have given contradictory results. No studies have yet evaluated these findings in the context of the new standard treatment of CML with imatinib [228]. Trials with DAC monotherapy in advanced CML during accelerated phase or blast phase have shown a high activity of this hypomethylating agent in this situation [229-231].

p15 is one of the most frequently hypermethylated genes in AML. The importance of p15 promoter methylation for the prognosis of AML remains controversial [232]. However, studies with a positive correlation between p15 methylation and a negative prognosis prevail [233]. One study suggested a positive correlation between promoter hypermethylation of E-cadherin and a negative outcome in AML as well as a more unfavorable prognosis in cases with simultaneous p15 CpG island methylation [115, 234]. A negative impact of p15 promoter hypermethylation could be observed especially in patients with APL [219]. A further gene silenced by promoter hypermethylation with a negative prognostic impact is the Wnt inhibitory factor-1. A poorer survival was found in patients with APL [235]. Contrary to the negative influence of promoter hypermethylation of those genes, ER methylation was associated with improved survival in adults with AML [232]. In a small study, promoter hypermethylation of the calcitonin gene was used as a marker for minimal residual disease detection in AML patients [27, 115]. When minimal residual disease, determined by MSP, persisted after end of treatment, bone marrow relapse occurred soon.

The concept of the CpG island methylator phenotype (CIMP) has been established particularly in ALL [236]. One trial analyzed 54 children with t(12;21)(p13;q22) positive B-ALL for the methylation status of 38 genes [26, 233]. In the group with the chromosomal abnormality, 89 % presented at least one hypermethylated gene. The group of patients with 0-2 of the 38 tested genes presenting hypermethylation (CIMP-) was compared with a group with more than 2 promoters with CpG island methylation (CIMP+). Remission rates did not differ between the groups of patients. Disease-free survival and overall survival at 9 years, however, were 92% and 100% for the CIMP- group and 33% and 73% for the CIMP+ group the differences being statistically highly significant. Comparable results were obtained in children with T-ALL. The prognostic impact of hypermethylation of inhibitors of the Wnt pathway could recently be confirmed in a general ALL patient population without recurring cytogenetic alterations [237, 238].

Epigenetic Alterations as Therapeutic Targets

Therapy with DNMT Inhibitors

Epigenetic changes in human cancer may be reversed by pharmacological inhibition of DNA methylation and histone deacetylation [84, 239, 240]. Figure 7 gives an overview about epigenetic factors contributing to the silencing of cancer-related genes and possible pharmacological intervention. The effects of therapeutic intervention with DNMT inhibitors on DNA methylation patterns are illustrated in figure 8.

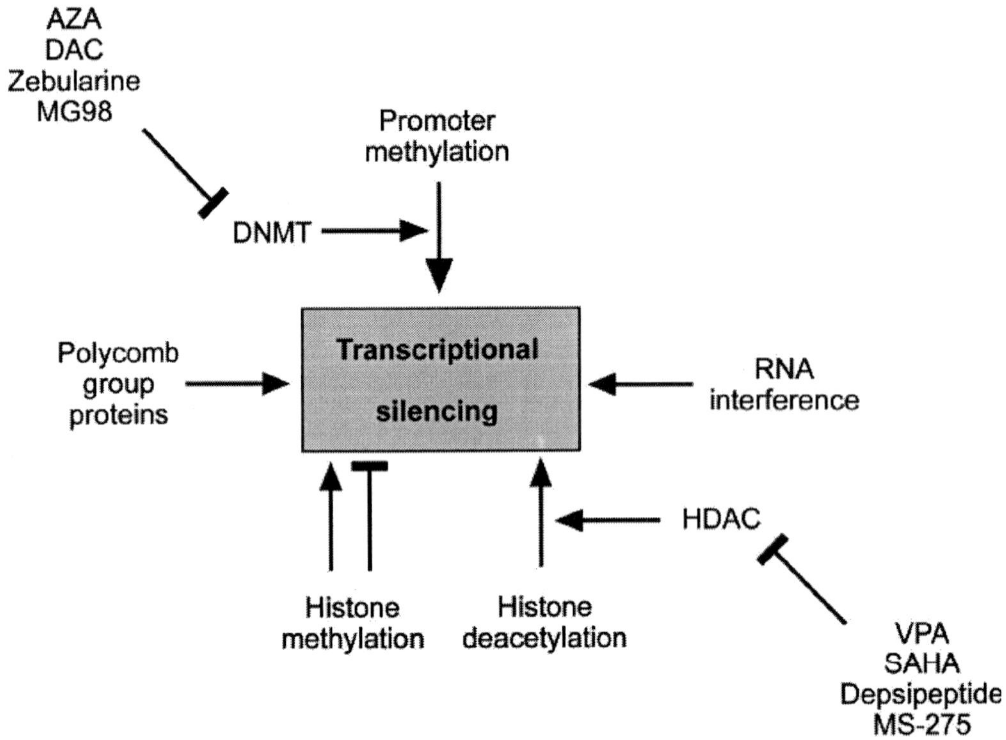

Figure 7. Epigenetic regulation of transcriptional activity and the influence of different agents used in clinical trials to modify DNA methylation patterns and histone acetylation status. Transcriptional silencing of tumor suppressor genes by epigenetic events is a fundamental mechanism in tumorigenesis. Epigenetic pathways influencing transcriptional silencing comprise promoter hypermethylation, histone acetylation and methylation, small inhibitors RNAs and polycomb group proteins. These mechanisms have multiple interactions with each other. Promoter hypermethylation and histone acetylation are at the moment the only factors in these interactions that can be influenced pharmacologically. Promoter hypermethylation is dependent on the activity of DNMTs that may be targeted by different azanucleosides (AZA, DAC, Zebularine) or the antisense nucleoside MG98. Deacetylation of histones may be reversed by different HDAC inhibitors such as VPA, SAHA, depsipeptide and MS-275.

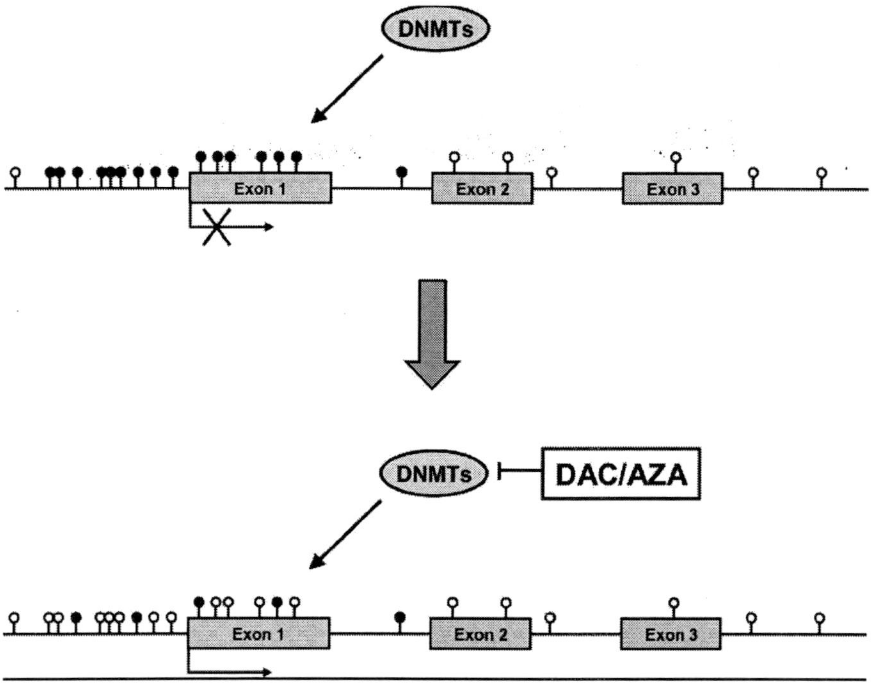

Figure 8. Treatment with demethylating agents induces global hypomethylation by inhibition of DNMTs, especially DNMT1, responsible for the maintenance of CpG methylation during mitosis. Demethylation may result in the reexpression of silenced genes and contribute to the reactivation of tumor suppressor genes.

Synthesized in the 1960s, both azanucleosides DAC and AZA were initially regarded as cytotoxic agents. DAC and AZA have similar mechanisms of action; DAC is incorporated into DNA only, while AZA is incorporated both into DNA and RNA. In early studies with the two drugs, intermediate to high doses were administered to patients with hematopoietic malignancies and solid tumors, often in combination with conventional chemotherapeutic agents [241]. However, the therapeutic index of DAC and AZA at those dose levels was limited by hematological and non-hematological toxicities, and the response rates were rather disappointing. For example, trials conducted with high dose azanucleosides in patients with refractory or relapsed AML resulted in remission rates of 24 % for AZA [242] and 12 % for DAC [243] and showed an unfavorable toxicity profile.

The demethylating properties of AZA have been first discovered in the 1980s by Jones and colleagues [244]. Based on the growing knowledge about gene silencing in association with promoter methylation, new treatment approaches with the prolonged administration of DNMT inhibitors at low doses were evaluated in vitro and also in clinical trials. In experimental settings, both DAC and AZA were shown to be capable of reactivating tumor suppressor genes silenced by promoter hypermethylation in association with regional DNA demethylation [186, 244-246]. These drugs incorporate into DNA as a cytosine analog to exert their effects but may also induce some non-specific toxic cellular effects, probably associated with DNA damage [247]. However, their best characterized mode of action is to induce DNA demethylation via irreversible inhibition of DNMTs. They act essentially on the DNMT1 enzyme that is responsible for the maintenance of the methylation status of the genome during replication. Thus, the administration of DAC and AZA leads to a sequential demethylation only in dividing cells during successive rounds of mitosis. Treatment with demethylating agents as inducers of differentiation must be considered as an approach that is different from conventional cytotoxic drugs. Differentiation arrest is a general phenomenon in cancer and may be influenced by DAC or AZA. However, the action of these drugs is not directed to specific genes or promoter regions. Further progress in the characterization of DNMTs will allow a more rational approach to drug design. Molecular models will enable the identification of more specific and selective molecules to obtain a targeted design of epigenetic drugs.

Phase II studies conducted with the administration of DAC at low doses in MDS have shown promising results with response rates of up to 76 % [248, 249]. Phase II dose finding studies for DAC in high risk MDS and chronic myelomonocytic leukemia (CMML) provided evidence for a dose of 20 mg/m²/d i.v. for 5 days as an optimal dose schedule. Other recent data from a phase I study with various hematopoietic malignancies indicate that DAC may be most effective when administered in a prolonged schedule at low doses for ten days, resulting in response rates of up to 65 % [229]. Interestingly, a high response rate of 64 % was especially observed by administration of low dose DAC in patients with high risk MDS subtypes [248]. Similar results were obtained in phase II studies with low dose AZA. In a first trial with 43 evaluable patients, 49 % responded with 12 % complete remission (CR), 25 % partial remission (PR) and 12 % with clinical benefit (CB) [250]. 39 % of patients responded with a transfusion independence following 7 days of subcutaneous administration of AZA in a phase II trial published by Gryn and colleagues [251].

Silverman and colleagues reported the first phase III study demonstrating that the clinical history of MDS may be changed using a non-intensive drug treatment with a positive impact on quality of life which played a pivotal role in the approval of AZA by the Food and Drug Administration [252]. In this randomized controlled phase III trial they compared the subcutaneous application of AZA to best supportive care. The overall response rate was 60 % in the AZA arm with a significant prolongation of median survival and a delay of time to leukemic transformation. Interestingly, a great percentage of responders showed only clinical benefit after the third or forth cycle of treatment, suggesting that repetitive exposure is important for initializing of the differentiation. In a phase III study using DAC, Kantarjian et al. reported complete remission rates of 9 %, partial remission of 8 % and hematological improvement in 13 % with a prolongation of the time to transformation in AML.

In another study evaluating the effects of DAC in CML, objective responses could be achieved in 28 % of the patients in blastic phase, in 55 % of the patients in accelerated phase, and in 63 % of the patients in chronic phase [253]. DAC was also applied to patients with imatinib-resistant CML (12 patients in chronic, 17 patients in accelerated and 6 patients in blastic phase). A hematological response rate of 54 % including 34 % CR and 20 % PR and/or hematological improvement were observed [229]. Major toxicity was myelosuppression with neutropenic fever in 23 % of treatment cycles. These encouraging results indicate significant anti-CML activity of DAC and may provide a basis for future studies combining demethylating agents with the tyrosine kinase inhibitor imatinib in BCR/ABL-positive CML.

Furthermore, two newly developed DNMT inhibitors are under clinical investigation. Zebularine is an azanucleoside that may be administered orally and has recently been introduced in clinical phase I trials [254-257]. MG98 is an antisense oligonucleotide of the DNMT1 gene [258, 259]. This phosphorothioate antisense oligodeoxynucleotide was shown to downregulate the DNMT1 protein in human cancer cells. Phase I results were obtained from patients with solid tumors with an acceptable toxicity profile. However, no phase II data are available at the moment.

In one trial using DAC in MDS, the methylation status of the cell cycle regulator p15 was analyzed as a surrogate marker for in vivo demethylating activity. Responding MDS patients appeared to have reversal of p15 hypermethylation and reestablishment of p15 protein expression in the bone marrow following DAC treatment [220]. However, it is not fully clear whether the DNA demethylation and gene reexpression reflects the return of normal cell populations or an acute response of the abnormal cells during initial therapy. Nevertheless, the results certainly are consistent with the idea that the clinical efficacy of demethylating agents may in fact be mediated by the reversal of gene silencing.

A summary of trials using demethylating agents in patients with hematopoietic malignancies is given in table 1.

Table 1. Clinical trials of DNMT inhibitors in adult patients with hematopoietic malignancies

Author	Disease	Dose schedule	n	CR/PR (%)
Trials with DAC				
Petti et al. 1993 [266]	Untreated AML	90-120 mg/m² tid for 3 days	12	25/8
Kantarjian et al. 1997 [243]	Relapsed/Refractory AML	90 mg/m² tid for 3 days	8	12/0
Kantarjian et al. 1997 [230]	Accelerated Phase CML	75-100 mg/m² bid for 5 days	17	6/12
Lübbert et al. 2005 [267]	AML elderly	15 mg/m² tid for 3 days	51	14/17
Sacchi et al. 1999 [231]	Blast crisis CML	50-100 mg/m² bid for 5 days	31	19/0
Issa et al. 2005 [268]	CML resistant to imatinib	15 mg/m² for 5 days	35	34/20
Wijermans et al. 2000 [248]	High risk MDS	45 mg/m² for 3 days	57	22/4
Kantarjian et al. 2007 [137]	MDS	10-20 mg/m² for 5-10 days	95	34/NR
Saba et al. 2004 [269]	MDS	45 mg/m² for 3 days	170	9/8
Trials with AZA				
Saiki et al. 1978 [242]	Relapsed/Refractory AML/ALL	300 mg/m² for 5 days	72	24/NR
Saiki et al. 1981 [270]	Relapsed/Refractory leukemia	150-300 mg/m² for 5-10 days	120	8/2
Trials with DAC				
Silverman et al. 1993 [250]	High risk MDS	75 mg/m² for 7 days	49	12/25
Gryn et al. 2002 [251]	High risk MDS	75 mg/m² for 7 days	48	NR/39 (CB)
Silverman et al. 2002 [252]	High risk MDS	75 mg/m² for 7 days	99	7/54 (PR + CB)

Combination Therapy with DNMT and HDAC Inhibitors

Owing to their mechanism of action and low toxicity, DNMT inhibitors have been investigated in combination with other anti-cancer agents. HDAC inhibitors as well as demethylating agents may contribute to the reexpression of epigenetically silenced genes in tumors. Enhanced reexpression of cancer-related genes has been shown to be achieved by treatment with a combination of HDAC and DNMT inhibitors. Direct interaction between gene silencing by promoter methylation and histone deacetylation is explained by HDAC recruitment to DNMTs in combination with methylcytosine-binding proteins [14, 260]. In vitro studies could demonstrate that HDAC inhibitors have the potential to synergize with demethylating agents for reactivation of epigenetically silenced genes when the DNMT inhibitor is given first to cause some initial degree of promoter demethylation [81, 203]. The optimal sequence of these agents to obtain a reexpression of epigenetically silenced genes is the exposure to a DNMT inhibitor at low doses for several days, followed by the use of a HDAC inhibitor [14, 81, 203, 240].

In a recent clinical study, patients with AML or MDS were treated with AZA subcutaneously for 10-14 days followed by a one week continuous infusion of the HDAC inhibitor phenylbutyrate (PB) [261]. Complete and partial responses could be observed in 50 % of patients treated by this regimen. Using this strategy, demethylation of the p15 promoter region and increases in histone acetylation could be demonstrated in vivo. An additional trial of the same study group was conducted with AZA for 5, 10 or 14 days subcutaneously followed by PB for 7 days as a continuous infusion [28]. Sequential bisulfite DNA sequencing showed a significant decrement in p15 methylation following treatment with AZA. A 50 % increase in global acetylation of histone H3 and H4 was already observed after administration of AZA alone. An additional gain in histone acetylation was demonstrated after the infusion of PB.

Furthermore, the combination of DAC (15 mg/m^2 for 10 days) and valproic acid (VPA) (20-50 mg/kg) was reported to induce CR in 22 % of pretreated patients with AML and MDS [262]. Global DNA hypomethylation, p15 reactivation and acetylation of histone H3 and H4 could also be observed in this patient group.

Combination treatment of DNMT inhibitors with cytotoxic drugs have been conducted in the 1990s using high doses of demethylating agents. DAC has been combined with daunorubicin, idarubicin and amsacrine, respectively. Remission rates were comparable to other chemotherapeutic regimens, but the relatively high hematological and gastrointestinal toxicity profile contributed to the abandoning of this approach [263, 264]. The combination of topoisomerase I inhibitors and cisplatin with either AZA or DAC has been conducted in solid tumors without improvement of remission rates but high treatment-associated toxicities [265].

A summary of clinical trials with combined treatment approaches involving demethylating agents and HDAC inhibitors in hematopoietic malignancies is shown in table 2.

Table 2. Clinical trials in adult patients with hematopoietic malignancies and treatment combination of DNMT inhibitors or HDAC inhibitors with other antineoplastic agents

Ref.	Agent combination	Disease	Dose schedule	n	CR/PR, SD (%)
Gore et al. 2004 [261]	AZA/PB	AML/MDS	AZA: 25-50 mg/m² for 10-14 days PB: 7 days at MTD	16	50 (CR + PR)
Garcia-Manero et al. 2006 [262]	DAC/VPA	AML/MDS	DAC: 15 mg/m²/d for 10 days VPA: 20-50 mg/kg	54	22 CR
Raffoux et al. 2005 [271]	AZA/VPA/ ATRA	AML	AZA: 75 mg/m²/d 7 days VPA: 20 mg/kg/d ATRA: 45 mg/m²/d day 8-28	5	40/40
Willemze et al. 1997 [263]	DAC with -Amsacrine or -Idarubicin	Acute leukemias	DAC: 125 mg/m² for 6 days Amsacrine: 120 mg/m² day 6,7 Idarubicin: 12 mg/m² days 5-7	30 33	27/NR 45/NR
Schwartsmann et al. 1997 [264]	DAC/ Daunorubicin	AML	DAC: 90 mg/m² for 5 days Daunorubicin: 50 mg/m² for 3 days	6	100
Gore et al. 2006 [28]	AZA/PB	AML/MDS	AZA: 25-75 mg/m² SC for 5-14 days PB: 375 mg/kg/m IV for 7 days	32	38 (CR + PR)

Conclusions and Future Perspectives

DNA methylation and post-transcriptional histone modifications interact in an epigenetic network that is critically involved in the regulation of chromatin structure and transcriptional activity. Both regional DNA hypermethylation and histone hypoacetylation have been recognized as promising therapeutic targets in hematopoietic malignancies and solid tumors. The clinical application of targeting the epigenome in cancer cells through demethylation and histone acetylation is an exiting novel pharmacologic concept in the treatment of malignant diseases. The modification of gene expression through epigenetic manipulation represents a new dimension in anti-cancer treatment: Administration of demethylating agents as inducers of differentiation must be considered as an approach different from conventional cytotoxic drugs. Based on the publication of the results of phase III trials in recent years, administration of DNMT inhibitors may now be considered as the standard treatment in patients with MDS.

Differentiation arrest is a general phenomenon in cancer and may be due to expression changes of many genes that can not specifically be influenced by DAC or AZA. However, the action of these drugs is not targeted to specific genes or their respective promoter regions. The phenomenon of RNA directed sequence specific DNA methylation has been described in C. elegans, plants, fungi, drosophila and very recently also to some extent in human cells. A number of obstacles such as non-specific responses to siRNA and yet undefined problems of histone alterations in conjunction with DNA methylation leave a wide-open area for translational research in this field.

So far, epigenetic therapies have been successful especially in AML and MDS. Clinical studies with the combined and sequential use of demethylating agents and HDAC inhibitors are currently ongoing in various hematopoietic malignancies. The development of more specific demethylating agents and HDAC inhibitors in the future will provide us with new powerful tools to target the deranged epigenome in cancer cells.

Acknowledgements

We thank Peter Glatte for expert assistance with the preparation of the figures. OG was supported by a grant from the Deutsche Krebshilfe.

References

[1] Hanahan D and Weinberg RA. The hallmarks of cancer. *Cell.* 2000; 100: 57-70.

[2] Vogelstein B and Kinzler KW. Cancer genes and the pathways they control. *Nat. Med.* 2004; 10: 789-799.

[3] Scaltriti M and Baselga J. The epidermal growth factor receptor pathway: a model for targeted therapy. *Clin. Cancer Res.* 2006; 12: 5268-5272.

[4] Bos JL. ras oncogenes in human cancer: a review. *Cancer Res.* 1989; 49: 4682-4689.

[5] Liggett WH, Jr. and Sidransky D. Role of the p16 tumor suppressor gene in cancer. *J. Clin. Oncol.* 1998; 16: 1197-1206.

[6] de Klein A, van Kessel AG, Grosveld G, Bartram CR, Hagemeijer A, Bootsma D, Spurr NK, Heisterkamp N, Groffen J, and Stephenson JR. A cellular oncogene is translocated to the Philadelphia chromosome in chronic myelocytic leukaemia. *Nature.* 1982; 300: 765-767.

[7] Kurzrock R, Gutterman JU, and Talpaz M. The molecular genetics of Philadelphia chromosome-positive leukemias. *N. Engl. J. Med.* 1988; 319: 990-998.

[8] Druker BJ. Perspectives on the development of a molecularly targeted agent. *Cancer Cell.* 2002; 1: 31-36.

[9] Gibbs JB. Mechanism-based target identification and drug discovery in cancer research. *Science.* 2000; 287: 1969-1973.

[10] Baselga J, Tripathy D, Mendelsohn J, Baughman S, Benz CC, Dantis L, Sklarin NT, Seidman AD, Hudis CA, Moore J, Rosen PP, Twaddell T, Henderson IC, and Norton L. Phase II study of weekly intravenous recombinant humanized anti-p185HER2 monoclonal antibody in patients with HER2/neu-overexpressing metastatic breast cancer. *J. Clin. Oncol.* 1996; 14: 737-744.

[11] Druker BJ, Sawyers CL, Kantarjian H, Resta DJ, Reese SF, Ford JM, Capdeville R, and Talpaz M. Activity of a specific inhibitor of the BCR-ABL tyrosine kinase in the blast crisis of chronic myeloid leukemia and acute lymphoblastic leukemia with the Philadelphia chromosome. *N. Engl. J. Med.* 2001; 344: 1038-1042.

[12] Druker BJ, Talpaz M, Resta DJ, Peng B, Buchdunger E, Ford JM, Lydon NB, Kantarjian H, Capdeville R, Ohno-Jones S, and Sawyers CL. Efficacy and safety of a specific inhibitor of the BCR-ABL tyrosine kinase in chronic myeloid leukemia. *N. Engl. J. Med.* 2001; 344: 1031-1037.

[13] Melnick A and Licht JD. Deconstructing a disease: RARalpha, its fusion partners, and their roles in the pathogenesis of acute promyelocytic leukemia. *Blood.* 1999; 93: 3167-3215.

[14] Egger G, Liang G, Aparicio A, and Jones PA. Epigenetics in human disease and prospects for epigenetic therapy. *Nature.* 2004; 429: 457-463.

[15] Esteller M. CpG island hypermethylation and tumor suppressor genes: a booming present, a brighter future. *Oncogene.* 2002; 21: 5427-5440.

[16] Herman JG and Baylin SB. Gene silencing in cancer in association with promoter hypermethylation. *N. Engl. J. Med.* 2003; 349: 2042-2054.

[17] Jones PA and Laird PW. Cancer epigenetics comes of age. *Nat. Genet.* 1999; 21: 163-167.

[18] Jones PA and Baylin SB. The fundamental role of epigenetic events in cancer. *Nat. Rev. Genet.* 2002; 3: 415-428.

[19] Richards EJ and Elgin SC. Epigenetic codes for heterochromatin formation and silencing: rounding up the usual suspects. *Cell.* 2002; 108: 489-500.

[20] Strahl BD and Allis CD. The language of covalent histone modifications. *Nature.* 2000; 403: 41-45.

[21] Kawasaki H and Taira K. Transcriptional gene silencing by short interfering RNAs. *Curr. Opin. Mol. Ther.* 2005; 7: 125-131.

[22] Sledz CA and Williams BR. RNA interference in biology and disease. *Blood.* 2005; 106: 787-794.

[23] Sparmann A and van Lohuizen M. Polycomb silencers control cell fate, development and cancer. *Nat. Rev. Cancer.* 2006; 6: 846-856.

[24] Corcoran M, Parker A, Orchard J, Davis Z, Wirtz M, Schmitz OJ, and Oscier D. ZAP-70 methylation status is associated with ZAP-70 expression status in chronic lymphocytic leukemia. *Haematologica.* 2005; 90: 1078-1088.

[25] Esteller M, Gaidano G, Goodman SN, Zagonel V, Capello D, Botto B, Rossi D, Gloghini A, Vitolo U, Carbone A, Baylin SB, and Herman JG. Hypermethylation of the DNA repair gene O(6)-methylguanine DNA methyltransferase and survival of patients with diffuse large B-cell lymphoma. *J. Natl. Cancer Inst.* 2002; 94: 26-32.

[26] Roman-Gomez J, Jimenez-Velasco A, Castillejo JA, Agirre X, Barrios M, Navarro G, Molina FJ, Calasanz MJ, Prosper F, Heiniger A, and Torres A. Promoter hypermethylation of cancer-related genes: a strong independent prognostic factor in acute lymphoblastic leukemia. *Blood.* 2004; 104: 2492-2498.

[27] Tang Y, Deng C, Du Q, Li G, Zeng Z, Zhang J, and Miu S. [Hypermethylation of the calcitonin gene as molecular genetic marker in detecting minimal residual disease of leukemia]. *Zhonghua Nei Ke Za Zhi.* 2001; 40: 226-228.

[28] Gore SD, Baylin S, Sugar E, Carraway H, Miller CB, Carducci M, Grever M, Galm O, Dauses T, Karp JE, Rudek MA, Zhao M, Smith BD, Manning J, Jiemjit A, Dover G, Mays A, Zwiebel J, Murgo A, Weng LJ, and Herman JG. Combined DNA methyltransferase and histone deacetylase inhibition in the treatment of myeloid neoplasms. *Cancer Res.* 2006; 66: 6361-6369.

[29] Holliday R and Grigg GW. DNA methylation and mutation. *Mutat Res.* 1993; 285: 61-67.

[30] Bird A. DNA methylation patterns and epigenetic memory. *Genes Dev.* 2002; 16: 6-21.

[31] Gardiner-Garden M and Frommer M. CpG islands in vertebrate genomes. *J. Mol. Biol.* 1987; 196: 261-282.

[32] Esteller M and Herman JG. Cancer as an epigenetic disease: DNA methylation and chromatin alterations in human tumours. *J. Pathol.* 2002; 196: 1-7.

[33] Constancia M, Pickard B, Kelsey G, and Reik W. Imprinting mechanisms. *Genome Res.* 1998; 8: 881-900.

[34] Robertson KD and Jones PA. DNA methylation: past, present and future directions. *Carcinogenesis.* 2000; 21: 461-467.

[35] Jaenisch R and Bird A. Epigenetic regulation of gene expression: how the genome integrates intrinsic and environmental signals. *Nat. Genet.* 2003; 33 Suppl: 245-254.

[36] Fischle W, Wang Y, and Allis CD. Histone and chromatin cross-talk. *Curr. Opin. Cell Biol.* 2003; 15: 172-183.

[37] Turner BM. Cellular memory and the histone code. *Cell.* 2002; 111: 285-291.

[38] Felsenfeld G and Groudine M. Controlling the double helix. *Nature.* 2003; 421: 448-453.

[39] Lachner M and Jenuwein T. The many faces of histone lysine methylation. *Curr. Opin. Cell Biol.* 2002; 14: 286-298.

[40] Jenuwein T and Allis CD. Translating the histone code. *Science.* 2001; 293: 1074-1080.

[41] Hake SB, Xiao A, and Allis CD. Linking the epigenetic 'language' of covalent histone modifications to cancer. *Br. J. Cancer.* 2004; 90: 761-769.

[42] Fire A, Xu S, Montgomery MK, Kostas SA, Driver SE, and Mello CC. Potent and specific genetic interference by double-stranded RNA in Caenorhabditis elegans. *Nature.* 1998; 391: 806-811.

[43] Chen CZ, Li L, Lodish HF, and Bartel DP. MicroRNAs modulate hematopoietic lineage differentiation. *Science.* 2004; 303: 83-86.

[44] Kim DH, Villeneuve LM, Morris KV, and Rossi JJ. Argonaute-1 directs siRNA-mediated transcriptional gene silencing in human cells. *Nat. Struct. Mol. Biol.* 2006; 13: 793-797.

[45] Ting AH, Schuebel KE, Herman JG, and Baylin SB. Short double-stranded RNA induces transcriptional gene silencing in human cancer cells in the absence of DNA methylation. *Nat. Genet.* 2005; 37: 906-910.

[46] Heidenreich O, Krauter J, Riehle H, Hadwiger P, John M, Heil G, Vornlocher HP, and Nordheim A. AML1/MTG8 oncogene suppression by small interfering RNAs supports myeloid differentiation of t(8;21)-positive leukemic cells. *Blood.* 2003; 101: 3157-3163.

[47] Scherr M, Battmer K, Winkler T, Heidenreich O, Ganser A, and Eder M. Specific inhibition of bcr-abl gene expression by small interfering RNA. *Blood.* 2003; 101: 1566-1569.

[48] Croonquist PA and Van Ness B. The polycomb group protein enhancer of zeste homolog 2 (EZH 2) is an oncogene that influences myeloma cell growth and the mutant ras phenotype. *Oncogene.* 2005; 24: 6269-6280.

[49] Laible G, Wolf A, Dorn R, Reuter G, Nislow C, Lebersorger A, Popkin D, Pillus L, and Jenuwein T. Mammalian homologues of the Polycomb-group gene Enhancer of zeste

mediate gene silencing in Drosophila heterochromatin and at S. cerevisiae telomeres. *Embo J.* 1997; 16: 3219-3232.

[50] van der Vlag J and Otte AP. Transcriptional repression mediated by the human polycomb-group protein EED involves histone deacetylation. *Nat. Genet.* 1999; 23: 474-478.

[51] Kuzmichev A, Jenuwein T, Tempst P, and Reinberg D. Different EZH2-containing complexes target methylation of histone H1 or nucleosomal histone H3. *Mol. Cell.* 2004; 14: 183-193.

[52] Vire E, Brenner C, Deplus R, Blanchon L, Fraga M, Didelot C, Morey L, Van Eynde A, Bernard D, Vanderwinden JM, Bollen M, Esteller M, Di Croce L, de Launoit Y, and Fuks F. The Polycomb group protein EZH2 directly controls DNA methylation. *Nature.* 2006; 439: 871-874.

[53] Cao R, Tsukada Y, and Zhang Y. Role of Bmi-1 and Ring1A in H2A ubiquitylation and Hox gene silencing. *Mol. Cell.* 2005; 20: 845-854.

[54] Cao R, Wang L, Wang H, Xia L, Erdjument-Bromage H, Tempst P, Jones RS, and Zhang Y. Role of histone H3 lysine 27 methylation in Polycomb-group silencing. *Science.* 2002; 298: 1039-1043.

[55] Li Z, Cao R, Wang M, Myers MP, Zhang Y, and Xu RM. Structure of a Bmi-1-Ring1B polycomb group ubiquitin ligase complex. *J. Biol. Chem.* 2006; 281: 20643-20649.

[56] Volpe TA, Kidner C, Hall IM, Teng G, Grewal SI, and Martienssen RA. Regulation of heterochromatic silencing and histone H3 lysine-9 methylation by RNAi. *Science.* 2002; 297: 1833-1837.

[57] van Kemenade FJ, Raaphorst FM, Blokzijl T, Fieret E, Hamer KM, Satijn DP, Otte AP, and Meijer CJ. Coexpression of BMI-1 and EZH2 polycomb-group proteins is associated with cycling cells and degree of malignancy in B-cell non-Hodgkin lymphoma. *Blood.* 2001; 97: 3896-3901.

[58] Bea S, Tort F, Pinyol M, Puig X, Hernandez L, Hernandez S, Fernandez PL, van Lohuizen M, Colomer D, and Campo E. BMI-1 gene amplification and overexpression in hematological malignancies occur mainly in mantle cell lymphomas. *Cancer Res.* 2001; 61: 2409-2412.

[59] Sawa M, Yamamoto K, Yokozawa T, Kiyoi H, Hishida A, Kajiguchi T, Seto M, Kohno A, Kitamura K, Itoh Y, Asou N, Hamajima N, Emi N, and Naoe T. BMI-1 is highly expressed in M0-subtype acute myeloid leukemia. *Int. J. Hematol.* 2005; 82: 42-47.

[60] Bachmann IM, Halvorsen OJ, Collett K, Stefansson IM, Straume O, Haukaas SA, Salvesen HB, Otte AP, and Akslen LA. EZH2 expression is associated with high proliferation rate and aggressive tumor subgroups in cutaneous melanoma and cancers of the endometrium, prostate, and breast. *J. Clin. Oncol.* 2006; 24: 268-273.

[61] Collett K, Eide GE, Arnes J, Stefansson IM, Eide J, Braaten A, Aas T, Otte AP, and Akslen LA. Expression of enhancer of zeste homologue 2 is significantly associated with increased tumor cell proliferation and is a marker of aggressive breast cancer. *Clin. Cancer Res.* 2006; 12: 1168-1174.

[62] Raaphorst FM, van Kemenade FJ, Blokzijl T, Fieret E, Hamer KM, Satijn DP, Otte AP, and Meijer CJ. Coexpression of BMI-1 and EZH2 polycomb group genes in Reed-Sternberg cells of Hodgkin's disease. *Am. J. Pathol.* 2000; 157: 709-715.

[63] Visser HP, Gunster MJ, Kluin-Nelemans HC, Manders EM, Raaphorst FM, Meijer CJ, Willemze R, and Otte AP. The Polycomb group protein EZH2 is upregulated in proliferating, cultured human mantle cell lymphoma. *Br. J. Haematol.* 2001; 112: 950-958.

[64] Weikert S, Christoph F, Kollermann J, Muller M, Schrader M, Miller K, and Krause H. Expression levels of the EZH2 polycomb transcriptional repressor correlate with aggressiveness and invasive potential of bladder carcinomas. *Int. J. Mol. Med.* 2005; 16: 349-353.

[65] Kuzmichev A, Margueron R, Vaquero A, Preissner TS, Scher M, Kirmizis A, Ouyang X, Brockdorff N, Abate-Shen C, Farnham P, and Reinberg D. Composition and histone substrates of polycomb repressive group complexes change during cellular differentiation. *Proc. Natl. Acad. Sci. U S A.* 2005; 102: 1859-1864.

[66] Lessard J and Sauvageau G. Bmi-1 determines the proliferative capacity of normal and leukaemic stem cells. *Nature.* 2003; 423: 255-260.

[67] Feinberg AP and Vogelstein B. Hypomethylation distinguishes genes of some human cancers from their normal counterparts. *Nature.* 1983; 301: 89-92.

[68] Feinberg AP, Gehrke CW, Kuo KC, and Ehrlich M. Reduced genomic 5-methylcytosine content in human colonic neoplasia. *Cancer Res.* 1988; 48: 1159-1161.

[69] Ehrlich M. DNA methylation in cancer: too much, but also too little. *Oncogene.* 2002; 21: 5400-5413.

[70] Gaudet F, Hodgson JG, Eden A, Jackson-Grusby L, Dausman J, Gray JW, Leonhardt H, and Jaenisch R. Induction of tumors in mice by genomic hypomethylation. *Science.* 2003; 300: 489-492.

[71] Baylin SB and Herman JG. DNA hypermethylation in tumorigenesis: epigenetics joins genetics. *Trends Genet.* 2000; 16: 168-174.

[72] Knudson AG. Two genetic hits (more or less) to cancer. *Nat. Rev. Cancer.* 2001; 1: 157-162.

[73] Okano M, Xie S, and Li E. Cloning and characterization of a family of novel mammalian DNA (cytosine-5) methyltransferases. *Nat. Genet.* 1998; 19: 219-220.

[74] Okano M, Bell DW, Haber DA, and Li E. DNA methyltransferases Dnmt3a and Dnmt3b are essential for de novo methylation and mammalian development. *Cell.* 1999; 99: 247-257.

[75] Fuks F, Burgers WA, Brehm A, Hughes-Davies L, and Kouzarides T. DNA methyltransferase Dnmt1 associates with histone deacetylase activity. *Nat. Genet.* 2000; 24: 88-91.

[76] Rountree MR, Bachman KE, and Baylin SB. DNMT1 binds HDAC2 and a new co-repressor, DMAP1, to form a complex at replication foci. *Nat. Genet.* 2000; 25: 269-277.

[77] Struhl K. Histone acetylation and transcriptional regulatory mechanisms. *Genes Dev.* 1998; 12: 599-606.

[78] Jones PL, Veenstra GJ, Wade PA, Vermaak D, Kass SU, Landsberger N, Strouboulis J, and Wolffe AP. Methylated DNA and MeCP2 recruit histone deacetylase to repress transcription. *Nat. Genet.* 1998; 19: 187-191.

[79] Baylin SB, Esteller M, Rountree MR, Bachman KE, Schuebel K, and Herman JG. Aberrant patterns of DNA methylation, chromatin formation and gene expression in cancer. *Hum. Mol. Genet.* 2001; 10: 687-692.

[80] Bachman KE, Park BH, Rhee I, Rajagopalan H, Herman JG, Baylin SB, Kinzler KW, and Vogelstein B. Histone modifications and silencing prior to DNA methylation of a tumor suppressor gene. *Cancer Cell.* 2003; 3: 89-95.

[81] Cameron EE, Bachman KE, Myohanen S, Herman JG, and Baylin SB. Synergy of demethylation and histone deacetylase inhibition in the re-expression of genes silenced in cancer. *Nat. Genet.* 1999; 21: 103-107.

[82] Esteller M, Corn PG, Baylin SB, and Herman JG. A gene hypermethylation profile of human cancer. *Cancer Res.* 2001; 61: 3225-3229.

[83] Esteller M. Profiling aberrant DNA methylation in hematologic neoplasms: a view from the tip of the iceberg. *Clin. Immunol.* 2003; 109: 80-88.

[84] Claus R and Lübbert M. Epigenetic targets in hematopoietic malignancies. *Oncogene.* 2003; 22: 6489-6496.

[85] Galm O and Esteller M. Beyond genetics-the emerging role of epigenetic changes in hematopoietic malignancies. *Int. J. Hematol.* 2004; 80: 120-127.

[86] Issa JP, Ottaviano YL, Celano P, Hamilton SR, Davidson NE, and Baylin SB. Methylation of the oestrogen receptor CpG island links ageing and neoplasia in human colon. *Nat. Genet.* 1994; 7: 536-540.

[87] Graff JR, Herman JG, Lapidus RG, Chopra H, Xu R, Jarrard DF, Isaacs WB, Pitha PM, Davidson NE, and Baylin SB. E-cadherin expression is silenced by DNA hypermethylation in human breast and prostate carcinomas. *Cancer Res.* 1995; 55: 5195-5199.

[88] Merlo A, Herman JG, Mao L, Lee DJ, Gabrielson E, Burger PC, Baylin SB, and Sidransky D. 5' CpG island methylation is associated with transcriptional silencing of the tumour suppressor p16/CDKN2/MTS1 in human cancers. *Nat. Med.* 1995; 1: 686-692.

[89] Esteller M, Corn PG, Urena JM, Gabrielson E, Baylin SB, and Herman JG. Inactivation of glutathione S-transferase P1 gene by promoter hypermethylation in human neoplasia. *Cancer Res.* 1998; 58: 4515-4518.

[90] Herman JG, Umar A, Polyak K, Graff JR, Ahuja N, Issa JP, Markowitz S, Willson JK, Hamilton SR, Kinzler KW, Kane MF, Kolodner RD, Vogelstein B, Kunkel TA, and Baylin SB. Incidence and functional consequences of hMLH1 promoter hypermethylation in colorectal carcinoma. *Proc. Natl. Acad. Sci. U S A.* 1998; 95: 6870-6875.

[91] Esteller M, Hamilton SR, Burger PC, Baylin SB, and Herman JG. Inactivation of the DNA repair gene O6-methylguanine-DNA methyltransferase by promoter hypermethylation is a common event in primary human neoplasia. *Cancer Res.* 1999; 59: 793-797.

[92] Corn PG, Kuerbitz SJ, van Noesel MM, Esteller M, Compitello N, Baylin SB, and Herman JG. Transcriptional silencing of the p73 gene in acute lymphoblastic leukemia and Burkitt's lymphoma is associated with 5' CpG island methylation. *Cancer Res.* 1999; 59: 3352-3356.

[93] Katzenellenbogen RA, Baylin SB, and Herman JG. Hypermethylation of the DAP-kinase CpG island is a common alteration in B-cell malignancies. *Blood.* 1999; 93: 4347-4353.

[94] Cameron EE, Baylin SB, and Herman JG. p15(INK4B) CpG island methylation in primary acute leukemia is heterogeneous and suggests density as a critical factor for transcriptional silencing. *Blood.* 1999; 94: 2445-2451.

[95] Virmani AK, Rathi A, Zochbauer-Muller S, Sacchi N, Fukuyama Y, Bryant D, Maitra A, Heda S, Fong KM, Thunnissen F, Minna JD, and Gazdar AF. Promoter methylation and silencing of the retinoic acid receptor-beta gene in lung carcinomas. *J. Natl. Cancer Inst.* 2000; 92: 1303-1307.

[96] Esteller M, Tortola S, Toyota M, Capella G, Peinado MA, Baylin SB, and Herman JG. Hypermethylation-associated inactivation of p14(ARF) is independent of p16(INK4a) methylation and p53 mutational status. *Cancer Res.* 2000; 60: 129-133.

[97] Yoshikawa H, Matsubara K, Qian GS, Jackson P, Groopman JD, Manning JE, Harris CC, and Herman JG. SOCS-1, a negative regulator of the JAK/STAT pathway, is silenced by methylation in human hepatocellular carcinoma and shows growth-suppression activity. *Nat. Genet.* 2001; 28: 29-35.

[98] Banelli B, Casciano I, Croce M, Di Vinci A, Gelvi I, Pagnan G, Brignole C, Allemanni G, Ferrini S, Ponzoni M, and Romani M. Expression and methylation of CASP8 in neuroblastoma: identification of a promoter region. *Nat. Med.* 2002; 8: 1333-1335.

[99] Esteller M, Guo M, Moreno V, Peinado MA, Capella G, Galm O, Baylin SB, and Herman JG. Hypermethylation-associated Inactivation of the Cellular Retinol-Binding-Protein 1 Gene in Human Cancer. *Cancer Res.* 2002; 62: 5902-5905.

[100] He B, You L, Uematsu K, Zang K, Xu Z, Lee AY, Costello JF, McCormick F, and Jablons DM. SOCS-3 is frequently silenced by hypermethylation and suppresses cell growth in human lung cancer. *Proc. Natl. Acad. Sci. U S A.* 2003; 100: 14133-14138.

[101] Galm O, Suzuki H, Akiyama Y, Esteller M, Brock MV, Osieka R, Baylin SB, and Herman JG. Inactivation of the tissue inhibitor of metalloproteinases-2 gene by promoter hypermethylation in lymphoid malignancies. *Oncogene.* 2005; 24: 4799-4805.

[102] Li Y, Nagai H, Ohno T, Yuge M, Hatano S, Ito E, Mori N, Saito H, and Kinoshita T. Aberrant DNA methylation of p57(KIP2) gene in the promoter region in lymphoid malignancies of B-cell phenotype. *Blood.* 2002; 100: 2572-2577.

[103] Rossi D, Capello D, Gloghini A, Franceschetti S, Paulli M, Bhatia K, Saglio G, Vitolo U, Pileri SA, Esteller M, Carbone A, and Gaidano G. Aberrant promoter methylation of multiple genes throughout the clinico-pathologic spectrum of B-cell neoplasia. *Haematologica.* 2004; 89: 154-164.

[104] Ng MH, Chung YF, Lo KW, Wickham NW, Lee JC, and Huang DP. Frequent hypermethylation of p16 and p15 genes in multiple myeloma. *Blood.* 1997; 89: 2500-2506.

[105] Gonzalez M, Mateos MV, Garcia-Sanz R, Balanzategui A, Lopez-Perez R, Chillon MC, Gonzalez D, Alaejos I, and San Miguel JF. De novo methylation of tumor suppressor gene p16/INK4a is a frequent finding in multiple myeloma patients at diagnosis. *Leukemia.* 2000; 14: 183-187.

[106] Guillerm G, Gyan E, Wolowiec D, Facon T, Avet-Loiseau H, Kuliczkowski K, Bauters F, Fenaux P, and Quesnel B. p16(INK4a) and p15(INK4b) gene methylations in plasma cells from monoclonal gammopathy of undetermined significance. *Blood.* 2001; 98: 244-246.

[107] Galm O, Yoshikawa H, Esteller M, Osieka R, and Herman JG. SOCS-1, a negative regulator of cytokine signaling, is frequently silenced by methylation in multiple myeloma. *Blood.* 2003; 101: 2784-2788.

[108] Ng MH, To KW, Lo KW, Chan S, Tsang KS, Cheng SH, and Ng HK. Frequent death-associated protein kinase promoter hypermethylation in multiple myeloma. *Clin. Cancer Res.* 2001; 7: 1724-1729.

[109] Seidl S, Ackermann J, Kaufmann H, Keck A, Nosslinger T, Zielinski CC, Drach J, and Zochbauer-Muller S. DNA-methylation analysis identifies the E-cadherin gene as a potential marker of disease progression in patients with monoclonal gammopathies. *Cancer.* 2004; 100: 2598-2606.

[110] Galm O, Wilop S, Reichelt J, Jost E, Gehbauer G, Herman JG, and Osieka R. DNA methylation changes in multiple myeloma. *Leukemia.* 2004; 18: 1687-1692.

[111] Nguyen TT, Mohrbacher AF, Tsai YC, Groffen J, Heisterkamp N, Nichols PW, Yu MC, Lubbert M, and Jones PA. Quantitative measure of c-abl and p15 methylation in chronic myelogenous leukemia: biological implications. *Blood.* 2000; 95: 2990-2992.

[112] Sun B, Jiang G, Zaydan MA, La Russa VF, Safah H, and Ehrlich M. ABL1 promoter methylation can exist independently of BCR-ABL transcription in chronic myeloid leukemia hematopoietic progenitors. *Cancer Res.* 2001; 61: 6931-6937.

[113] Garcia-Manero G, Daniel J, Smith TL, Kornblau SM, Lee MS, Kantarjian HM, and Issa JP. DNA methylation of multiple promoter-associated CpG islands in adult acute lymphocytic leukemia. *Clin. Cancer Res.* 2002; 8: 2217-2224.

[114] Gutierrez MI, Siraj AK, Bhargava M, Ozbek U, Banavali S, Chaudhary MA, El Solh H, and Bhatia K. Concurrent methylation of multiple genes in childhood ALL: Correlation with phenotype and molecular subgroup. *Leukemia.* 2003; 17: 1845-1850.

[115] Melki JR, Vincent PC, and Clark SJ. Concurrent DNA hypermethylation of multiple genes in acute myeloid leukemia. *Cancer Res.* 1999; 59: 3730-3740.

[116] Toyota M, Kopecky KJ, Toyota MO, Jair KW, Willman CL, and Issa JP. Methylation profiling in acute myeloid leukemia. *Blood.* 2001; 97: 2823-2829.

[117] Chen CY, Tsay W, Tang JL, Shen HL, Lin SW, Huang SY, Yao M, Chen YC, Shen MC, Wang CH, and Tien HF. SOCS1 methylation in patients with newly diagnosed acute myeloid leukemia. *Genes Chromosomes. Cancer* .2003; 37: 300-305.

[118] Galm O, Wilop S, Luders C, Jost E, Gehbauer G, Herman JG, and Osieka R. Clinical implications of aberrant DNA methylation patterns in acute myelogenous leukemia. *Ann. Hematol.* 2005: 1-8.

[119] Shapiro GI. Cyclin-dependent kinase pathways as targets for cancer treatment. *J. Clin. Oncol.* 2006; 24: 1770-1783.

[120] Macaluso M, Montanari M, Cinti C, and Giordano A. Modulation of cell cycle components by epigenetic and genetic events. *Semin. Oncol.* 2005; 32: 452-457.

[121] Jen J, Harper JW, Bigner SH, Bigner DD, Papadopoulos N, Markowitz S, Willson JK, Kinzler KW, and Vogelstein B. Deletion of p16 and p15 genes in brain tumors. *Cancer Res.* 1994; 54: 6353-6358.

[122] Cairns P, Polascik TJ, Eby Y, Tokino K, Califano J, Merlo A, Mao L, Herath J, Jenkins R, Westra W, and et al. Frequency of homozygous deletion at p16/CDKN2 in primary human tumours. *Nat. Genet.* 1995; 11: 210-212.

[123] Spike BT and Macleod KF. The Rb tumor suppressor in stress responses and hematopoietic homeostasis. *Cell Cycle.* 2005; 4: 42-45.

[124] Wang JC, Chen W, Nallusamy S, Chen C, and Novetsky AD. Hypermethylation of the P15INK4b and P16INK4a in agnogenic myeloid metaplasia (AMM) and AMM in leukaemic transformation. *Br. J. Haematol.* 2002; 116: 582-586.

[125] Lowe SW, Cepero E, and Evan G. Intrinsic tumour suppression. *Nature.* 2004; 432: 307-315.

[126] Harris SL and Levine AJ. The p53 pathway: positive and negative feedback loops. *Oncogene.* 2005; 24: 2899-2908.

[127] Sharpless NE. INK4a/ARF: a multifunctional tumor suppressor locus. *Mutat. Res.* 2005; 576: 22-38.

[128] Sugimoto M, Tahara H, Ide T, and Furuichi Y. Steps involved in immortalization and tumorigenesis in human B-lymphoblastoid cell lines transformed by Epstein-Barr virus. *Cancer Res.* 2004; 64: 3361-3364.

[129] Sherr CJ and McCormick F. The RB and p53 pathways in cancer. *Cancer Cell.* 2002; 2: 103-112.

[130] Yamasaki L. Role of the RB tumor suppressor in cancer. *Cancer Treat Res.* 2003; 115: 209-239.

[131] Ferreira R, Naguibneva I, Pritchard LL, Ait-Si-Ali S, and Harel-Bellan A. The Rb/chromatin connection and epigenetic control: opinion. *Oncogene.* 2001; 20: 3128-3133.

[132] Dominguez G, Silva J, Garcia JM, Silva JM, Rodriguez R, Munoz C, Chacon I, Sanchez R, Carballido J, Colas A, Espana P, and Bonilla F. Prevalence of aberrant methylation of p14ARF over p16INK4a in some human primary tumors. *Mutat. Res.* 2003; 530: 9-17.

[133] Ruas M and Peters G. The p16INK4a/CDKN2A tumor suppressor and its relatives. *Biochim. Biophys. Acta.* 1998; 1378: F115-177.

[134] Ohta M, Nagai H, Shimizu M, Rasio D, Berd D, Mastrangelo M, Singh AD, Shields JA, Shields CL, Croce CM, and et al. Rarity of somatic and germline mutations of the cyclin-dependent kinase 4 inhibitor gene, CDK4I, in melanoma. *Cancer Res.* 1994; 54: 5269-5272.

[135] Drexler HG. Review of alterations of the cyclin-dependent kinase inhibitor INK4 family genes p15, p16, p18 and p19 in human leukemia-lymphoma cells. *Leukemia.* 1998; 12: 845-859.

[136] Faderl S, Kantarjian HM, Manshouri T, Chan CY, Pierce S, Hays KJ, Cortes J, Thomas D, Estrov Z, and Albitar M. The prognostic significance of p16INK4a/p14ARF and p15INK4b deletions in adult acute lymphoblastic leukemia. *Clin. Cancer Res.* 1999; 5: 1855-1861.

[137] Kantarjian H, Oki Y, Garcia-Manero G, Huang X, O'Brien S, Cortes J, Faderl S, Bueso-Ramos C, Ravandi F, Estrov Z, Ferrajoli A, Wierda W, Shan J, Davis J, Giles F, Saba HI, and Issa JP. Results of a randomized study of 3 schedules of low-dose decitabine in higher-risk myelodysplastic syndrome and chronic myelomonocytic leukemia. *Blood.* 2007; 109: 52-57.

[138] Bueso-Ramos C, Xu Y, McDonnell TJ, Brisbay S, Pierce S, Kantarjian H, Rosner G, and Garcia-Manero G. Protein expression of a triad of frequently methylated genes, p73, p57Kip2, and p15, has prognostic value in adult acute lymphocytic leukemia independently of its methylation status. *J. Clin. Oncol.* 2005; 23: 3932-3939.

[139] Hatada I and Mukai T. Genomic imprinting and Beckwith-Wiedemann syndrome. *Histol. Histopathol.* 2000; 15: 309-312.

[140] Nagy E, Beck Z, Kiss A, Csoma E, Telek B, Konya J, Olah E, Rak K, and Toth FD. Frequent methylation of p16INK4A and p14ARF genes implicated in the evolution of chronic myeloid leukaemia from its chronic to accelerated phase. *Eur. J. Cancer.* 2003; 39: 2298-2305.

[141] Roman-Gomez J, Castillejo JA, Jimenez A, Gonzalez MG, Moreno F, Rodriguez Mdel C, Barrios M, Maldonado J, and Torres A. 5' CpG island hypermethylation is associated with transcriptional silencing of the p21(CIP1/WAF1/SDI1) gene and confers poor prognosis in acute lymphoblastic leukemia. *Blood.* 2002; 99: 2291-2296.

[142] Danial NN and Korsmeyer SJ. Cell death: critical control points. *Cell.* 2004; 116: 205-219.

[143] Igney FH and Krammer PH. Death and anti-death: tumour resistance to apoptosis. *Nat. Rev. Cancer.* 2002; 2: 277-288.

[144] Vousden KH and Lu X. Live or let die: the cell's response to p53. *Nat. Rev. Cancer.* 2002; 2: 594-604.

[145] Nagata S. Apoptosis by death factor. *Cell.* 1997; 88: 355-365.

[146] Riedl SJ and Shi Y. Molecular mechanisms of caspase regulation during apoptosis. *Nat. Rev. Mol. Cell Biol.* 2004; 5: 897-907.

[147] Chaudhari BR, Murphy RF, and Agrawal DK. Following the TRAIL to apoptosis. *Immunol. Res.* 2006; 35: 249-262.

[148] Kaufmann SH and Steensma DP. On the TRAIL of a new therapy for leukemia. *Leukemia.* 2005; 19: 2195-2202.

[149] Panner A, Parsa AT, and Pieper RO. Use of APO2L/TRAIL with mTOR inhibitors in the treatment of glioblastoma multiforme. *Expert Rev. Anticancer Ther.* 2006; 6: 1313-1322.

[150] Zhang L and Fang B. Mechanisms of resistance to TRAIL-induced apoptosis in cancer. *Cancer Gene Ther.* 2005; 12: 228-237.

[151] Roos WP and Kaina B. DNA damage-induced cell death by apoptosis. *Trends Mol. Med.* 2006; 12: 440-450.

[152] Wu X and Deng Y. Bax and BH3-domain-only proteins in p53-mediated apoptosis. *Front Biosci.* 2002; 7: d151-156.

[153] Gutierrez MI, Siraj AK, Ibrahim MM, Hussain A, and Bhatia K. Childhood and adult ALL: differences in epigenetic lesions associated with cell cycle genes. *Am. J. Hematol.* 2005; 80: 158-160.

[154] Agirre X, Novo FJ, Calasanz MJ, Larrayoz MJ, Lahortiga I, Valganon M, Garcia-Delgado M, and Vizmanos JL. TP53 is frequently altered by methylation, mutation, and/or deletion in acute lymphoblastic leukaemia. *Mol. Carcinog.* 2003; 38: 201-208.

[155] Guinn BA and Mills KI. p53 mutations, methylation and genomic instability in the progression of chronic myeloid leukaemia. *Leuk. Lymphoma.* 1997; 26: 211-226.

[156] Chim CS, Fung TK, Wong KF, Lau JS, and Liang R. Frequent DAP kinase but not p14 or Apaf-1 hypermethylation in B-cell chronic lymphocytic leukemia. *J. Hum. Genet.* 2006; 51: 832-838.

[157] Kawano S, Miller CW, Gombart AF, Bartram CR, Matsuo Y, Asou H, Sakashita A, Said J, Tatsumi E, and Koeffler HP. Loss of p73 gene expression in leukemias/lymphomas due to hypermethylation. *Blood.* 1999; 94: 1113-1120.

[158] Pluta A, Nyman U, Joseph B, Robak T, Zhivotovsky B, and Smolewski P. The role of p73 in hematological malignancies. *Leukemia.* 2006; 20: 757-766.

[159] Bialik S and Kimchi A. DAP-kinase as a target for drug design in cancer and diseases associated with accelerated cell death. *Semin. Cancer Biol.* 2004; 14: 283-294.

[160] Kaufmann WK. Cell cycle checkpoints and DNA repair preserve the stability of the human genome. *Cancer Metastasis Rev.* 1995; 14: 31-41.

[161] Wong P, Verselis SJ, Garber JE, Schneider K, DiGianni L, Stockwell DH, Li FP, and Syngal S. Prevalence of early onset colorectal cancer in 397 patients with classic Li-Fraumeni syndrome. *Gastroenterology.* 2006; 130: 73-79.

[162] Lynch JP and Hoops TC. The genetic pathogenesis of colorectal cancer. *Hematol. Oncol. Clin. North Am.* 2002; 16: 775-810.

[163] Yu Q. Restoring p53-mediated apoptosis in cancer cells: new opportunities for cancer therapy. *Drug Resist. Updat.* 2006; 9: 19-25.

[164] Esteller M, Garcia-Foncillas J, Andion E, Goodman SN, Hidalgo OF, Vanaclocha V, Baylin SB, and Herman JG. Inactivation of the DNA-repair gene MGMT and the clinical response of gliomas to alkylating agents. *N. Engl. J. Med.* 2000; 343: 1350-1354.

[165] Brandes AA, Tosoni A, Cavallo G, Reni M, Franceschi E, Bonaldi L, Bertorelle R, Gardiman M, Ghimenton C, Iuzzolino P, Pession A, Blatt V, and Ermani M. Correlations between O6-methylguanine DNA methyltransferase promoter methylation status, 1p and 19q deletions, and response to temozolomide in anaplastic and recurrent oligodendroglioma: a prospective GICNO study. *J. Clin. Oncol.* 2006; 24: 4746-4753.

[166] Martin P, Santon A, Garcia-Cosio M, and Bellas C. hMLH1 and MGMT inactivation as a mechanism of tumorigenesis in monoclonal gammopathies. *Mod. Pathol.* 2006; 19: 914-921.

[167] Lenz G, Hutter G, Hiddemann W, and Dreyling M. Promoter methylation and expression of DNA repair genes hMLH1 and MGMT in acute myeloid leukemia. *Ann. Hematol.* 2004; 83: 628-633.

[168] Seedhouse CH, Das-Gupta EP, and Russell NH. Methylation of the hMLH1 promoter and its association with microsatellite instability in acute myeloid leukemia. *Leukemia.* 2003; 17: 83-88.

[169] Fulop Z, Csernus B, Timar B, Szepesi A, and Matolcsy A. Microsatellite instability and hMLH1 promoter hypermethylation in Richter's transformation of chronic lymphocytic leukemia. *Leukemia.* 2003; 17: 411-415.

[170] Bergers G and Benjamin LE. Tumorigenesis and the angiogenic switch. *Nat. Rev. Cancer.* 2003; 3: 401-410.

[171] Carmeliet P, Dor Y, Herbert JM, Fukumura D, Brusselmans K, Dewerchin M, Neeman M, Bono F, Abramovitch R, Maxwell P, Koch CJ, Ratcliffe P, Moons L, Jain RK, Collen D, and Keshert E. Role of HIF-1alpha in hypoxia-mediated apoptosis, cell proliferation and tumour angiogenesis. *Nature.* 1998; 394: 485-490.

[172] Ferrara N and Kerbel RS. Angiogenesis as a therapeutic target. *Nature.* 2005; 438: 967-974.

[173] Chantrain CF, Henriet P, Jodele S, Emonard H, Feron O, Courtoy PJ, DeClerck YA, and Marbaix E. Mechanisms of pericyte recruitment in tumour angiogenesis: a new role for metalloproteinases. *Eur. J. Cancer* .2006; 42: 310-318.

[174] Baluk P, Hashizume H, and McDonald DM. Cellular abnormalities of blood vessels as targets in cancer. *Curr. Opin. Genet. Dev.* 2005; 15: 102-111.

[175] Lee CG, Link H, Baluk P, Homer RJ, Chapoval S, Bhandari V, Kang MJ, Cohn L, Kim YK, McDonald DM, and Elias JA. Vascular endothelial growth factor (VEGF) induces remodeling and enhances TH2-mediated sensitization and inflammation in the lung. *Nat. Med.* 2004; 10: 1095-1103.

[176] Ergun S, Tilki D, Oliveira-Ferrer L, Schuch G, and Kilic N. Significance of vascular stabilization for tumor growth and metastasis. *Cancer Lett.* 2006; 238: 180-187.

[177] List AF. Vascular endothelial growth factor signaling pathway as an emerging target in hematologic malignancies. *Oncologist.* 2001; 6: 24-31.

[178] Yu Q. The dynamic roles of angiopoietins in tumor angiogenesis. *Future Oncol.* 2005; 1: 475-484.

[179] Kobayashi H and Lin PC. Angiopoietin/Tie2 signaling, tumor angiogenesis and inflammatory diseases. *Front. Biosci.* 2005; 10: 666-674.

[180] Plank MJ, Sleeman BD, and Jones PF. The role of the angiopoietins in tumour angiogenesis. *Growth Factors.* 2004; 22: 1-11.

[181] Hajitou A, Sounni NE, Devy L, Grignet-Debrus C, Lewalle JM, Li H, Deroanne CF, Lu H, Colige A, Nusgens BV, Frankenne F, Maron A, Yeh P, Perricaudet M, Chang Y, Soria C, Calberg-Bacq CM, Foidart JM, and Noel A. Down-regulation of vascular endothelial growth factor by tissue inhibitor of metalloproteinase-2: effect on in vivo mammary tumor growth and angiogenesis. *Cancer Res.* 2001; 61: 3450-3457.

[182] Vergani V, Garofalo A, Bani MR, Borsotti P, Parker MP, Drudis T, Mazzarol G, Viale G, Giavazzi R, Stetler-Stevenson WG, and Taraboletti G. Inhibition of matrix metalloproteinases by over-expression of tissue inhibitor of metalloproteinase-2 inhibits the growth of experimental hemangiomas. *Int. J. Cancer.* 2001; 91: 241-247.

[183] Seo DW, Li H, Guedez L, Wingfield PT, Diaz T, Salloum R, Wei BY, and Stetler-Stevenson WG. TIMP-2 mediated inhibition of angiogenesis: an MMP-independent mechanism. *Cell.* 2003; 114: 171-180.

[184] Hellebrekers DM, Castermans K, Vire E, Dings RP, Hoebers NT, Mayo KH, Oude Egbrink MG, Molema G, Fuks F, van Engeland M, and Griffioen AW. Epigenetic

regulation of tumor endothelial cell anergy: silencing of intercellular adhesion molecule-1 by histone modifications. *Cancer Res.* 2006; 66: 10770-10777.

[185] Hellebrekers DM, Jair KW, Vire E, Eguchi S, Hoebers NT, Fraga MF, Esteller M, Fuks F, Baylin SB, van Engeland M, and Griffioen AW. Angiostatic activity of DNA methyltransferase inhibitors. *Mol. Cancer Ther.* 2006; 5: 467-475.

[186] Bachman KE, Herman JG, Corn PG, Merlo A, Costello JF, Cavenee WK, Baylin SB, and Graff JR. Methylation-associated silencing of the tissue inhibitor of metalloproteinase-3 gene suggest a suppressor role in kidney, brain, and other human cancers. *Cancer Res.* 1999; 59: 798-802.

[187] Albini A, Melchiori A, Santi L, Liotta LA, Brown PD, and Stetler-Stevenson WG. Tumor cell invasion inhibited by TIMP-2. *J. Natl. Cancer Inst.* 1991; 83: 775-779.

[188] Brock MV, Gou M, Akiyama Y, Muller A, Wu TT, Montgomery E, Deasel M, Germonpre P, Rubinson L, Heitmiller RF, Yang SC, Forastiere AA, Baylin SB, and Herman JG. Prognostic importance of promoter hypermethylation of multiple genes in esophageal adenocarcinoma. *Clin. Cancer Res.* 2003; 9: 2912-2919.

[189] Jankowski JA, Bedford FK, and Kim YS. Changes in gene structure and regulation of E-cadherin during epithelial development, differentiation, and disease. *Prog. Nucleic. Acid. Res. Mol. Biol.* 1997; 57: 187-215.

[190] Nass SJ, Herman JG, Gabrielson E, Iversen PW, Parl FF, Davidson NE, and Graff JR. Aberrant methylation of the estrogen receptor and E-cadherin 5' CpG islands increases with malignant progression in human breast cancer. *Cancer Res.* 2000; 60: 4346-4348.

[191] Corn PG, Heath EI, Heitmiller R, Fogt F, Forastiere AA, Herman JG, and Wu TT. Frequent hypermethylation of the 5' CpG island of E-cadherin in esophageal adenocarcinoma. *Clin. Cancer Res.* 2001; 7: 2765-2769.

[192] Melki JR, Vincent PC, Brown RD, and Clark SJ. Hypermethylation of E-cadherin in leukemia. *Blood.* 2000; 95: 3208-3213.

[193] Imada K and Leonard WJ. The Jak-STAT pathway. *Mol. Immunol.* 2000; 37: 1-11.

[194] Baxter EJ, Scott LM, Campbell PJ, East C, Fourouclas N, Swanton S, Vassiliou GS, Bench AJ, Boyd EM, Curtin N, Scott MA, Erber WN, and Green AR. Acquired mutation of the tyrosine kinase JAK2 in human myeloproliferative disorders. *Lancet.* 2005; 365: 1054-1061.

[195] James C, Ugo V, Le Couedic JP, Staerk J, Delhommeau F, Lacout C, Garcon L, Raslova H, Berger R, Bennaceur-Griscelli A, Villeval JL, Constantinescu SN, Casadevall N, and Vainchenker W. A unique clonal JAK2 mutation leading to constitutive signalling causes polycythaemia vera. *Nature.* 2005; 434: 1144-1148.

[196] Alexander WS, Starr R, Metcalf D, Nicholson SE, Farley A, Elefanty AG, Brysha M, Kile BT, Richardson R, Baca M, Zhang JG, Willson TA, Viney EM, Sprigg NS, Rakar S, Corbin J, Mifsud S, DiRago L, Cary D, Nicola NA, and Hilton DJ. Suppressors of cytokine signaling (SOCS): negative regulators of signal transduction. *J. Leukoc. Biol.* 1999; 66: 588-592.

[197] Naka T, Fujimoto M, and Kishimoto T. Negative regulation of cytokine signaling: STAT-induced STAT inhibitor. *Trends Biochem. Sci.* 1999; 24: 394-398.

[198] Wu C, Guan Q, Wang Y, Zhao ZJ, and Zhou GW. SHP-1 suppresses cancer cell growth by promoting degradation of JAK kinases. *J. Cell Biochem.* 2003; 90: 1026-1037.

[199] Melzner I, Bucur AJ, Bruderlein S, Dorsch K, Hasel C, Barth TF, Leithauser F, and Moller P. Biallelic mutation of SOCS-1 impairs JAK2 degradation and sustains phospho-JAK2 action in the MedB-1 mediastinal lymphoma line. *Blood.* 2005; 105: 2535-2542.

[200] Chim CS, Wong KY, Loong F, and Srivastava G. SOCS1 and SHP1 hypermethylation in mantle cell lymphoma and follicular lymphoma: implications for epigenetic activation of the Jak/STAT pathway. *Leukemia.* 2004; 18: 356-358.

[201] Jost E, do Ó N, Maintz C, Jousten P, Habets L, Wilop S, Herman J, Osieka R, and Galm O. Epigenetic alterations complement mutation of JAK2 tyrosine kinase in patients with BCR/ABL-negative myeloproliferative disorders. *Leukemia.* 2007; in press.

[202] Chim CS, Fung TK, Cheung WC, Liang R, and Kwong YL. SOCS1 and SHP1 hypermethylation in multiple myeloma: implications for epigenetic activation of the Jak/STAT pathway. *Blood.* 2004; 103: 4630-4635.

[203] Suzuki H, Gabrielson E, Chen W, Anbazhagan R, van Engeland M, Weijenberg MP, Herman JG, and Baylin SB. A genomic screen for genes upregulated by demethylation and histone deacetylase inhibition in human colorectal cancer. *Nat. Genet.* 2002; 31: 141-149.

[204] Caldwell GM, Jones C, Gensberg K, Jan S, Hardy RG, Byrd P, Chughtai S, Wallis Y, Matthews GM, and Morton DG. The Wnt antagonist sFRP1 in colorectal tumorigenesis. *Cancer Res.* 2004; 64: 883-888.

[205] Fukui T, Kondo M, Ito G, Maeda O, Sato N, Yoshioka H, Yokoi K, Ueda Y, Shimokata K, and Sekido Y. Transcriptional silencing of secreted frizzled related protein 1 (SFRP 1) by promoter hypermethylation in non-small-cell lung cancer. *Oncogene.* 2005; 24: 6323-6327.

[206] He B, Lee AY, Dadfarmay S, You L, Xu Z, Reguart N, Mazieres J, Mikami I, McCormick F, and Jablons DM. Secreted frizzled-related protein 4 is silenced by hypermethylation and induces apoptosis in beta-catenin-deficient human mesothelioma cells. *Cancer Res.* 2005; 65: 743-748.

[207] Lee AY, He B, You L, Dadfarmay S, Xu Z, Mazieres J, Mikami I, McCormick F, and Jablons DM. Expression of the secreted frizzled-related protein gene family is downregulated in human mesothelioma. *Oncogene.* 2004; 23: 6672-6676.

[208] Liu TH, Raval A, Chen SS, Matkovic JJ, Byrd JC, and Plass C. CpG island methylation and expression of the secreted frizzled-related protein gene family in chronic lymphocytic leukemia. *Cancer Res.* 2006; 66: 653-658.

[209] Asimakopoulos FA, Shteper PJ, Krichevsky S, Fibach E, Polliack A, Rachmilewitz E, Ben-Neriah Y, and Ben-Yehuda D. ABL1 methylation is a distinct molecular event associated with clonal evolution of chronic myeloid leukemia. *Blood.* 1999; 94: 2452-2460.

[210] Jones LC, Tefferi A, Idos GE, Kumagai T, Hofmann WK, and Koeffler HP. RARbeta2 is a candidate tumor suppressor gene in myelofibrosis with myeloid metaplasia. *Oncogene.* 2004; 23: 7846-7853.

[211] Agrelo R, Cheng WH, Setien F, Ropero S, Espada J, Fraga MF, Herranz M, Paz MF, Sanchez-Cespedes M, Artiga MJ, Guerrero D, Castells A, von Kobbe C, Bohr VA, and

Esteller M. Epigenetic inactivation of the premature aging Werner syndrome gene in human cancer. *Proc. Natl. Acad. Sci. U S A.* 2006; 103: 8822-8827.

[212] Lim JH, Kim SP, Gabrielson E, Park YB, Park JW, and Kwon TK. Activation of human cancer/testis antigen gene, XAGE-1, in tumor cells is correlated with CpG island hypomethylation. *Int. J. Cancer.* 2005; 116: 200-206.

[213] Maio M, Coral S, Fratta E, Altomonte M, and Sigalotti L. Epigenetic targets for immune intervention in human malignancies. *Oncogene.* 2003; 22: 6484-6488.

[214] Sigalotti L, Coral S, Fratta E, Lamaj E, Danielli R, Di Giacomo AM, Altomonte M, and Maio M. Epigenetic modulation of solid tumors as a novel approach for cancer immunotherapy. *Semin. Oncol.* 2005; 32: 473-478.

[215] Venter JC, Adams MD, Myers EW, Li PW, Mural RJ, Sutton GG, Smith HO, Yandell M, Evans CA, Holt RA, Gocayne JD, Amanatides P, Ballew RM, Huson DH, Wortman JR, Zhang Q, Kodira CD, Zheng XH, Chen L, Skupski M, Subramanian G, Thomas PD, Zhang J, Gabor Miklos GL, Nelson C, Broder S, Clark AG, Nadeau J, McKusick VA, Zinder N, Levine AJ, Roberts RJ, Simon M, Slayman C, Hunkapiller M, Bolanos R, Delcher A, Dew I, Fasulo D, Flanigan M, Florea L, Halpern A, Hannenhalli S, Kravitz S, Levy S, Mobarry C, Reinert K, Remington K, Abu-Threideh J, Beasley E, Biddick K, Bonazzi V, Brandon R, Cargill M, Chandramouliswaran I, Charlab R, Chaturvedi K, Deng Z, Di Francesco V, Dunn P, Eilbeck K, Evangelista C, Gabrielian AE, Gan W, Ge W, Gong F, Gu Z, Guan P, Heiman TJ, Higgins ME, Ji RR, Ke Z, Ketchum KA, Lai Z, Lei Y, Li Z, Li J, Liang Y, Lin X, Lu F, Merkulov GV, Milshina N, Moore HM, Naik AK, Narayan VA, Neelam B, Nusskern D, Rusch DB, Salzberg S, Shao W, Shue B, Sun J, Wang Z, Wang A, Wang X, Wang J, Wei M, Wides R, Xiao C, Yan C, Yao A, Ye J, Zhan M, Zhang W, Zhang H, Zhao Q, Zheng L, Zhong F, Zhong W, Zhu S, Zhao S, Gilbert D, Baumhueter S, Spier G, Carter C, Cravchik A, Woodage T, Ali F, An H, Awe A, Baldwin D, Baden H, Barnstead M, Barrow I, Beeson K, Busam D, Carver A, Center A, Cheng ML, Curry L, Danaher S, Davenport L, Desilets R, Dietz S, Dodson K, Doup L, Ferriera S, Garg N, Gluecksmann A, Hart B, Haynes J, Haynes C, Heiner C, Hladun S, Hostin D, Houck J, Howland T, Ibegwam C, Johnson J, Kalush F, Kline L, Koduru S, Love A, Mann F, May D, McCawley S, McIntosh T, McMullen I, Moy M, Moy L, Murphy B, Nelson K, Pfannkoch C, Pratts E, Puri V, Qureshi H, Reardon M, Rodriguez R, Rogers YH, Romblad D, Ruhfel B, Scott R, Sitter C, Smallwood M, Stewart E, Strong R, Suh E, Thomas R, Tint NN, Tse S, Vech C, Wang G, Wetter J, Williams S, Williams M, Windsor S, Winn-Deen E, Wolfe K, Zaveri J, Zaveri K, Abril JF, Guigo R, Campbell MJ, Sjolander KV, Karlak B, Kejariwal A, Mi H, Lazareva B, Hatton T, Narechania A, Diemer K, Muruganujan A, Guo N, Sato S, Bafna V, Istrail S, Lippert R, Schwartz R, Walenz B, Yooseph S, Allen D, Basu A, Baxendale J, Blick L, Caminha M, Carnes-Stine J, Caulk P, Chiang YH, Coyne M, Dahlke C, Mays A, Dombroski M, Donnelly M, Ely D, Esparham S, Fosler C, Gire H, Glanowski S, Glasser K, Glodek A, Gorokhov M, Graham K, Gropman B, Harris M, Heil J, Henderson S, Hoover J, Jennings D, Jordan C, Jordan J, Kasha J, Kagan L, Kraft C, Levitsky A, Lewis M, Liu X, Lopez J, Ma D, Majoros W, McDaniel J, Murphy S, Newman M, Nguyen T, Nguyen N, Nodell M, Pan S, Peck J, Peterson M, Rowe W, Sanders R, Scott J, Simpson M, Smith T, Sprague A, Stockwell T, Turner R,

Venter E, Wang M, Wen M, Wu D, Wu M, Xia A, Zandieh A and Zhu X. The sequence of the human genome. *Science.* 2001; 291: 1304-1351.

[216] Lander ES, Linton LM, Birren B, Nusbaum C, Zody MC, Baldwin J, Devon K, Dewar K, Doyle M, FitzHugh W, Funke R, Gage D, Harris K, Heaford A, Howland J, Kann L, Lehoczky J, LeVine R, McEwan P, McKernan K, Meldrim J, Mesirov JP, Miranda C, Morris W, Naylor J, Raymond C, Rosetti M, Santos R, Sheridan A, Sougnez C, Stange-Thomann N, Stojanovic N, Subramanian A, Wyman D, Rogers J, Sulston J, Ainscough R, Beck S, Bentley D, Burton J, Clee C, Carter N, Coulson A, Deadman R, Deloukas P, Dunham A, Dunham I, Durbin R, French L, Grafham D, Gregory S, Hubbard T, Humphray S, Hunt A, Jones M, Lloyd C, McMurray A, Matthews L, Mercer S, Milne S, Mullikin JC, Mungall A, Plumb R, Ross M, Shownkeen R, Sims S, Waterston RH, Wilson RK, Hillier LW, McPherson JD, Marra MA, Mardis ER, Fulton LA, Chinwalla AT, Pepin KH, Gish WR, Chissoe SL, Wendl MC, Delehaunty KD, Miner TL, Delehaunty A, Kramer JB, Cook LL, Fulton RS, Johnson DL, Minx PJ, Clifton SW, Hawkins T, Branscomb E, Predki P, Richardson P, Wenning S, Slezak T, Doggett N, Cheng JF, Olsen A, Lucas S, Elkin C, Uberbacher E, Frazier M, Gibbs RA, Muzny DM, Scherer SE, Bouck JB, Sodergren EJ, Worley KC, Rives CM, Gorrell JH, Metzker ML, Naylor SL, Kucherlapati RS, Nelson DL, Weinstock GM, Sakaki Y, Fujiyama A, Hattori M, Yada T, Toyoda A, Itoh T, Kawagoe C, Watanabe H, Totoki Y, Taylor T, Weissenbach J, Heilig R, Saurin W, Artiguenave F, Brottier P, Bruls T, Pelletier E, Robert C, Wincker P, Smith DR, Doucette-Stamm L, Rubenfield M, Weinstock K, Lee HM, Dubois J, Rosenthal A, Platzer M, Nyakatura G, Taudien S, Rump A, Yang H, Yu J, Wang J, Huang G, Gu J, Hood L, Rowen L, Madan A, Qin S, Davis RW, Federspiel NA, Abola AP, Proctor MJ, Myers RM, Schmutz J, Dickson M, Grimwood J, Cox DR, Olson MV, Kaul R, Shimizu N, Kawasaki K, Minoshima S, Evans GA, Athanasiou M, Schultz R, Roe BA, Chen F, Pan H, Ramser J, Lehrach H, Reinhardt R, McCombie WR, de la Bastide M, Dedhia N, Blocker H, Hornischer K, Nordsiek G, Agarwala R, Aravind L, Bailey JA, Bateman A, Batzoglou S, Birney E, Bork P, Brown DG, Burge CB, Cerutti L, Chen HC, Church D, Clamp M, Copley RR, Doerks T, Eddy SR, Eichler EE, Furey TS, Galagan J, Gilbert JG, Harmon C, Hayashizaki Y, Haussler D, Hermjakob H, Hokamp K, Jang W, Johnson LS, Jones TA, Kasif S, Kaspryzk A, Kennedy S, Kent WJ, Kitts P, Koonin EV, Korf I, Kulp D, Lancet D, Lowe TM, McLysaght A, Mikkelsen T, Moran JV, Mulder N, Pollara VJ, Ponting CP, Schuler G, Schultz J, Slater G, Smit AF, Stupka E, Szustakowski J, Thierry-Mieg D, Thierry-Mieg J, Wagner L, Wallis J, Wheeler R, Williams A, Wolf YI, Wolfe KH, Yang SP, Yeh RF, Collins F, Guyer MS, Peterson J, Felsenfeld A, Wetterstrand KA, Patrinos A, Morgan MJ, de Jong P, Catanese JJ, Osoegawa K, Shizuya H, Choi S and Chen YJ. Initial sequencing and analysis of the human genome. *Nature* 2001; 409: 860-921.

[217] Fraga MF and Esteller M. DNA methylation: a profile of methods and applications. *Biotechniques.* 2002; 33: 632, 634, 636-649.

[218] Ushijima T. Detection and interpretation of altered methylation patterns in cancer cells. *Nat. Rev. Cancer.* 2005; 5: 223-231.

[219] Chim CS, Liang R, Tam CY, and Kwong YL. Methylation of p15 and p16 Genes in Acute Promyelocytic Leukemia: Potential Diagnostic and Prognostic Significance. *J. Clin. Oncol.* 2001; 19: 2033-2040.

[220] Daskalakis M, Nguyen TT, Nguyen C, Guldberg P, Kohler G, Wijermans P, Jones PA, and Lübbert M. Demethylation of a hypermethylated P15/INK4B gene in patients with myelodysplastic syndrome by 5-Aza-2'-deoxycytidine (decitabine) treatment. *Blood.* 2002; 100: 2957-2964.

[221] Xiong Z and Laird PW. COBRA: a sensitive and quantitative DNA methylation assay. *Nucleic Acids Res.* 1997; 25: 2532-2534.

[222] Kim H, Kwon YM, Kim JS, Lee H, Park JH, Shim YM, Han J, Park J, and Kim DH. Tumor-specific methylation in bronchial lavage for the early detection of non-small-cell lung cancer. *J. Clin. Oncol.* 2004; 22: 2363-2370.

[223] Petko Z, Ghiassi M, Shuber A, Gorham J, Smalley W, Washington MK, Schultenover S, Gautam S, Markowitz SD, and Grady WM. Aberrantly methylated CDKN2A, MGMT, and MLH1 in colon polyps and in fecal DNA from patients with colorectal polyps. *Clin. Cancer Res.* 2005; 11: 1203-1209.

[224] Cavalli LR, Urban CA, Dai D, de Assis S, Tavares DC, Rone JD, Bleggi-Torres LF, Lima RS, Cavalli IJ, Issa JP, and Haddad BR. Genetic and epigenetic alterations in sentinel lymph nodes metastatic lesions compared to their corresponding primary breast tumors. *Cancer Genet. Cytogenet.* 2003; 146: 33-40.

[225] Hiraga J, Kinoshita T, Ohno T, Mori N, Ohashi H, Fukami S, Noda A, Ichikawa A, and Naoe T. Promoter hypermethylation of the DNA-repair gene O6-methylguanine-DNA methyltransferase and p53 mutation in diffuse large B-cell lymphoma. *Int. J. Hematol.* 2006; 84: 248-255.

[226] Takada S, Morita K, Hayashi K, Matsushima T, Sawamura M, Murakami H, and Nojima Y. Methylation status of fragile histidine triad (FHIT) gene and its clinical impact on prognosis of patients with multiple myeloma. *Eur. J. Haematol.* 2005; 75: 505-510.

[227] Quesnel B, Guillerm G, Vereecque R, Wattel E, Preudhomme C, Bauters F, Vanrumbeke M, and Fenaux P. Methylation of the p15(INK4b) gene in myelodysplastic syndromes is frequent and acquired during disease progression. *Blood.* 1998; 91: 2985-2990.

[228] Deininger M, Buchdunger E, and Druker BJ. The development of imatinib as a therapeutic agent for chronic myeloid leukemia. *Blood.* 2005; 105: 2640-2653.

[229] Issa JP, Gharibyan V, Cortes J, Jelinek J, Morris G, Verstovsek S, Talpaz M, Garcia-Manero G, and Kantarjian HM. Phase II study of low-dose decitabine in patients with chronic myelogenous leukemia resistant to imatinib mesylate. *J. Clin. Oncol.* 2005; 23: 3948-3956.

[230] Kantarjian HM, O'Brien SM, Keating M, Beran M, Estey E, Giralt S, Kornblau S, Rios MB, de Vos D, and Talpaz M. Results of decitabine therapy in the accelerated and blastic phases of chronic myelogenous leukemia. *Leukemia.* 1997; 11: 1617-1620.

[231] Sacchi S, Kantarjian HM, O'Brien S, Cortes J, Rios MB, Giles FJ, Beran M, Koller CA, Keating MJ, and Talpaz M. Chronic myelogenous leukemia in nonlymphoid blastic

phase: analysis of the results of first salvage therapy with three different treatment approaches for 162 patients. *Cancer.* 1999; 86: 2632-2641.

[232] Issa JP, Baylin SB, and Herman JG. DNA methylation changes in hematologic malignancies: biologic and clinical implications. *Leukemia* .1997; 11 Suppl 1: S7-11.

[233] Wong IH, Ng MH, Huang DP, and Lee JC. Aberrant p15 promoter methylation in adult and childhood acute leukemias of nearly all morphologic subtypes: potential prognostic implications. *Blood.* 2000; 95: 1942-1949.

[234] Corn PG, Smith BD, Ruckdeschel ES, Douglas D, Baylin SB, and Herman JG. E-cadherin expression is silenced by 5' CpG island methylation in acute leukemia. *Clin. Cancer Res.* 2000; 6: 4243-4248.

[235] Chim CS, Chan WW, Pang A, and Kwong YL. Preferential methylation of Wnt inhibitory factor-1 in acute promyelocytic leukemia: an independent poor prognostic factor. *Leukemia.* 2006; 20: 907-909.

[236] Issa JP. CpG island methylator phenotype in cancer. *Nat. Rev. Cancer.* 2004; 4: 988-993.

[237] Roman-Gomez J, Cordeu L, Agirre X, Jimenez-Velasco A, San Jose-Eneriz E, Garate L, Calasanz MJ, Heiniger A, Torres A, and Prosper F. Epigenetic regulation of WNT signaling pathway in acute lymphoblastic leukemia. *Blood.* 2006.

[238] Roman-Gomez J, Jimenez-Velasco A, Agirre X, Castillejo JA, Navarro G, Barrios M, Andreu EJ, Prosper F, Heiniger A, and Torres A. Transcriptional silencing of the Dickkopfs-3 (Dkk-3) gene by CpG hypermethylation in acute lymphoblastic leukaemia. *Br. J. Cancer.* 2004; 91: 707-713.

[239] Gilbert J, Gore SD, Herman JG, and Carducci MA. The clinical application of targeting cancer through histone acetylation and hypomethylation. *Clin. Cancer Res.* 2004; 10: 4589-4596.

[240] Garcia-Manero G and Gore SD. Future directions for the use of hypomethylating agents. *Semin. Hematol.* 2005; 42: S50-59.

[241] Santini V, Kantarjian HM, and Issa JP. Changes in DNA methylation in neoplasia: pathophysiology and therapeutic implications. *Ann. Intern. Med.* 2001; 134: 573-586.

[242] Saiki JH, McCredie KB, Vietti TJ, Hewlett JS, Morrison FS, Costanzi JJ, Stuckey WJ, Whitecar J, and Hoogstraten B. 5-azacytidine in acute leukemia. *Cancer.* 1978; 42: 2111-2114.

[243] Kantarjian HM, O'Brien SM, Estey E, Giralt S, Beran M, Rios MB, Keating M, de Vos D, and Talpaz M. Decitabine studies in chronic and acute myelogenous leukemia. *Leukemia.* 1997; 11 Suppl 1: S35-36.

[244] Jones PA and Taylor SM. Cellular differentiation, cytidine analogs and DNA methylation. *Cell.* 1980; 20: 85-93.

[245] Galm O, Rountree MR, Bachman KE, Jair KW, Baylin SB, and Herman JG. Enzymatic regional methylation assay: a novel method to quantify regional CpG methylation density. *Genome Res* .2002; 12: 153-157.

[246] Paz MF, Fraga MF, Avila S, Guo M, Pollan M, Herman JG, and Esteller M. A systematic profile of DNA methylation in human cancer cell lines. *Cancer Res.* 2003; 63: 1114-1121.

[247] Issa JP. Decitabine. *Curr. Opin. Oncol.* 2003; 15: 446-451.

[248] Wijermans P, Lubbert M, Verhoef G, Bosly A, Ravoet C, Andre M, and Ferrant A. Low-dose 5-aza-2'-deoxycytidine, a DNA hypomethylating agent, for the treatment of high-risk myelodysplastic syndrome: a multicenter phase II study in elderly patients. *J. Clin. Oncol.* 2000; 18: 956-962.

[249] Kantarjian H, Oki Y, Garcia-Manero G, Huang X, O'Brien S, Cortes J, Faderl S, Bueso-Ramos C, Ravandi F, Estrov Z, Ferrajoli A, Wierda W, Shan J, Davis J, Giles F, Saba HI, and Issa JP. Results of a randomized study of three schedules of low-dose decitabine in higher risk myelodysplastic syndrome and chronic myelomonocytic leukemia. *Blood.* 2006.

[250] Silverman LR, Holland JF, Weinberg RS, Alter BP, Davis RB, Ellison RR, Demakos EP, Cornell CJ, Jr., Carey RW, Schiffer C, and et al. Effects of treatment with 5-azacytidine on the in vivo and in vitro hematopoiesis in patients with myelodysplastic syndromes. *Leukemia.* 1993; 7 Suppl 1: 21-29.

[251] Gryn J, Zeigler ZR, Shadduck RK, Lister J, Raymond JM, Sbeitan I, Srodes C, Meisner D, and Evans C. Treatment of myelodysplastic syndromes with 5-azacytidine. *Leuk. Res.* 2002; 26: 893-897.

[252] Silverman LR, Demakos EP, Peterson BL, Kornblith AB, Holland JC, Odchimar-Reissig R, Stone RM, Nelson D, Powell BL, DeCastro CM, Ellerton J, Larson RA, Schiffer CA, and Holland JF. Randomized controlled trial of azacitidine in patients with the myelodysplastic syndrome: a study of the cancer and leukemia group B. *J. Clin. Oncol.* 2002; 20: 2429-2440.

[253] Kantarjian HM, O'Brien S, Cortes J, Giles FJ, Faderl S, Issa JP, Garcia-Manero G, Rios MB, Shan J, Andreeff M, Keating M, and Talpaz M. Results of decitabine (5-aza-2'deoxycytidine) therapy in 130 patients with chronic myelogenous leukemia. *Cancer.* 2003; 98: 522-528.

[254] Zhou L, Cheng X, Connolly BA, Dickman MJ, Hurd PJ, and Hornby DP. Zebularine: a novel DNA methylation inhibitor that forms a covalent complex with DNA methyltransferases. *J. Mol. Biol.* 2002; 321: 591-599.

[255] Cheng JC, Matsen CB, Gonzales FA, Ye W, Greer S, Marquez VE, Jones PA, and Selker EU. Inhibition of DNA methylation and reactivation of silenced genes by zebularine. *J. Natl. Cancer Inst.* 2003; 95: 399-409.

[256] Cheng JC, Yoo CB, Weisenberger DJ, Chuang J, Wozniak C, Liang G, Marquez VE, Greer S, Orntoft TF, Thykjaer T, and Jones PA. Preferential response of cancer cells to zebularine. *Cancer Cell.* 2004; 6: 151-158.

[257] Holleran JL, Parise RA, Joseph E, Eiseman JL, Covey JM, Glaze ER, Lyubimov AV, Chen YF, D'Argenio DZ, and Egorin MJ. Plasma pharmacokinetics, oral bioavailability, and interspecies scaling of the DNA methyltransferase inhibitor, zebularine. *Clin. Cancer Res.* 2005; 11: 3862-3868.

[258] Fournel M, Sapieha P, Beaulieu N, Besterman JM, and MacLeod AR. Down-regulation of human DNA-(cytosine-5) methyltransferase induces cell cycle regulators p16(ink4A) and p21(WAF/Cip1) by distinct mechanisms. *J. Biol. Chem.* 1999; 274: 24250-24256.

[259] Stewart DJ, Donehower RC, Eisenhauer EA, Wainman N, Shah AK, Bonfils C, MacLeod AR, Besterman JM, and Reid GK. A phase I pharmacokinetic and

pharmacodynamic study of the DNA methyltransferase 1 inhibitor MG98 administered twice weekly. *Ann. Oncol.* 2003; 14: 766-774.

[260] Rountree MR, Bachman KE, Herman JG, and Baylin SB. DNA methylation, chromatin inheritance, and cancer. *Oncogene.* 2001; 20: 3156-3165.

[261] Gore SD, Baylin SB, Dauses T, Grever MR, Jiemijt A, Manning J, Miller CB, Murgo AJ, Zwiebel JA, and Herman JG. Changes in promoter methylation and gene expression in patients with MDS and MDS-AML treated with 5-azacytidine and sodium phenylbutyrate [abstract]. *Blood.* 2004; 104: 469a.

[262] Garcia-Manero G, Kantarjian HM, Sanchez-Gonzalez B, Yang H, Rosner G, Verstovsek S, Rytting M, Wierda WG, Ravandi F, Koller C, Xiao L, Faderl S, Estrov Z, Cortes J, O'Brien S, Estey E, Bueso-Ramos C, Fiorentino J, Jabbour E, and Issa JP. Phase 1/2 study of the combination of 5-aza-2'-deoxycytidine with valproic acid in patients with leukemia. *Blood.* 2006; 108: 3271-3279.

[263] Willemze R, Suciu S, Archimbaud E, Muus P, Stryckmans P, Louwagie EA, Berneman Z, Tjean M, Wijermans P, Dohner H, Jehn U, Labar B, Jaksic B, Dardenne M, and Zittoun R. A randomized phase II study on the effects of 5-Aza-2'-deoxycytidine combined with either amsacrine or idarubicin in patients with relapsed acute leukemia: an EORTC Leukemia Cooperative Group phase II study (06893). *Leukemia.* 1997; 11 Suppl 1: S24-27.

[264] Schwartsmann G, Fernandes MS, Schaan MD, Moschen M, Gerhardt LM, Di Leone L, Loitzembauer B, and Kalakun L. Decitabine (5-Aza-2'-deoxycytidine; DAC) plus daunorubicin as a first line treatment in patients with acute myeloid leukemia: preliminary observations. *Leukemia.* 1997; 11 Suppl 1: S28-31.

[265] Schwartsmann G, Schunemann H, Gorini CN, Filho AF, Garbino C, Sabini G, Muse I, DiLeone L, and Mans DR. A phase I trial of cisplatin plus decitabine, a new DNA-hypomethylating agent, in patients with advanced solid tumors and a follow-up early phase II evaluation in patients with inoperable non-small cell lung cancer. *Invest. New Drugs.* 2000; 18: 83-91.

[266] Petti MC, Mandelli F, Zagonel V, De Gregoris C, Merola MC, Latagliata R, Gattei V, Fazi P, Monfardini S, and Pinto A. Pilot study of 5-aza-2'-deoxycytidine (Decitabine) in the treatment of poor prognosis acute myelogenous leukemia patients: preliminary results. *Leukemia.* 1993; 7 Suppl 1: 36-41.

[267] Lübbert M, Rueter B, Schnid M, Knipp S, Germing U, Dobbelstein C, Eimermacher H, Deschler B, Claus R, Schlenk R, Ganser A, Doehner H, Schmoor C, and Henss H. Continued Low-dose Decitabine (DAC) is an active First-Line Treatment of older AML Patients: First results of a Multicenter Phase II Study. *Blood.* 2005; 106: 1852a.

[268] Issa JP, Gharibyan V, Jelinek J, Morris G, Verstovsek S, Talpaz M, Garcia-Manero G, and Kantarjian HM. Hypomethylation Dynamics Following Decitabine Therapy in Chronic Myelogenous Leukemia. *Blood.* 2004; 104: 2956a.

[269] Saba H, Rosenfeld C, Issa JP, DiPersio J, Raza A, Klimek V, Slack J, de Castro C, Mettinger K, and Kantarjian HM. First Report of the Phase III North American Trial of Decitabine in Advanced Myelodysplastic Syndrome (MDS). *Blood.* 2004; 104: 67a.

[270] Saiki JH, Bodey GP, Hewlett JS, Amare M, Morrison FS, Wilson HE, and Linman JW. Effect of schedule on activity and toxicity of 5-azacytidine in acute leukemia: a Southwest Oncology Group Study. *Cancer.* 1981; 47: 1739-1742.

[271] Raffoux E, de Labarthe A, Gardin C, Daniel MT, Chomienne C, Fenaux P, Degos L, and Dombret H. Sequential Treatment with 5-Azacytidine, Valproic Acid (VPA), and All-Trans Retinoic Acid (ATRA) in Patients with High-Risk Acute Myeloid Leukemia (AML). *Blood.* 2005; 106: 2798a.

In: DNA Methylation Research Trends
Editor: Taku B. Kobayashi, pp. 183-208
ISBN: 978-1-60021-721-0

Chapter VII

DNA Methylation and Tumor Suppression

***Iftikhar Qayumi*[1] *and Muhammad Ashraf*[2]**
[1] Womens Medical College Abbottabad, Pakistan
[2] Department of Plant Sciences; Quaid-I-Azam University Islamabad, Pakistan

Abstract

DNA methylation is currently the only known natural base modification of the DNA molecule occurring under the control of a group of nuclear enzymes called DNA methylases, the principal one of these being DNA methyltransferase 1 (DNMT1). Methylation most commonly involves conversion of the base cytosine to 5'-methylcytosine (5'-MeC) and is a fundamental 'epigenetic' pathway involved in 'gene silencing'. Such events occur preferentially at specific loci, designated CpG areas (cytosine-guanine pairs or doublets) located throughout DNA coding areas. Runs of CpG doublets called CpG islands occur in the promoter regions of a number of important genes, though these remain unmethylated under most normal circumstances.

Along with regulating such important events like daughter strand methylation, genomic imprinting of paternal and maternal strands/allele selection and X-chromosome inactivation in females, methylation is also a key regulator of tumor promoter or tumor suppressor genes by virtue of its ability to methylate (hence 'silence') or demethylate (hence 'activate') these genes. Under normal conditions, this delicate and exquisitely balanced system helps to maintain normal mitosis and interphase; however abnormal methylation has been documented as a key aberration in a wide variety of malignancies, by far the commoner pattern being hypermethylation of tumor suppressor genes (such as p15, p16, p53, etc.) accompanied by global genomic hypomethylation. Most of the hypermethylation involves the CpG islands of the promoter regions of tumor suppressor genes, so that these genes are rendered functionally inactive or 'hypoactive'. This abnormal methylation pattern confers upon the cell a certain degree of abnormal 'unstable' genomic configuration that leads initially to the phenotype of the 'transformed' cell and later on to the fully malignant cell after several abnormal mitotic cycles. Studies of normal and abnormal methylation patterns have thus become central to the approach towards informed studies of the progression of a normal cell to a malignant one and also the progressive acquisition of greater malignant properties by neoplastic cells.

This chapter will attempt to provide a comprehensive overview of the processes of normal and abnormal DNA methylation, the genes involved in methylation processes and the roles of the major tumor suppressor genes in malignant transformation and progression, as they become progressively deranged in both structure and function (by abnormal methylation-related changes) during oncogenesis. It will also provide a view of current therapeutic strategies based on insights into methylation-related epigenetic pathways that can be exploited to therapeutic advantage for effective tumor control in cancer patients. It remains as yet for future researchers to develop those strategies that offer the promise of adopting a preventive approach to the cancer problem, particularly for high-risk individuals and for the population in general.

Introduction

Our understanding of DNA, the exquisitely preserved 'protein-making mould' of all known life forms on earth, is undergoing quantum changes with each passing day. Gone are the dogmas of 'one gene, one protein' and the concept of millions of genes packed inside tiny cell nuclei. Though some research work has been anti-climactic from the pedantic viewpoints, most of modern-day concepts in molecular genetics stand well in the face of logical criticism and strengthen what we know about the strong conservative drives inherent in nature and all natural systems. That a gene should be adaptable to environmental changes, thus having the inbuilt or acquired property of being able to synthesize a number of messenger RNA molecules, and hence proteins, by alternate splicing is but to be expected by natural scientists, as it fits neatly into their earlier observations about the workings of nature.

Among these relatively newer discoveries and concepts is one of an 'epigenetic' control of gene functioning. Previous concepts of changes in gene function relied on the 'mutation model' wherein any change in the base sequence of a gene could bestow on it no change, a beneficial change or a deleterious change. This gave rise to a near-dogma that a genome would continue unchanged unless a mutation brought about some change, which would cause the corresponding cell to gain a 'genetic advantage' or a 'genetic disadvantage' that determined its survival as well as the future fate of its progeny. Such changes were believed responsible for the 'evolutionary drives' that brought about genetic changes ensuring 'survival of the fittest' as the deleterious mutations would gradually be weeded out.

The epigenetic model was derived from early observations in bacteria and other lower life forms where it was observed that DNA introduced into bacterial cells underwent a particular modification called DNA methylation, wherein preferentially cytosine bases (and infrequently adenine in lower organisms) of the foreign DNA were methylated by additions of methyl groups to these bases. DNA segments that underwent successful methylation of a specific pattern were incorporated into host DNA, while segments undergoing methylation in other patterns or remaining unmethylated were degraded as foreign DNA, i.e., rejected. Such specific concerns for conservation of both self and foreign DNA in primitive life forms developed the notion that the process of DNA methylation was a fundamental biochemical process of much importance to the preservation of native or host DNA.

Further research spurred by these landmark discoveries led to the conclusion that DNA methylation was indeed a fundamental process found in all cells that maintained functioning genomes, and that this process played a crucial role in the process of orchestrating the genetic components (genes) selected for transcriptional activity or for silencing of genes not required immediately or on a long term basis for transcription. That such a sophisticated, yet simple, mechanism existed as an inbuilt system of modulation of gene activity was startling, yet fascinating. The grounds were then established for a flurry of research activity on the topic of DNA methylation, starting in the 1980s and continuing to date, as more and more significant discoveries continue to be made about the role of DNA methylation in a myriad of cellular processes.

DNA Methylation

DNA methylation is the only known physiological covalent modification of the DNA molecule and is accomplished by transfer of a methyl group from S-Adenosyl Methionine (SAM) to the 5' position of Cytosine. It is believed to be an almost universal biological phenomenon found in both prokaryotes and eukaryotes, with the notable exception of some species such as *Drosophilae melanogaster.*[1]

In lower life forms, both Adenine and Cytosine bases can be methylated, whereas in higher organisms, the predominant base is Cytosine. Methylation of bases is carried out by enzymes identified and labelled as DNA Methyl Transferases or (DNA-MTases); these were initially isolated in bacteria such as *E. coli* and later in other life forms as well.[2] For bacteria in particular, DNA methylation is a means of neutralizing foreign inserted strands of DNA, so that the bacteria are protected from undesirable 'active' foreign genes inserted into them by bacteriophages, viruses, or other means. This protection is offered by the fact that endogenous DNA has its built-in methylation pattern that is unique to the species or even the micro organism and differs significantly from other patterns of methylated DNA. Thus Restriction Endonucleases are able to recognize foreign DNA and proceed to fragment it.[2]

As an offshoot, such an activity by native cells offers them an innate protection against viral invasion, being an example of a built-in antiviral activity.[3] This has important implications for malignant transformation, as it may imply that perturbations in this antiviral protection allow viral genomes to incorporate into cellular genomes and/or takeover the cellular machinery propelling it towards malignant transformation. Thus it may be that defective DNA methylation-dependent antiviral programs are the first or initiating cellular events that allow further progression to a malignant state. Alternately successful 'oncogenic' viruses may have acquired the ability to escape from cellular antiviral programs or to modify them to their advantage so as to enable a successful invasion.

The unique methylation pattern given to each individual DNA acts as a 'fingerprint' for the self-recognition of an organism's own DNA and is maintained by high fidelity methylation of the daughter strand after DNA replication, with the original methylated strand acting as the template.[4] This event happens soon after replication so much so that even the Okazaki fragments are already methylated. Such methylation is given the name of *'Maintenance Methylation'* and is essential to species survival. The first maintenance methyltransferase was identified in the mouse and given the name of DNA Methyltransferase 1 or DNMT1.[5] Later it was identified in humans and was shown to have a high homology to the mouse enzyme.[6] It was soon established that the enzyme is essential for normal embryonic development in the mouse and that total homozygous knockout of mouse DNMT1 was lethal for the embryo.[7] It was shown that DNMT1 resides in the replication complex and carries out the methylation of the newly replicated daughter strand immediately by using the original methylated strand as the template.[8]

DNMT1 has a relative mass of 190 kDa, contains 1620 amino acids[9] and is present in all cell types. It has a C-terminal domain of about 500 amino acids that has a close homology to bacterial restriction methyltransferases;[10] the N-terminal domain consists of over 1000 amino acids and has multiple regulatory properties.[11] The large N-terminal domain has multiple domains that provide functions such as the co-ordination of replication and methylation during S-phase, and the partial suppression of *de novo* methylation.[9] The human DNMT1 gene has been mapped to chromosome 19 by hybrid cell panel experiments and to 19p13.2-13.3 by FISH techniques.

Another DNA methyltransferase was discovered and named DNMT2;[12] however to date this enzyme has no known activity of either of the two types of DNA methylation in eukaryotic cells.[13] Although the exact biological role of any member of the Dnmt2 family in any species remains to be demonstrated, a postulated role in some aspects of centromere function is a possibility.[9] It has been mapped to chromosome 10p.

Yet another type of DNA methylation was discovered that involved the methylation of new DNA sites for the purpose of differentiation of embryonic or adult tissues;[14,15] it was given the name of '*de novo methylation*' to differentiate it from maintenance methylation. De Novo methylation was significantly different from maintenance methylation and was shown to be carried out by another set of methyltransferases called DNMT3a and DNMT3b.[16,17] These enzymes have no maintenance methylation activity, but are essential for proper developmental processes in mouse embryonic stem cells. DNMT3a knockout mice are born alive but become runted and die at about 4 weeks of age;[18] DNMT3b knockout mice are not viable and mutant embryos show numerous developmental defects and growth impairment.[7]

DNMT3a has been mapped to chromosome 2p, while DNMT3b is located on chromosome 20q; a homologue DNMT3L maps to chromosome 21q in the critical region associated with Down's syndrome, but whether its increased expression has any role to play in Down's syndrome is not known.[9]

DNA methylation thus refers to a key process that acts as a chemical control over gene expression in all cells.[19-22] Several pioneering studies linked the activity status of DNA and genes to the amount of methylation that they had undergone. Although all aspects of this key regulatory process are not yet fully understood, it is universally accepted that addition of methyl groups (methylation) to a gene will decrease its expression, 'silencing' it, as it were; removal of methyl groups (demethylation) from a gene allows it to be expressed and display its biological activity. In humans, this process involves the base Cytosine, so that the methylcytosine content of DNA or a gene is an index of expression of its activity. The same gene can be made more or less functional by respectively hypo- or hyper- methylating it. It has been discovered that although all the cytosine bases can be methylated, preferentially bases forming CG doublet sequences are methylated to act as regulators of gene expression. Thus the CG sequence of a gene plays a determining role in allowing it to be regulated by the action of methyltransferases. These CG doublets are referred to as CpG (p for *pair*) denoting the pairing sequence and that C is 5' (proximal) to G. A CpG island is defined as a sequence with a G+C content of greater than 60% and ratio of CpG to GpC of at least 0.6.[23] Approximately 70 - 75% of CpG areas are methylated in human DNA, whereas the total methylated DNA ranges from 2 to 7% with an average of 5%.[19,20]

Methylating and demethylating functions of the enzyme DNA methyltransferase, or the conditions required for directing the reaction in a specific direction have not been fully elucidated as yet. Nevertheless it is established that whenever the enzyme encounters a CpG island, it stalls at this site, and then carries out the reaction, as required. It is pertinent to mention that the methyl group attached to position 5 of cytosine sticks out in the major groove of DNA and is easily recognized by scanning enzymes. Factors determining the original methylation of DNA have not been determined, although it is believed to be a highly specific pattern, unique for each individual and perhaps even for each tissue. Initially determined in the zygote, at the time of recombination, it is stable and passed on to all daughter cells. There are regional differences in tissues arising as a consequence of gene expression specific to the tissue, but the basic pattern is unique to the individual.

CpG Islands

Methylation of DNA by the methyltransferases may occur at any Cytosine base converting it to 5' Methyl Cytosine (5MeC); however it was found that methylation of Cytosine bases occurs preferentially in areas where the bases are present proximal to Guanine bases so that a CG pair forms. Such dinucleotide pairs are called CpG pairs or doublets denoting that Cytosine and Guanine form a pair and that Cytosine is proximal to Guanine.[19] CpG pairs may occur as isolated pairs randomly in DNA or may occur in runs of pairs – when this occurs, these areas are called CpG islands. A CpG island is defined as an area with G+C content greater than 60% and the ratio of CpG to GpC of at least 0.60.[23] Such islands have been found more often in 5' proximal promoter regions of genes and remain unmethylated under most normal conditions in accordance with their 'activity statuses'. The majority of genes that contain CpG islands in them are usually 'housekeeping' genes and have a broad pattern of tissue expression. It is estimated that approximately 2 – 10% (average 5%) of the human genome consists of CpG islands; as they are found predominantly in the promoter areas of genes, the promoters are likely to be regulated by methylation. Whereas approximately 70 – 80% of the CpG doublets are in a state of methylation,[1, 24] the CpG islands are generally unmethylated[9, 25-28] with the exception of malignant cells where promoter regions of tumor suppressor genes are found methylated and even hypermethylated with resulting 'gene silencing'.[29, 30]

Methylation of CpG areas confers on them a state of 'epigenetic-induced' silencing of the gene in question; the reasons for this are not entirely clear, but it is believed that methylation of cytosines prevents attachment of ligands that are required normally for gene activation. Methylation thus appears to be a biological phenomenon of great importance in eukaryotes, for it provides the main 'epigenetic' regulatory mechanism for control of gene expression.

The methyltransferases have the ability to recognize CpG sequences in DNA, perhaps because the 5MeC tends to stick out in the major DNA groove. Exactly how this is accomplished is not known, however it is observed that whenever the enzymes encounter a CpG pair, they stall at this site and then carry out methylation or demethylation, as the case may be.

It is by now fairly well established that promoter regions containing CpG islands have to be in a non-methylated state for them to transcribe, though the mere presence of non methylated CpG islands does not mean an actively transcribing gene.[31] This implies that one of the basic prerequisites for genes to be able to transcribe is that they should be in a non-methylated state. It also implies that methylated genes would be inactive. Methylation confers, in addition to other properties, a condensed chromatin state that makes transcription factors inaccessible to promoter regions;[32] the cooperative activity of histone deacetylation mediated through DNMT1 contributes to this chromatin condensation.

Methylation Related Gene Functions

Although it has not been established that all gene expression is controlled by methylation, this process is believed to be involved in a number of important cellular processes:

a. Methylation of relevant genes establishes the specific genetic identity of an individual by recombination events in the zygote.[19,20]
b. Genetic imprinting, which is the inheritance of either maternal or paternal traits or selection of an allele pair, also involves methylation-specific selection events.[19,20]
c. The inactivation of the second X-chromosome in the female through the process of Lyonization and formation of the Barr body, is a methylation-specific process, in that the Barr body is inactive because its genes are methylated in contrast to the unmethylated active X.[19,20]
d. Use of a cytosine analog 5'-azacytidine has been shown to alter methylation patterns of genes and their expression. This drug has a Nitrogen atom at position 5 and hence cannot be methylated. When incorporated in place of cytosine in DNA, it causes genes to remain unmethylated and has been shown to increase their expression levels. Interestingly, the drug has been shown to affect differentiation, so that fibroblasts have been made to differentiate to muscle cells, implying changes in the types of genes expressed.[33, 34]
e. Studies on malignant cells have shown alterations of the methylation patterns, so that the tumor suppressor genes have been shown to be hypermethylated and hence inactive in a large number of tumors.[35-45] There are reports also of hypomethylated tumor promoter genes.[46]
f. Studies on prokaryotic and eukaryotic systems have shown that methylation of even a single base in a gene may profoundly affect the gene expression.[19, 20]

g. Methylation of the newly synthesized DNA daughter strands occurs soon afterwards through enzymatic action. This is done by recognition of the pattern of methylation on the original (parent) strand and then methylating the daughter strand. Experiments in which the daughter strands have been allowed to remain unmethylated have shown different patterns of gene expression from these strands.[19, 20]
h. Mismatch repair of DNA strands or other repairs involving removal of bases and incorporation of new ones will show defective gene expression, unless they have been remethylated according to the original correct pattern. Systems in which methylation is defective will accumulate errors over periods of time. [19, 20]
i. Methylation patterns, being specific to any individual DNA, offer a means of recognizing foreign DNA if it enters the cell. It can be cleaved by specific Restriction Endonucleases, which recognize the CpG islands. [19, 20]

It is thus being realized that DNA methylation and demethylation are important events underlying the normal process of gene expression. It is believed to be one of the major mechanisms controlling gene expression. The enzymatic mechanisms have not yet been fully elucidated, nor is it known whether the same enzyme carries out methylation and demethylation, or that DNA Demethylases exist as well though some studies have shown the existence of possible demethylases.[47]

DNA Methylation and Histone Acetylation

Whereas DNA methylation serves to regulate gene expression, it is by no means the only epigenetic pathway to do so. It has been seen that the eight core histone proteins (H2A, H2B, H3 and H4) function not only to provide a convenient framework or physical scaffold for binding the DNA or chromatin thread, they also undergo chemical processes of acetylation and deacetylation of their N-terminal tails exposed on the nucleosome surface, which has important effects on chromatin remodeling and gene expression. Acetylation has been seen to occur preferentially of the lysine residues. Part of this regulation of gene expression is believed to be due to conformational changes related to acetylation so that the chromatin areas become more accessible to ligands such as transcription factors after histones are maintained in the acetylated state. Acetylation of the lysine residues at the N terminus of histone proteins neutralizes the positive charges of the basic histone proteins thereby reducing the affinity between histones and DNA. This loosening of histone-DNA binding allows RNA polymerase and transcription factors to access the promoter region of active genes. Therefore, in most cases, histone acetylation enhances transcription while histone deacetylation represses transcription by determining the level of compaction of nucleosome structure and the coiled state of DNA.[48]

It has been over thirty years that a strong correlation between increased levels of histone acetylation and elevated levels of gene expression was reported. Acetylation of histone tails correlates with transcriptional activity in many genes. The importance of histone acetylation in the regulation of transcription is reinforced by the observations that histone deacetylation causes repression of gene expression. Histone deacetylases often form complexes that bind transcriptional repressors.[49, 50]

Histone acetylation is catalyzed by histone acetyltransferases (HATs) and histone deacetylation is catalyzed by histone deacetylases (HDs or HDACs). Several different forms of HATs and HDs have been identified. Among them, *CBP/p300* is probably the most important, since it can interact with numerous transcription regulators. Histone acetylation is reversible; HATs transfer the acetyl moiety from acetyl coenzyme A to the lysine residues while HDACs remove the acetyl groups re-establishing positive charges of histones. At least six human HDAC enzymes exist; HDAC1 was first purified using an affinity matrix based on the deacetylase inhibitor trapoxin.[51]

Acetylation of lysine residues on histones H3 and H4 correlates with active or open chromatin, which allow various transcription factors access to the promoters of target genes. By contrast, deacetylation of lysine residues by HDACs results in chromatin compaction and inactivation of genes.[52]

It has been demonstrated that a link exists between DNA methylation and histone deacetylation, so that as DNA is methylated histone proteins are deacetylated, thus forming a simultaneous operating system for chemical and physical 'gene silencing'. Acetylated histones are associated with unmethylated DNA and are nearly absent from methylated DNA regions. This association appears to be local and does not propagate along the DNA,[53] thereby restricting gene activation to sites specified for transcriptional activity. It is widely accepted that histone modification and DNA methylation are intricately interrelated, working together to determine the status of gene expression and to decide cell fate.[54]

The process of DNA methylation involves not only 5' cytosine methylation at CpG areas or islands of promoter regions, but during this, it also binds additional gene repressor mechanisms such as methyl-(CpG)-binding proteins (MBPs) as a part of a large complex that also contains HDACs.[55, 56] This complex then ensures that histone deacetylation takes place concomitantly with DNA methylation to ensure control of transcriptional activity by making transcriptional genes inaccessible and inactive. During DNA demethylation, the reverse of these events ensures that transcriptional genes are accessible and active.

Histones, in particular the lysine residues are also amenable to methylation by a group of methylases called Histone Lysine Methyltransferases (HMTs). Histone lysine methylation can result in either activation or repression, depending on which residue is involved. Thus histone H3 lysine 4 (H3-K4) methylation is a well-known 'active' marker whereas examples of 'Inactive' markers include methylation of H3-K98 and trimethylation of H4-K209; up to three methyl groups per lysine residue can be present.[52]

DNA Methylation and Malignancy

With the advent of modern medicine just over a century ago, newer techniques and technologies promised tremendous insights into the biology of life, both normal and abnormal, so that researchers became optimistic of finding cures for even the more dreaded diseases, including cancer. Fundamental landmark researches in this field can be categorized as:

1. The understanding that cancer is based on an uncontrolled proliferation of body cells that results in the production of an obvious mass or tumor that is clinically evident.
2. The knowledge that fundamental cellular processes of mitosis and differentiation underlie the altered biology of cancer cells, so that cancer cells can be differentiated from normal cells by qualitative and quantitative abnormalities of these fundamental cellular processes.
3. The demonstration that the ultimate abnormal driving forces in a cancer cell reside at the level of the cell genome, even though other cell parts such as the membrane and various receptors play additional or supportive roles; Cancer is thus currently classified as a 'genetic' disease, though not in the sense of the classical genetic diseases that show Mendelian patterns of inheritance.

Even though the field of genetics, particularly molecular genetics, is relatively recent seeing significant advances only in the last 30 odd years or so, it has enabled powerful techniques to be applied to medical research, and cancer genetics has remained a key area spurring large volumes of research work in the last few decades. Painstaking and elaborate research work has enabled a faint and early glimpse of the vastly complex genetic array that allows normal cell function, and has yet to provide a coherent and comprehensible image of the abnormalities that underlie neoplastic transformation of normal cells to malignant ones.

The process of malignant transformation of somatic cells is believed to be a prime example of dysregulated malfunction of a number of fundamental growth regulating genes[19,20] These genes are classified as: a) Growth Promoter Genes and b) Growth Suppressor Genes. The normal working of these two groups in tandem and in coordination determine normal cell growth and division as required, for all body processes from the zygote onwards. One of the key aspects of normal cell mitosis is the genetic (and epigenetic) knowledge of the number of mitoses required for any particular situation, after which the process (of mitosis) is automatically switched off. This latter function is also presumably under the control of Growth Suppressor Genes. *A priori*, it can be said that the zygote is genetically programmed to undergo an (almost) infinite number of cell divisions, and this process (of mitosis) is secondarily controlled or regulated by a number of growth regulatory genes, which direct it to undergo a finite number of mitoses, as needed, along with cellular differentiation, resulting in a directed formation of all the embryonic, and later on, the adult body parts. This basic system of regulated or controlled mitosis is kept as an inherent genetic system in all mitotic cells of the body in adult life, in order to maintain tissue renewal, turnover and repair phenomena.

Neoplastic cells are characterized by losses of a number of key regulatory functions determined by genes in both the growth promoter and growth suppressor groups. There are numerous examples of "overfunctioning" of Growth Promoter Genes and "underfunctioning" of Growth Suppressor Genes" in a large number of malignant tumors. In fact, these observations were crucial to the development of the key concepts of "Oncogenes and Anti-Oncogenes"[19, 20] that are the two determinants of cell growth control, loss of which is a fundamental aspect of the uncontrolled mitosis displayed by malignant cells. More emphasis is now being laid on the factors determining the "state of balance" between oncogenes and anti-oncogenes, as certain imbalances are crucial pre-requisites to the initiation and promotion of malignancy in somatic cells.

Oncogenes and Anti-Oncogenes

These two groups of genes are found in all mitotic cells – however the designation "proto-oncogene" is applied to the normal counterpart and "oncogene" to the transformed or malignant one. The transformation process (whether physical, chemical, viral or spontaneous) induces a change, loosely referred to as a "mutation", in the structure and/or function of the proto-oncogene, converting it to an oncogene. The fundamental aspect of this change is the dominant or uncontrolled activity of the oncogene (tumor promoter) out of synchronization with the growth (or tumor) suppressor or anti-oncogene. The mechanism of this change may be a suppression or loss of activity of the anti-oncogene or a direct 'continuous' activation of the oncogene beyond the controlling ability of the anti-oncogene. The consensus of research findings so far points to a larger role for loss of tumor suppressor gene activity in the majority of tumors studied. Well known examples of 'altered' tumor suppressor genes are: p53, p73, p15, p16 and Rb genes.[35-45] Common examples of implicated tumor promoter genes are: *ras, myc, erb* and *neu*. These abnormalities are found in epithelial, soft tissue and haematologic malignancies.

The traditional approach to understanding the molecular basis of neoplasia is to look for mutations in the DNA of cancer cells, particularly in the genes affecting cellular growth (tumor promoter and tumor suppressor genes). A large number of possible candidate genes and specific mutations have also been discovered and described by researchers over the last few decades. However, so far, none of these lesions could be categorically said to be the underlying cause or molecular basis of the uncontrolled cell proliferation that is the hallmark of malignantly transformed cells.

Since the 1980s a new line of research has been gradually emerging based on research work on the genes of lower life forms such as bacteria and yeasts; this has been extended to humans in the last decade, with promising, though early, results. Experiments on bacteria showed the existence of special areas in the base sequence of DNA, containing cytosine bases coupled with guanine bases to form CG pairs or doublets; these areas were designated as CpG to denote that cytosine paired with guanine and was proximal to guanine.[2, 19] These areas were found in the regulatory regions of genes or scattered in specific spots in the DNA. In bacteria it was observed that each species had its own specific pattern of CpG pair configuration, so much so that these areas could be used for bacterial taxonomy. More interestingly, however, was the finding that the cytosine bases in these CpG areas were different from other cytosine bases of DNA, in that they had a methyl group attached to them; they were in fact 5'- methylcytosines. Although all the cytosines were not methylated, the pattern of methylation was also very specific for each strain of bacteria.[19,20,43,44]

The concept of DNA methylation was further developed and elaborated with further experiments, which indicated that this process was continuous and ongoing, so that certain CpG areas, particularly those related to gene function could be either methylated or demethylated at various times. This brought about the concept of gene regulation being performed through the process of cytosine methylation and demethylation. This concept was further supported by research work, showing that hypermethylated genes were in effect 'switched off' while hypomethylated genes were 'switched on'. Thus the process of CpG methylation / demethylation was indeed shown to be a regulator of gene function.

It was further shown that this function of gene methylation / demethylation was conducted through an enzyme which could add or remove methyl groups from the cytosines of the CpG islands. This enzyme, initially identified in mice was named DNA Methyltransferase or DNMT1 for short.[5] Further work showed the existence of further DNMT enzymes called DNMT2 and DNMT3, including subclasses.[12,16,17] However to date, DNMT1 is believed to be the major enzyme involved in gene regulation via CpG methylation / demethylation.

The relationship of DNA methylation status to cell function was made strikingly clear in the case of malignant cells, when it was shown that in a vast majority of tumor types, the major tumor suppressor genes were hypermethylated (hence rendered inactive), while the corresponding tumor promoter genes were hypomethylated (hence overactive). Such work gave rise to the concept of an 'epigenetic' change in specific genes as a forerunner of malignant change, and gave the first concept of cancer as a disease of genetic regulation rather than of genetic mutation. Thus the genes were normal, but their activity status was changed epigenetically to cause a 'dysregulation' underlying malignant transformation.

Although the relationship of DNA methylation status to carcinogenesis is still being actively studied, it has been shown that the DNMT1 gene itself is also variously affected in malignant cells. Reports indicate that this gene may also be hypomethylated[57,58] or hypermethylated,[59] hence unable to carry out its normal methylation / demethylation function. It is presumed that such abnormalities are the forerunners, if not the actual biochemical / molecular changes that cause malignant transformation.

Methylation Patterns in Leukemia and Lymphoma

Abnormal methylation patterns are being increasingly recognized in a number of common malignancies, and the group of haematological malignancies is no exception.[60]

Hypermethylation with resultant silencing of p53 and its homologous p73 (tumor suppressors) gene expression has been reported in acute lymphoblastic leukaemia, Burkitt's lymphoma, non-Hodgkin's lymphoma, acute myelogenous leukaemia and multiple myeloma.[42-44] It is pertinent that there were no mutations or other changes in the gene in question and the hypermethylation was considered the sole explanation for the observations.

Hypermethylation of the DAP-Kinase (Death Associated Protein Kinase) CpG island has also been reported from a number of B-cell malignancies.[61] DAP-Kinase is important for inducing gamma-interferon induced apoptosis and its loss of function could confer immortality on affected cell lines. In one series, all Burkitt's lymphoma and 84% of B-cell non-Hodgkin's lymphoma samples showed hypermethylation of this island.

Unmethylated MDR1 (Multi Drug Resistance 1) gene has been observed in haematopoeitic malignancies.[62-64] The expression of this gene causes malignant cell lines to become resistant to chemotherapy.

Hypermethylated HIC1 (a tumor suppressor) gene has been observed in leukaemias.[65] This gene is located on chromosome 17p13.3.

Global hypomethylation of DNA has been observed in EBV (Epstein-Barr virus) transformed cells.[65]

Silencing of p15 (a protein kinase inhibitor) gene by altered hypermethylated pattern has been described in acute myeloid leukaemia and other haematologic malignancies.[36-40]

Similar silencing has been described for the p16 gene in multiple myeloma, plasmacytoma and other haematologic malignancies.[36-40]

The hypermethylated sites are usually promoter regions of genes known to be involved at some point in the process of growth control or cell division (tumor suppressor genes). In addition, the hypermethylation increases with progression of disease or with recurrences, accelerated phases or blast crises.

There is evidence of increased DNA methyltransferase activity in acute leukaemia as detected by RT-PCR on mRNA from malignant cells.[66]

Chromosomes from leukaemic patients have been assessed for the presence of methylcytosine by means of monoclonal fluorescent anti-methylcytosine antibody techniques.[67] There was definite evidence of hypermethylation as judged by increased immunostaining for methylated cytosine in malignant cells as compared to normal controls.

Further evidence links methylated cytosines to the later step of mutation. Methylcytosine is unstable and undergoes spontaneous deamination, after which a keto group is added to it and it is converted to Thymine.[19,68] Proofreading enzymes recognize this base change and the complementary base is changed from G to A. Thus a TA mutation occurs. In this way, the presence of hypermethylated CpG islands may act as a means to inducing DNA mutations as a further or final step in carcinogenesis.

Based on research work conducted so far, it seems relevant to consider altered (or 'permanently altered') DNA methylation pattern as an indicator of the early changes of malignant transformation. Perhaps such changes are the forerunners of the more permanent malignant mutations (base changes) that make malignancy inaccessible or refractory to control or therapeutic procedures.

DNA Methylation and Cancer

There are convincing reports of altered methylation patterns in the genomes of malignant or transformed cells. Generally it is agreed that there is a global hypomethylation in malignant cells,[69-71] accompanied by regional hypermethylation1,[72,73] and in most cases also an increased DNA methyltransferase activity.[65,74,75]

Global hypomethylation in malignant cells is believed to contribute towards a general 'activation' of a previously repressed genome thus providing a 'soil of turbulence' for future cell transformation. However it has also important contributions to cell division, as preferential hypomethylation of tumor promoter genes have been described;[76] hypomethylation also increases the mutation rates of cells.[77]

Regional hypermethylation has been described essentially in the context of hypermethylated promoter regions of tumor suppressor genes, so that their inactivation could form a basis for the uncontrolled proliferative drive of cells required for transformation and malignancy.[78-82] This aspect has perhaps been the most widely studied in relation to methylation-induced abnormalities in the epigenetic regulation of cancer genes. A study describes the carcinogenic mechanism of Arsenic and concludes that Arsenic-induced hypermethylation of p53 promoter regions is the primary mechanism for cellular transformation in lung cancer cells.[83]

Increased DNA methyltransferase activity of tumor cells has also received a good deal of attention, having been described in a wide variety of malignant tissues. It has been shown that the DNMT1 gene is itself hypomethylated in malignant cells, thus providing a basis for increased DNA methyltransferase activity in tumor cells.[57,58] The finding of increased DNA methyltransferase activity is essential to any concept of an epigenetic dysregulation of DNA methylation as an underlying basis of neoplastic transformation.

An additional basis for transformation is provided not by an epigenetic change, but by the effects of methylation of the Cytosine bases themselves. It has been established that 5MeC acts as a 'hot spot' for further mutations, so that it is changed to Thymine by spontaneous deamination and the subsequent addition of a keto group.[20,67,84] In one study it was found that 28% of mutations at the human p53 gene locus were due to C to T transitions at CpG dinucleotides.[85]

The frequency of p16 methylation in bronchial epithelial cells (BECs) was documented among people who never smoked and compared to smokers with preneoplastic lesions (hyperplasia, metaplasia, carcinoma in situ) and patients with squamous cell carcinoma of the lung.[86] Whereas never-smoked had no case of p16 methylation, an increasing frequency of p16 methylation was found in smokers with premalignant lesions and in squamous cell carcinoma patients ($p<0.05$ for linear trend from normal to carcinoma in situ). By analogy with a similar progression in cases of colonic carcinomas from benign colonic lesions,[75] this indicates a progressive state of hypermethylation of tumor suppressor genes that ultimately leads to an irreversible state of malignant transformation.

Oncogenes and DNA Methylation

The formation of oncogenes from proto-oncogenes is fundamental to the concept of a genetically derived origin of malignant cells. The mechanism of oncogene formation has also been studied from the viewpoints of either a genetic mutation or an epigenetic dysregulation of both tumor promoter and tumor suppressor gene systems. Exactly which system operates in individual tumors is the target of modern day molecular genetic research.

The concept of DNA methylation abnormalities however serves as a unifying link in the otherwise widening gulf between these two schools of thought. Here we have not only an established epigenetic mechanism of transformation, but as can be seen from C to T mutations induced by 5MeC, also a plausible bridge between the epigenetic and genetic domains. It has been shown that the rate of spontaneous deamination at 5MeC:G base pairs exceeds that of C:G base pairs by more than 21-fold.[87] It remains to be determined exactly what the conditions are that cause 5MeC to undergo mutations in some cells and not in others. One line of research points to the relative amounts of the methyl donor S-Adenosyl

Methionine (SAM), DNMT1 activity and 5MeC in cells as the final determinants of mutation and oncogene formation. In this scenario, increased DNMT1 activity brought about by any mechanism leads further to C to T transitions in 5MeC related to the levels of SAM in cells. The deamination of cytosine is inhibited by normal levels of SAM and enhanced by low SAM levels.[88,89] However it has been shown that DNMT1 can by itself cause C to T mutations irrespective of SAM levels and thus is an inherent mutagen as well.[89]

Tumor Protein 53 (TP53) Gene

The TP53 gene encodes a 53 kDa protein hence its name. TP53 gene, located on chromosome 17 is one of the earliest 'oncogene' to have been studied. Initially it was observed that increased p53 expression correlated with cellular transformation and carcinogenesis in as many as 50-70% of human cancers. P53 was thus categorized as a tumor promoter oncogene.[19] Later studies revealed that the p53 found at high levels in tumor cells was a mutated form of the wild type p53 and that the wild type p53 was in fact a tumor suppressor gene.[90] It thus became clear that a tumor suppressor gene had been inactivated and the mutated form was being expressed in tumor cell lines. Many studies have confirmed this finding and increased TP53 expression is now considered one of the main markers of tumor cells.

The role of p53 in cellular biology has been highlighted by a number of elegant experimental studies. It is now clear that normal p53 acts as the main cell cycle check point control at the G1 to S transition point. Evidence has accumulated to the effect that p53 checkpoint control is mediated by recognition of cells with altered DNA (mutations or unrepaired damages) and the prevention of these cells from entering the S phase. Such cells are later made to undergo apoptosis as well by p53 mediated mechanisms. The checkpoint thus serves as an efficient filter that allows only normal cells to undergo DNA replication and subsequent cell division. Exactly how the recognition of altered cells is mediated is unknown; however mutated p53 is unable to perform this vital function. In this situation, altered or damaged cells can undergo mitosis and accumulate further errors with each cell division – a scenario of prime requirement for the accumulated errors found in cancer cells. There is evidence that p53 also contributes to a checkpoint at G2 to M,[91-93] so that faulty post replication DNA is not allowed to proceed to mitosis. Such a vast control over the quality of cellular DNA has earned p53 the name of 'Guardian of the genome'.[94-96]

The role of p53 has also been studied extensively in human leukaemias and lymphomas. Despite some conflicting studies, the evidence points to a role for p53 mutations in the early stages of cell transformation in these malignancies. High levels of TP53 expression have been found in non Hodgkin's[97-103] and Hodgkin's Lymphomas[103-105] as well as in Multiple Myelomas[106-108] with implications not only for tumorigenesis but also for disease prognosis[109,110]

Recent work has demonstrated a direct interaction between DNMT1 and the p53 gene in transcriptional repression resulting in silencing of promoter genes.[111] In this case, p53 stimulated DNMT1 expression and methylation. It was demonstrated that p53 bound the strongest to amino acids 1081–1409 of DNMT1 showing a close physical association between these two. This binding was interpreted as leading to *de novo* methylation and gene silencing; moreover a crucial role for p53 in gene silencing by means of DNA methylation was demonstrated, as $p53^{-/-}$ and $DNMT1^{-/-}$ cells were unable to cause gene silencing. It was apparent that p53 acts in this capacity only in the presence of DNMT1 and the formation of a p53-DNMT1 complex is essential to promoter-level gene silencing. It appears that normal or wild type (WT) p53 performs this crucial 'epigenetic' regulation of gene expression in normal cells, preferentially on damaged DNA, whereas this activity is lost in the mutant p53.

Implications for Therapeutic Exploitation

Both DNA methylation and histone acetylation are targets of 'epigenetic therapy' of neoplasia, perhaps because of the understanding that unlike mutations, these processes are reversible in most tumors types if not accompanied by mutated genes.

Initial therapies were based on the use of DNMT1 antagonists (inhibitors) such as 5-aza-2'-deoxycytidine, a cytosine analogue that could get incorporated into DNA and not allow methylation of its already occupied 5' position. This drug thus had a demethylating effect and was used successfully to demethylate the hypermethylated tumor suppressor genes in a mouse model of neoplasia.[112] Its use in routine for human neoplasia has been limited because of strong cytotoxic side effects if used in therapeutic concentrations.

Similarly HDAC inhibitors have been employed to block histone deacetylation thus allowing repressed tumor suppressor genes to become reactivated. Even though their exact mode of action and even sites of action have not been satisfactorily identified at present, their anti-tumour effects have continued research directed towards finding a suitable drug(s). The efficacy of these drugs provides additional benefits in the use of combined DNMT1 and HDAC inhibitors so that lower doses of both drugs are possible with fewer cytotoxic side effects.

DNA Methylation Inhibitors

Two classes of these drugs have been utilized: The Nucleoside analogues and the Non Nucleoside analogues.

The nucleoside analogues include: 5-azacytidine, 5-aza-2'-deoxycytidine, 5-fluoro-2'-deoxycytidine, 5,6-Dihydro-5-azacytidine and Zebularine. These drugs have a modified cytosine ring attached to either a ribose or a deoxyribose molecule. They get incorporated into DNA and there inhibit DNA methylation. The RNA analogues ultimately get incorporated into RNA where their exact mode of action is not currently understood. Though most of these drugs have been used for haematological malignancies, current trials are underway in solid tumors such as those of the breast, colon, kidney and urinary bladder. An advantage of epigenetic-inhibiting drugs is that they have been shown to be effective in low doses thus offsetting the toxicities associated with them.

The non nucleoside analogues include short chain fatty acids such as butyrate and valproic acid, and the hydroxamic acid compounds. Since these drugs are few in number, there have been few clinical trials with them, but research into newer drugs is expected to yield more compounds. These drugs hold great promise because they do not need to get incorporated into DNA, rather act by binding directly to the DNMT1 enzyme in its catalytic regions and render it inactive[113,114]

Inhibition of DNMT can also be accomplished by the use of antisense oligodeoxynucleotides (ODNs). These relatively short synthetic nucleic acids are specifically designed to hybridize to a specific mRNA sequence so that hybridization can block mRNA translation and cause mRNA degradation.[115] Antisense ODNs directed against DNMT mRNA have caused a decrease in DNMT mRNA and protein levels; they also cause demethylation,[116] normalization of gene expression and antitumor activity in mice.[117]

Histone Deacetylase Inhibitors

There are in general four classes of these drugs: short chain fatty acids, hydroxamic acids, cyclic tetrapeptides and benzamides. Although each group has different active sites, they all inhibit the deacetylation of histones by HDACs, hence rendering histone and chromatin in an acetylated and active configuration, so that inactivated tumor suppressor genes are rendered active.

HDAC inhibitors have been shown to induce cyclin-dependent kinase (CDK) inhibitors, such as *p21*, which are responsive to p53 in causing cell-cycle arrest in G1 and G2 phases.[118] Activation of the death-receptor-mediated and intrinsic apoptotic pathways is also mediated by HDAC inhibitors[119,120] involving factors such as nuclear factor-κB (NF-κB)91, c-JNK92 and BCL293. Furthermore, the expression of genes involved in the inhibition of angiogenesis[121,122] and metastasis[123] are also affected. It is possible that hyperacetylation of histones occurs causing genomic instability, which ultimately triggers the cell-cycle checkpoint[124] HDAC inhibitors also lead to acetylation of tumour suppressor proteins and transcription factors such as *p53*[125] and NF-κB.[126]

A particularly potent hydroxamic acid derivative, Trichostan A (TSA) derived from *Streptomyces*, is a potent inducer of cellular differentiation and cell cycle arrest in addition to, and perhaps related to, its powerful anti-HDAC activity.[127] In experimental mouse models, TSA has been shown to possess synergism when used in combination with 5-aza-2'-deoxycytidine (5-Aza-CdR) causing tumor regression and return of suppressed gene activity.[128] Such synergistic actions are being researched because the benefits of a two-pronged attack and lower dosages would provide immense benefits in controlling the growth of tumor cells.

Epigenetic therapeutic agents can also be used in conjunction with older modes of cancer therapy such as chemotherapy, radiotherapy and even immunotherapy. Current protocols indicate that initial epigenetic therapy sensitizes cancer cells to other forms of therapy particularly radiotherapy, may be helpful in reverting cancer drug-resistant genotypes and improve the chances of successful outcomes with immunotherapy. Restoration of other genes such as DNA repair genes, apoptotic genes, etc. may also be possible in future with more selective and selective epigenetic drugs. Thus we may see a future where the cancer therapist is armed with a cocktail; or series of epigenetic drugs, each acting at specific genes, which by a specific sequence of reactivating or suppressing may allow a normal cell genotype and phenotype to be reverted.

Certain disadvantages currently restrict the general use of epigenetic drugs. Among these are a lack of specificity for tumor cells and a short duration of action. These drugs have effects on all body cells and may cause normal cells to change their functional and structural patterns with unforeseen effects. This lack of specificity is in part due to precise biochemical identification of tumor cells so that epigenetic drugs can be supplied only to tumor cells and currently applies to other forms of chemotherapy as well. This concern for disadvantageous effects on normal cells is obviated somewhat by the fact that most CpG islands remain unmethylated in normal cells, thus a more specific targeting of tumor cells where these islands are methylated, is a distinct possibility. The transient nature of action of epigenetic drugs implies that these drugs would have to be used in conjunction with other more permanently effective drugs, so that the epigenetic strategy would enable tumor cells to become more sensitive to other effective drugs.

We are seeing an era where epigenetic therapy is in its infancy, and though promising, yet has a long way to go before it can be said to be safe and effective for patients suffering from neoplastic diseases. A lot more research is needed to more precisely outline the targets of epigenetic drug action in normal and tumor cells before this mode of therapy can be said to fulfill the stringent demands for an ideal therapeutic approach to tumor control.

Conclusion

A vast amount of directed research has outlined DNA methylation as an essential epigenetic mechanism for regulating the coordinated mitotic and differentiation activities of normal body cells right from the zygote through embryogenesis and the adult life of essentially all organisms. It thus represents perhaps one of the fundamental regulatory systems found in all cells and tissues. That a relatively simple biochemical process could have such profound implications for all life on earth has not only baffled biologists, but has also provided the impetus for research into this still not completely understood phenomenon. It is conceivable that life would not have existed in its complexity and diversity if DNA could not be regulated by this basic yet powerful and versatile system for controlling gene expression. In this context, the presence of CpG doublets and islands on DNA represents a giant evolutionary leap that made it possible for DNA to express its full potential as a source of the complex information and coding to produce the vast variety of proteins required for all life on earth.

No doubt future research will uncover the role played by aberrations in this regulatory system as the underlying pathology in a wide variety of human diseases; in the present context, however, the role of aberrant methylation in setting a cell to embark upon the initial biochemical pathway that would later on transform it to a neoplastic cell is to be highlighted. The intricate balance that exists between the tumor promoter and tumor suppressor genes is so devastatingly upset by methylation-induced silencing of tumor suppressor genes that it can progress towards irreversible mutations, previously thought to be the hallmark of a malignant cell. It is now apparent that before the actual mutation takes place, the tumor suppressor genes remain for an indefinite time in a state of reversible transformation, which if exploited, can stop its progression to malignancy. Though this currently remains the dream of epigenetic researchers and cancer therapists, it offers the most promise of a biochemical 'cure' or prevention of transformation of normal cells to the malignant state. Towards this preventative approach, future researchers must address themselves, if they are to understand fully the biological and biochemical pathways of malignant transformation.

Intriguing is the concept that de novo methylation acts in the manner of an antiviral program as it cleaves inserted foreign DNA and methylates it to the patterns found in host or native DNA. This implies the tremendously important role of maintenance of a 'self' configuration of DNA, presumably as 'DNA foreignness' of certain types is likely to bring about fundamental changes in gene expression that could be detrimental to the normal structure and function of the cell. It is tempting to postulate that cells that can be transformed to malignant genotypes are somehow defective in this function of maintaining 'self' DNA, so that ingested DNA may be a causal agent in initiating genomic changes in such cells. The abnormal methylation patterns thus introduced into these cells would be stably passed on to daughter cells and in time may enable transformation through other additional aberrations incrementally.

The plausible approach to such a situation would be to elucidate the precise 'methylome' of each individual, and then to develop systems (perhaps sensitized primed immune cells) that would recognize any aberrant methylome in somatic cells and eliminate aberrantly methylated cells before they get the chance to develop into malignant genotypes. It may be possible also to develop tailor-made epigenetic drugs based on specific methylome analysis of individual tumors, so that the aberrant methylation patterns and configurations found in these tumors can be reversed, much like a tape played backwards in time till the normal cellular methylome is reestablished.

As to whether a preventive approach towards methylation-induced cell transformation is possible or not will ultimately depend on the elucidation of factors that maintain the cell in a tilt towards tumor suppression or a regulated control on cell proliferation under intrinsic and extrinsic proliferative influences. This will require precise knowledge of the role played by methylation towards tumor suppression and the plausibility that methylation pathways can be strengthened by biochemical means to maintain a tilt towards controlled cell proliferation. At present not much is known about these events and more research should be directed towards the tumor promoter and tumor suppressor genetic systems.

To sum up, epigenetic methylation and acetylation pathways represent fundamental gene regulatory systems, of immense importance to both normal and abnormal cell structure and function. Hopefully future research will enable therapeutic control of the currently known and yet to be discovered diseases that occur due to aberrations in these pathways.

References

[1] Bird A, Tate P, Nan X, Campoy J, Meehan R, Cross S et al. Studies of DNA methylation in animals. *J. Cell Sci. Suppl.* 1995; 19: 37-9.

[2] Noyer-Weidner M, Trautner TA. Methylation of DNA in prokaryotes. *EXS.* 1993; 64: 39-108.

[3] Jahner D, Stuhlmann H, Stewart CL, Harbers K, Lohler J, Simon I et al. De novo methylation and expression of retroviral genomes during mouse embryogenesis. *Nature.* 1982;298(5875):623-8.

[4] Araujo FD, Knox JD, Szyf M, Price GB, Zannis-Hadjopoulos M. Concurrent replication and methylation at mammalian origins of replication. *Mol. Cell Biol.* 1998;18: 3475-82.

[5] Bestor T, Laudano A, Mattaliano R, Ingram V. Cloning and sequencing of a cDNA encoding DNA methyltransferase of mouse cells. The carboxy-terminal domain of the mammalian enzymes is related to bacterial restriction endonucleases. *J. Mol. Biol.* 1988; 203: 971-83.

[6] Yen RW, Vertino PM, Nelkin BD, Yu JJ, el-Deiry W, Cumaraswamy A et al. Isolation and characterization of the cDNA encoding human DNA methyltransferase. *Nucleic Acids Res.* 1992; 20:2287-91.

[7] Li E, Bestor TH, Jaenisch R. Targeted mutation of the DNA methyltransferase gene results in embryonic lethality. *Cell.* 1992; 69: 915-26.

[8] Leonhardt H, Page AW, Weier HU, Bestor TH. A targeting sequence directs DNA methyltransferase to sites of DNA replication in mammalian nuclei. *Cell.* 1992; 71: 865-73.

[9] Bestor TH. The DNA methyltransferases of mammals. *Human Molecular Genetics.* 2000; 9(16): 2395-2402.

[10] Bestor T, Laudano A, Mattaliano R, Ingram V. Cloning and sequencing of a cDNA encoding DNA methyltransferase of mouse cells. The carboxy-terminal domain of the mammalian enzymes is related to bacterial restriction endonucleases. *J. Mol. Biol.* 1988; 203: 971-83.

[11] Bestor TH, Verdine GL. DNA methyltransferases. *Curr. Opin. Cell Biol.* 1994; 6(3):380-9.

[12] Yoder JA, Bestor TH. A candidate mammalian DNA methyltransferase related to pmt1p of fission yeast. *Hum. Mol. Genet.* 1998; 7: 279-84.

[13] Okano M, Xie S, Li E. DNMT2 is not required for de novo and maintenance methylation of viral DNA in embryonic stem cells. *Nucleic Acids Res.* 1998; 26: 2536-40.

[14] Monk M. Changes in DNA methylation during mouse embryonic development in relation to X-chromosome activity and imprinting. *Philos Trans. R. Soc. Lond. Biol. Sci.* 1990; 326: 299-312.

[15] Razin A, Cedar H. DNA methylation and embryogenesis. *EXS.* 1993; 64: 343-57.

[16] Robertson KD, Uzvolgyi E, Liang G, Talmadge C, Sumegi J, Gonzalez FA, Jones PA. The human DNA methyltransferases (DNMTs) 1, 3a and 3b: coordinate mRNA expression in normal tissues and over expression in tumors. *Nucleic Acids Res.* 1999; 27: 2291-8.

[17] Xie S, Wang Z, Okano M, Nogami M, Li Y, He WW, Okumura K, Li E. Cloning, expression and chromosome locations of the human DNMT3 gene family. *Gene.* 1999; 236(1):87-95.

[18] Okano M, Bell DW, Haber DA, Li W. DNA methyltransferases DNMT3a and DNMT3b are essential for de novo methylation and mammalian development. *Cell.* 1999; 99: 247-57.

[19] Watson JD, Gilman M, Witkowski J, Zoller M. Recombinant DNA. 2nd ed., W. H. Freeman and Co. USA. 1992.

[20] Lewin B. Genes V. Oxford University Press. UK. 1994.

[21] Siegfried Z, Eden S, Mendelsohn M, Feng X, Tsuberi BZ, Cedar H. DNA methylation represses transcription in vivo. *Nat. Genet.* 1999; 22(2):203-6.

[22] Ramchandani S, Bhattacharya SK, Cervoni N; Szyf M. DNA methylation is a reversible biological signal. *Proc. Natl. Acad. Sci. USA.* 1999; 25; 96(11):6107-12.

[23] Baylin SB, Herman JG, Graff JR, Vertino PM, Issa JP. Alterations in DNA methylation: a fundamental aspect of neoplasia. *Adv. Cancer Res.* 1998; 72:141-96.

[24] Antequara F, Bird A. Number of CpG islands and genes in human and mouse. *Proc. Natl. Acad. Sci. USA.* 1993; 90: 11995-9.

[25] Bird AP. CpG-rich islands and the function of DNA methylation. *Nature.* 1986; 321: 209-13.

[26] Bird A. CpG islands as gene markers in the vertebrate nucleus. *Trends Genet.* 1987; 3: 342-6.

[27] Gardiner-Garden M, Frommer M. CpG islands in vertebrate genomes. *J. Mol. Biol.* 1987; 196: 261-82.

[28] Larsen F, Gundersen G, Lopez R, Prydz H. CpG islands as gene markers in the human genome. *Genomics.* 1992; 13: 1095-1107.

[29] Di Croce L et al. Methyltransferase recruitment and DNA hypermethylation of target promoters by an oncogenic transcription factor. *Science.* 2002; 295: 1079-82.

[30] Santoro R, Li J, Grummt I. The nuclear remodelling complex NoRC mediates chromatin formation and silencing of ribosomal gene transcription. *Nat. Genet.* 2002; 32: 393-6.

[31] Tazi J, Bird A. Alternative chromatin structure of at CpG islands. *Cell.* 1990; 60: 909-20.

[32] Kass SU, Landsberger N, Wolfe AP. DNA methylation directs a time-dependent repression of transcription initiation. *Curr. Biol.* 1997; 7: 157-65.

[33] Kees UR, Avramis V. Biochemical pharmacology and DNA methylation studies of arabinosyl 5-azacytidine and 5,6-dihydro-5-azacytidine in two human leukemia cell lines PER-145 and PER-163. *Anticancer Drugs.* 1995; 6(2):303-10.

[34] Moller MB, Ino Y, Gerdes AM, Skjodt K, Louis DN, Pedersen NT. Aberrations of the p53 pathway components p53, MDM2 and CDKN2A appear independent in diffuse large B cell lymphoma. *Leukemia.* 1999; 13(3):453-9.

[35] Herman JG. Hypermethylation of tumor suppressor genes in cancer. *Semin. Canc. Biol.* 1999; 9: 359-67.

[36] Batova A, Diccianni MB, Yu JC, Nobori T, Link MP, Pullen J, Yu AL. Frequent and selective methylation of p15 and deletion of both p15 and p16 in T-cell acute lymphoblastic leukemia. *Cancer Res.* 1997; 57(5):832-6.

[37] Herman JG, Civin CI, Issa JP, Collector MI, Sharkis SJ, Baylin SB. Distinct patterns of inactivation of p15INK4B and p16INK4A characterize the major types of hematological malignancies. *Cancer Res.* 1997; 57(5):837-41.

[38] Dodge JE, List AF, Futscher BW. Selective variegated methylation of the p15 CpG island in acute myeloid leukemia. *Int. J. Cancer.* 1998; 23;78(5):561-7.

[39] Drexler HG. Review of alterations of the cyclin-dependent kinase inhibitor INK4 family genes p15, p16, p18 and p19 in human leukemia-lymphoma cells. *Leukemia.* 1998; 12(6):845-59.

[40] Martinez-Delgado B, Robledo M, Arranz E, Osorio A, Garcia MJ, Echezarreta G, Rivas C, Benitez J. Hypermethylation of p15/ink4b/MTS2 gene is differentially implicated among non-Hodgkin's lymphomas. *Leukemia.* 1998; 12(6):937-41.

[41] Aggerholm A, Guldberg P, Hokland M, Hokland P. Extensive intra- and inter-individual heterogeneity of p15INK4B methylation in acute myeloid leukemia. *Cancer Res.* 1999; 59(2):436-41.

[42] Corn PG, Kuerbitz SJ, van Noesel MM, Esteller M, Compitello N, Baylin SB, Herman JG. Transcriptional silencing of the p73 gene in acute lymphoblastic leukemia and Burkitt's lymphoma is associated with 5' CpG island methylation. *Cancer Res.* 1999; 59(14):3352-6.

[43] Kawano S, Miller CW, Gombart AF, Bartram CR, Matsuo Y, Asou H, Sakashita A, Said J, Tatsumi E, Koeffler HP. Loss of p73 gene expression in leukemias/ lymphomas due to hypermethylation. *Blood.* 1999; 94(3):1113-20.

[44] Moller MB, Ino Y, Gerdes AM, Skjodt K, Louis DN, Pedersen NT. Aberrations of the p53 pathway components p53, MDM2 and CDKN2A appear independent in diffuse large B cell lymphoma. *Leukemia.* 1999; 13(3):453-9.

[45] Nakamura M, Sugita K, Inukai T, Goi K, Iijima K, Tezuka T et al. p16/MTS1/INK4A gene is frequently inactivated by hypermethylation in childhood acute lymphoblastic leukemia with 11q23 translocation. *Leukemia.* 1999; 13(6):884-90.

[46] Crossen PE, Morrison MJ. Methylation status of the 3rd exon of the c-MYC oncogene in B-cell malignancies. *Leuk. Res.* 1999; 23(3):251-3.

[47] Oue N, Kuraoka K, Kuniyasu H, Yokozaki H, Wakikawa A, Matsusaki K et al. DNA methylation status of hMLH1, p16(INK4a), and CDH1 is not associated with mRNA expression levels of DNA methyltransferase and DNA demethylase in gastric carcinomas. *Oncol. Rep.* 2001; 8(5):1085-9.

[48] Liu H, Wang J, Epner EM. Cyclin D1 activation in B-cell malignancy: association with changes in histone acetylation, DNA methylation, and RNA polymerase II binding to both promoter and distal sequences. *Blood.* 2004; 104(8); 2505-13.

[49] Razin A. CpG methylation, chromatin structure and gene silencing—a three-way connection. *EMBO J.* 1998; 17: 4905-8.

[50] Herman JG, Baylin SB. Gene silencing in cancer in association with promoter hypermethylation. *N. Engl. J. Med.* 2003; 349: 2042-54.

[51] Taunton J, Hassig CA, Schreiber SL. A mammalian histone deacetylase related to the yeast transcriptional regulator Rpd3p. *Science.* 1996; 27 2: 408-11.

[52] Yoo CB, Jones PA. Epigenetic therapy of cancer: past, present and future. *Nature Reviews.* 2006; 5: 37-50.

[53] Irvine RA, Lin IG, Hsieh C-L. DNA methylation has a local effect on transcription and histone acetylation. *Molecular and Cellular Biology.* 2002; 22(19): 6689-96.

[54] Hashimshony T, Zhang J, Keshet I, Bustin M, Cedar H. The role of DNA methylation in setting up chromatin structure during development. *Nature Genet.* 2003; 34: 187–92.

[55] Fujita N, Takebayashi S, Okumura K, Kudo S, Chiba T, Saya H et al. Methylation-mediated transcriptional silencing in euchromatin by methyl-CpG binding protein MBD1 isoforms. *Mol. Cell Biol.* 1999; 19: 6415–26.

[56] Hendrich B, Abbott C, McQueen H, Chambers D, Cross S, Bird A. Genomic structure and chromosomal mapping of the murine and human Mbd1, Mbd2, Mbd3, and Mbd4 genes. *Mamm. Genome.* 1999; 10: 906–12.

[57] Butler TL, Kay PH, Jacobsen PF. Hypomethylation of cytosine 5-methyltransferase in human neoplasms. *Anticancer Res.* 2000; 20: 1435-8.

[58] Qayum I, Ashraf M. Increased DNA methyltransferase 1 (DNMT1) gene expression in human lymphomas by fluorescence in situ hybridization. *J. Ayub. Med. Coll.* 2004; 16(4): 1-6.

[59] Kimura F, Seifert HH, Florl AR, Santourlidis S, Steinhoff C, Swiatkowski S et al. Decrease of DNA methyltransferase 1 expression relative to cell proliferation in transitional cell carcinoma. *Int. J. Cancer.* 2003; 104(5):568-78.

[60] Issa JP, Baylin SB, Herman JG. DNA methylation changes in hematologic malignancies: biologic and clinical implications. Leukemia 1997; 11(Suppl 1): S7-11.

[61] Katzenellenbogen RA, Baylin SB, Herman JG. Hypermethylation of the DAP-kinase CpG island is a common alteration in B-cell malignancies. *Blood.* 1999; 93(12):4347-53.

[62] Bernardino J, Gerbault-Seureau M, Zafrani B, Dericke Y, Boudou E, Magdelenat H et al. Homogeneously staining regions in 223 breast carcinomas: cytogenetic and clinicopathological correlations. *Br. J. Cancer.* 1998; 78(9):1214-8.

[63] Fryxell KB, McGee SB, Simoneaux DK, Willman CL, Cornwell MM. Methylation analysis of the human multidrug resistance 1 gene in normal and leukemic hematopoietic cells. *Leukemia.* 1999; 13(6):910-7.

[64] Kusaba H, Nakayama M, Harada T, Nomoto M, Kohno K, Kuwano M, Wada M. Association of 5' CpG demethylation and altered chromatin structure in the promoter region with transcriptional activation of the multidrug resistance 1 gene in human cancer cells. *Eur. J. Biochem.* 1999; 262(3):924-32.

[65] Melki JR, Vincent PC, Clark SJ. Cancer-specific region of hypermethylation identified within the HIC1 putative tumour suppressor gene in acute myeloid leukaemia. *Leukemia.* 1999; 13(6):877-83.

[66] Melki JR, Warnecke P, Vincent PC, Clark SJ. Increased DNA methyltransferase expression in leukaemia. *Leukemia.* 1998; 12(3):311-6.

[67] Bensaada M, Kiefer H, Tachdjian G, Lapierre JM, Cacheux V, Niveleau A et al. Altered patterns of DNA methylation on chromosomes from leukemia cell lines: identification of 5-methylcytosines by indirect immunodetection. *Cancer Genet. Cytogenet.* 1998; 103(2):101-9.

[68] Lutsenko E, Bhagwat AS. Principal causes of hot spots for cytosine to thymine mutations at sites of cytosine methylation in growing cells. A model, its experimental support and implications. *Mutat. Res.* 1999; 437(1):11-20.

[69] Lapeyre JN, Becker FF. 5-Methylcytosine content of nuclear DNA during chemical hepatocarcinogenesis and in carcinomas which result. *Biochem. Biophys. Res. Commun.* 1979; 87: 698-705.

[70] Gama-Sosa MA, Slagel VA, Trewyn RW, Oxenhandler R, Kuo KC, Gehrke CW, Ehrlich M. The 5-Methylcytosine content of DNA from human tumors. *Nucleic Acids Res.* 1983; 11: 6883-94.

[71] Feinberg AP, Gehrke CW, Kuo KC, Ehrlich M. Reduced genomic 5-methylcytosine content in human colonic neoplasia. *Cancer Res.* 1988; 48: 1159-61.

[72] Baylin SB, Hoppener JW, deBustros A, Steenbergh PH, Lips CJ, Nelkin BD. DNA methylation patterns of the calcitonin gene in human lung cancers and lymphomas. *Cancer Res.* 1986; 46: 2917-22.

[73] Baylin SB, Fearon ER, Vogelstein B, deBustros A, Sharkis SJ, Burke PJ et al. Hypermethylation of the 5' region of the calcitonin gene is a property of human lymphoid and acute myeloid malignancies. *Blood.* 1987; 70: 412-7.

[74] Kautiainen TL, Jones PA. DNA methyltransferase levels in tumorigenic and non tumorigenic cells in culture. *J. Biol. Chem.* 1986; 261: 1594-8.

[75] Issa JP, Vertino PM, Wu J, Sazawal S, Celano P, Nelkin BD et al. Increased cytosine DNA-methyltransferase activity during cancer colon progression. *J. Natl. Cancer Inst.* 1993; 85: 1235-40.

[76] Feinberg AP, Vogelstein B. Hypomethylation of ras oncogenes in primary human cancers. *Biochem. Biophys. Res. Commun.* 1983; 111: 47-54.

[77] Chen RZ, Pettersson U, Beard C, Jackson-Grusby L, Jaenisch R. DNA hypomethylation leads to elevated mutation rates. *Nature.* 1998; 395:89-93.

[78] Gonzalez-Zulueta M, Bender CM, Yang AS, Nguyen TD, Beart RW, Van Tornout JM et al. Methylation of the CpG island of the p16/CDKN2 tumor suppressor gene in normal and transformed human tissues correlates with gene silencing. *Cancer Res.* 1995; 55:4531-5.

[79] Hiltunen MO, Alhonen L, Koistinaho J, Myohanen S, Paakkonen M, Marin S et al. Hypermethylation of the APC (adenomatous polyposis coli) gene promoter region in human colorectal carcinoma. *Int. J. Cancer.* 1997; 70:644-8.

[80] Antequera F, Boyes J, Bird A. High levels of de novo methylation and altered chromatin structure at CpG islands in cell lines. *Cell.* 1990; 62(3):503-14.

[81] Baylin SB, Makos M, Wu JJ, Yen RW, de Bustros A, Vertino P et al. Abnormal patterns of DNA methylation in human neoplasia: potential consequences for tumor progression. *Cancer Cells.* 1991; 3: 383-90.

[82] Herman JG. Hypermethylation of tumor suppressor genes in cancer. *Semin. Canc. Biol.* 1999; 9: 359-67.

[83] Mass MJ, Wang L. Arsenic alters cytosine methylation patterns of the promoter of the tumor suppressor gene p53 in human lung cells: a model for a mechanism of carcinogenesis. *Mutat. Res.* 1997; 386(3):263-77.

[84] Warnecke PM, Bestor TH. Cytosine methylation and human cancer. *Curr. Opin. Oncol.* 2000; 12: 68-73.

[85] Hainaut P, Hernandez T, Robinson A, Rodriguez-Tome P, Flores T, Hollstein M et al. IARC database of p53 gene mutations in human tumors and cell lines: updated compilation, revised formats and new visualization tools. *Nucleic Acids Res.* 1998; 26: 205-13.

[86] Belinsky SA. Silencing of genes by promoter hypermethylation: key event in rodent and human lung cancer. *Carcinogenesis.* 2005; 26(9):1481-1487.

[87] Zhang X, Mathews CK. Effect of DNA cytosine methylation upon deamination-induced mutagenesis in a natural target sequence in duplex DNA. *J. Biol. Chem.* 1994; 269: 7066-9.

[88] Shen JC, Rideout III WM, Jones PA. High frequency mutagenesis by a DNA methyltransferase. *Cell.* 1992; 71: 1073-80.

[89] Yebra MJ, Bhagwat AS. A cytosine methyltransferase converts 5-methylcytosine in DNA to thymine. *Biochemistry.* 1995; 34(45):14752-7.

[90] Finlay CA, Hinds PW, Levine AJ. The p53 proto-oncogene can act as a suppressor of transformation. *Cell.* 1989; 57: 1083-93.

[91] Taylor WR, Agarwal ML, Agarwal A, Stacey DW, Stark GR. p53 inhibits entry into mitosis when DNA synthesis is blocked. *Oncogene.* 1999; 18(2):283-95.

[92] Innocente SA, Abrahamson JL, Cogswell JP, Lee JM. p53 regulates a G2 checkpoint through cyclin B1. *Proc. Natl. Acad. Sci. USA.* 1999; 96(5):2147-52.

[93] Park M, Chae HD, Yun J, Jung M, Kim YS, Kim SH et al. Constitutive activation of cyclin B1-associated cdc2 kinase overrides p53-mediated G2-M arrest. *Cancer Res.* 2000; 60(3):542-45.

[94] Schwartz D, Rotter V. (1998). p53-dependent cell cycle control: response to genotoxic stress. *Semin. Cancer Biol.* 8(5):325-36.

[95] Albrechtsen N, Dornreiter I, Grosse F, Kim E, Wiesmuller L, Deppert W. Maintenance of genomic integrity by p53: complementary roles for activated and non-activated p53. *Oncogene.* 1999; 18(53):7706-17.

[96] Janus F, Albrechtsen N, Dornreiter I, Wiesmuller L, Grosse F, Deppert W. The dual role model for p53 in maintaining genomic integrity. *Cell Mol. Life Sci.* 1999; 55(1):12-27.

[97] Chen Y, Xiang Z, Li H, Yang N, Zhang H. P53 gene mutations in non-Hodgkin's lymphoma. *J. Tongji. Med. Univ.* 1999; 19(1):27-30.

[98] Klumb CE, de Resende LM, Tajara EH, Bertelli EC, Rumjanek VM, Maia RC. p53 gene analysis in childhood B non-Hodgkin's lymphoma. *Sao Paulo Med. J.* 2001, 119(6):212-5.

[99] Llanos M, Alvarez-Arguelles H, Aleman R, Oramas J, Diaz-Flores L, Batista N. Prognostic significance of Ki-67 nuclear proliferative antigen, bcl-2 protein, and p53 expression in follicular and diffuse large B-cell lymphoma. *Med. Oncol.* 2001;18(1):15-22.

[100] Naresh KN, Banavali SD, Bhatia KG, Magrath I, Soman CS, Advani SH. Expression of P53 and bcl-2 proteins in T-cell lymphoblastic lymphoma: prognostic implications. *Leuk. Lymphoma.* 2002; 43(2):333-7.

[101] Hatta Y, Koeffler HP. Role of tumor suppressor genes in the development of adult T cell leukemia/lymphoma (ATLL). *Leukemia.* 2002; 16(6):1069-85.

[102] Moller MB, Nielsen O, Pedersen NT. Frequent alteration of MDM2 and p53 in the molecular progression of recurring non-Hodgkin's lymphoma. *Histopathology.* 2002; 41(4):322-30.

[103] Qayum I, Ashraf M. DNA methyltransferase 1 (DNMT1) gene activity in human lymphomas correlates with aberrant p53 gene expression. *J. Ayub. Med. Coll. Abbottabad.* 2006;18(1):1-6.

[104] Garcia JF, Camacho FI, Morente M, Fraga M, Montalban C, Alvaro T et al. Hodgkin and Reed-Sternberg cells harbor alterations in the major tumor suppressor pathways and cell-cycle checkpoints: analyses using tissue microarrays. *Blood.* 2003; 101(2):681-9.

[105] Qi ZL, Zhao T, Zhou XH, Zhang JH, Han XQ, Zhu MG. Expressions of latent membrane protein 1, p53 and bcl-2 proteins and their significance in Hodgkin's lymphoma. *Di Yi Jun Yi Da Xue Xue Bao.* 2003; 23(3):225-7.

[106] Nickenig C, Lang NK, Schoch C, Hiddemann W, Haferlach T. New insights into the biology of multiple myeloma using a combination of May-Grunwald-Giemsa staining and fluorescence in situ hybridization techniques at the single cell level. *Ann. Hematol.* 2001; 80(11):662-8.

[107] Pruneri G, Carboni N, Baldini L, Intini D, Colombi M, Bertolini F et al. Cell cycle regulators in multiple myeloma: prognostic implications of p53 nuclear accumulation. *Hum. Pathol.* 2003; 34(1):41-7.

[108] Ortega MM, Melo MB, De Souza CA, Lorand-Metze I, Costa FF, Lima CS. A possible role of the P53 gene deletion as a prognostic factor in multiple myeloma. *Ann. Hematol.* 2003; 82(7):405-9.

[109] Ichikawa A. Prognostic and predictive significance of p53 mutation in aggressive B-cell lymphoma. *Int. J. Hematol.* 2000; 71(3):211-20.

[110] Stokke T, Galteland E, Holte H, Smedshammer L, Suo Z, Smeland EB et al. Oncogenic aberrations in the p53 pathway are associated with a high S phase fraction and poor patient survival in B-cell Non-Hodgkin's lymphoma. *Int. J. Cancer.* 2000; 89(4):313-24.

[111] Esteve PO, Chin HG, Pradhan S. Human maintenance DNA (cytosine-5)-methyltransferase and p53 modulate expression of p53-repressed promoters. *Proc. Natl. Acad. Sci. USA.* 2005; 102(4):1000-5.

[112] Laird PW, Jackson-Grusby L, Fazeli A, Dickinson SL, Jung WE, Li E et al. Suppression of intestinal neoplasia by DNA hypomethylation. *Cell.* 1995; 81:197–205.

[113] Brueckner B, Boy RG, Siedlecki P, Musch T, Kliem HC, Zielenkiewicz P et al. Epigenetic reactivation of tumor suppressor genes by a novel small-molecule inhibitor of human DNA methyltransferases. *Cancer Res.* 2005; 65: 6305–11.

[114] Schuebel K, Baylin S. In living color: DNA methyltransferase caught in the act. *Nature Methods.* 2005; 2: 736–8.

[115] Crooke ST. Potential roles of antisense technology in cancer chemotherapy. *Oncogene.* 2000; 19: 6651–9.

[116] MacLeod AR, Szyf M. Expression of antisense to DNA methyltransferase mRNA induces DNA demethylation and inhibits tumorigenesis. *J. Biol. Chem.* 1995; 270: 8037–43.

[117] Ramchandani S, MacLeod AR, Pinard M et al. Inhibition of tumorigenesis by a cytosine-DNA, methyltransferase, antisense oligodeoxynucleotide. *Proc. Natl. Acad. Sci. USA.* 1997; 94: 684–9.

[118] Rocchi P, Tonelli R, Camerin C, Purgato S, Fronza R, Bianucci F et al. p21Waf1/Cip1 is a common target induced by short-chain fatty acid HDAC inhibitors (valproic acid, tributyrin and sodium butyrate) in neuroblastoma cells. *Oncol. Rep.* 2005; 13: 1139–44.

[119] Nebbioso A, Clarke N, Voltz E, Germain E, Ambrosino C, Bontempo P et al. Tumor-selective action of HDAC inhibitors involves TRAIL induction in acute myeloid leukemia cells. *Nature Med.* 2005; 11: 77–84.

[120] Peart MJ, Smyth GK, van Laar RK, Bowtell DD, Richon VM, Marks PA et al. Identification and functional significance of genes regulated by structurally different histone deacetylase inhibitors. *Proc. Natl. Acad. Sci. USA.* 2005; 102: 3697–3702.

[121] Michaelis M, Michaelis UR, Fleming I, Suhan T, Cinatl J, Blaheta RA et al. Valproic acid inhibits angiogenesis in vitro and in vivo. *Mol. Pharmacol.* 2004; 65: 520–7.

[122] Hellebrekers DMEI, Jair K-W, Viré E, Eguchi S, Hoebers NTH, Fraga MF et al. Angiostatic activity of DNA methyltransferase inhibitors. *Mol. Cancer Ther.* 2006; 5: 467-75.

[123] Joseph J, Mudduluru G, Antony S, Vashistha S, Ajitkumar P, Somasundaram K. Expression profiling of sodium butyrate (NaB)-treated cells: identification of regulation of genes related to cytokine signaling and cancer metastasis by NaB. *Oncogene.* 2004; 23: 6304–15.

[124] Qiu L, Burgess A, Fairlie DP, Leonard H, Parsons PG, Gabrielli BG. Histone deacetylase inhibitors trigger a G2 checkpoint in normal cells that is defective in tumor cells. *Mol. Biol. Cell.* 2000; 11, 2069–83.

[125] Gu W, Roeder RG Activation of p53 sequence specific DNA binding by acetylation of the p53 C-terminal domain. *Cell.* 1997; 90: 595–606.

[126] Chen LF, Greene WC. Regulation of distinct biological activities of the NF-κB transcription factor complex by acetylation. *J. Mol. Med.* 2003; 81: 549–57.

[127] Yoshida M, Kijima M, Akita M, Beppu T. Potent and specific inhibition of mammalian histone deacetylase both in vivo and in vitro by trichostatin A. *J. Biol. Chem.* 1990; 265: 17174–79.

[128] Cameron EE, Bachman KE, Myohanen S, Herman JG, Baylin SB. Synergy of demethylation and histone deacetylase inhibition in the re-expression of genes silenced in cancer. *Nature Genet.* 1999; 21: 103–7.

In: DNA Methylation Research Trends
Editor: Taku B. Kobayashi, pp. 209-236
ISBN: 978-1-60021-721-0

Chapter VIII

Methylation and Loss of Heterozygosity in Cancer Are Linked by siRNA?

Coral Wynter**, *Marcelo Alfonzo and Itala Lippo de Becemberg
Seccion de Biomembranas, Instituto de Medicina Experimental, Catedra de Patologia General y Fisiopatologia, Facultad de Medicina, Universidad Central de Venezuela, Apdo 50587 Sabana Grande, Caracas, Venezuela

Abstract

The recent discoveries of the RNA-mediated interference system in cells could explain all of the known features of human carcinogenesis, including methylation. A key, novel idea, proposed here, is that the cell has the ability to recognise a mutated protein and/or mRNA. Secondly, the cell can generate its own short interfering RNA (siRNA) using an RNA polymerase to destroy mutated mRNA, even when only a single base pair in the gene has mutated. The anti-sense strand of the short RNA molecule (called sicRNA), targets the mutated mRNA of an oncogene or a tumour suppressor. The resulting double stranded RNA, using the RNA-induced silencing complex (RISC) in the cytoplasm dices the mutated mRNA. During cell mitosis, the sicRNA complex can move into the nucleus to target the mutated gene. The sicRNA triggers the assembly of protein complexes leading to epigenetic modification of the promoter site of the mutant gene. In some instances, instead of methylation, the homologous DNA is degraded, leading to loss of heterozygosity (LOH). The factors controlling these two actions are unknown but the result is gene silencing by methylation or physical destruction of the mutant gene. The cell survives dependent on the functioning of the single, wild-type allele. A pretumor begins when CpG Methylation and/or LOH disrupt a key microRNA (miRNA) specific for that tissue. This produces a change in the profile of correct proteins for that tissue, resulting in dedifferentiation. The variability of the cancer response may depend on which specific miRNA is being transcribed at the time the sicRNA enters the nucleus. The driving force of carcinogenesis is the sequential loss of active miRNAs.

* Coral Wynter: Email: c.wynter@mailbox.uq.edu.au; Fax: 058 212 662 8877

Introduction

Cancer has been studied over the last 30 years by the examination of hundreds, if not thousands, of different genes, with more still being discovered. Familial cancers, which develop in an individual usually before the age of 55 years, due to an inherited mutation in one allele of a key gene, have led to many important discoveries of the myriad of pathways that occur in spontaneous cancers due to old age. Each gene has a role in controlling cell division, apoptosis, maintenance of the integrity of the genome or is located at a DNA damage checkpoint. It is considered highly significant if only 15-20% of all cancers in one organ have a mutation in one particular gene under examination. One allele of a gene could have hundreds of different mutations, sometimes concentrated in hotspots, usually involving only one base change, but the result inevitably led to a cancer. Thus the number of different oncogenic phenotypes in one tissue has been confusing, defying simplistic classifications.

The complex cellular ramifications of carcinogenesis resulting in a complete restructuring of the genome such as chromosomal disintegration, fragmentation and multiple copies; methylation of specific promoter sites of genes, possibly up to 1,000 changes in one cell; demethylation of mostly repeated sequences; the disappearance of segments of genes, known as loss of heterozygosity (LOH); dedifferentiation and loss of all tissue specificity; has all been very puzzling. How did the cell survive and actually thrive with such massive changes in the nucleus? Faced with this enormous variability, and a vast amount of data that made little sense, the question as to what mechanism generated such chaos was studiously avoided, as there was no possible answer. Now the recent discoveries of RNA interference of mRNA (RNAi) and the role of short interfering RNAs (siRNAs) as well as that of microRNAs (miRNAs) in controlling mRNA viability and protein translation in the cytoplasm begin to provide an answer to the cancer puzzle (Filipowicz, 2005, Hall et al, 2005;Wynter, 2006).

One of the ironies of scientific research is that RNAi was discovered as a result of a purely commercial project to create a purple flower for sale on the lucrative, international market. The project to create a purple petunia flower by introducing a chimeric chalcone synthase gene using the anti-sense RNA was a commercial failure as the only product was a white flower (Napoli et al, 1990). Other biotech laboratories tried to make a fortune by being the first to patent a black rose. However, these futile attempts led to a major scientific breakthrough, which we are only now beginning to fully realise. Closer investigation of why the petunia was white instead of a fashionable purple showed that short double stranded RNA could knock out genes with a similar sequence. The mechanism was more fully described in the nematode worm *Caenorhabditis elegans* some years later (Fire et al, 1998) for which they gained the Nobel Prize for Medicine in 2006. In 2001 it was finally demonstrated that a 21-23 nucleotide double stranded RNA (dsRNA) acted as a trigger for RNAi in mammalian cells (Elbashir et al, 2001; Caplen et al, 2001). RNAi was not just a peculiar property of plants but was most likely an ancient defence mechanism against viruses and transposons that also operated in humans.

Mechanism of RNAi

In order to understand cancer in the context of massive hypermethylation and hypomethylation of the genome as well as the process of LOH, it is first of all necessary to describe RNAi and its ramifications. Briefly, the mechanism of RNAi is a process whereby sequence specific, post-transcriptional gene silencing can be initiated by a short dsRNA, which can be either a miRNA or a siRNA. This can happen in two ways, either by post-transcriptional cleavage of mRNA through extensive complementarity in the coding region or the 3′untranslated region (UTR) or by translational repression of the mRNA again in the 3'UTR (Fig 1). Perfect homology induces mRNA degradation, but high complementarity induces repression of translation. Once the antisense strand of dsRNA or the guide strand is assembled onto a RNA-protein complex, called the RNA-induced silencing complex (RISC), it is the RISC that cleaves the targeted mRNA. Groups of mRNAs that form part of a metabolic pathway can be regulated at one time using a miRNA, which identifies it as a pleiotropic regulator of gene expression.

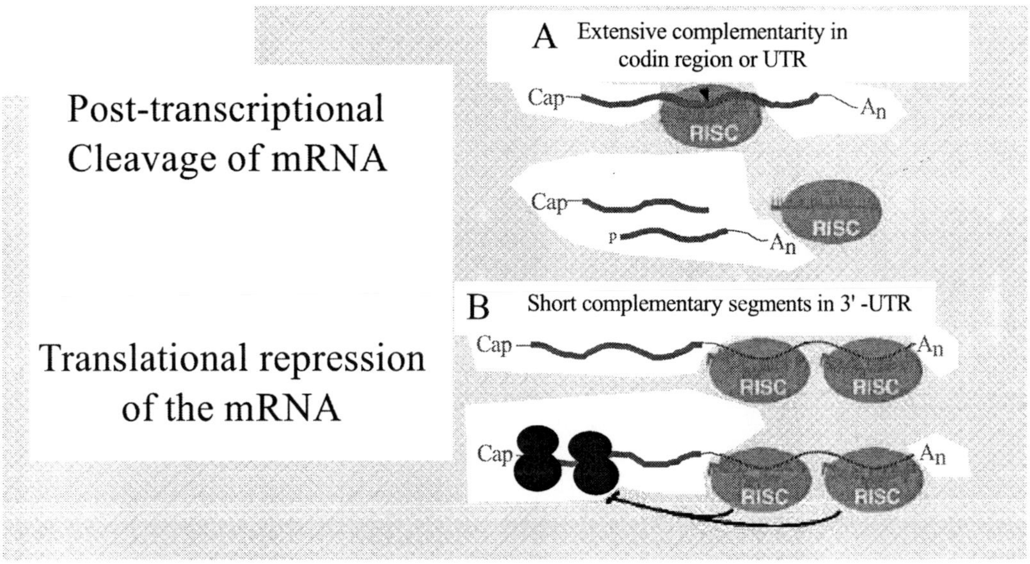

Figure 1. Degradation of mRNA or translational silencing by siRNA. A. The siRNA loaded onto the RNA-protein complex, called the RNA-induced silencing complex (RISC), can target the complementary strand of a mRNA. The RISC cleaves the mRNA at a position 10 nucleotides from the 5' end of the siRNA. B. The siRNA and RISC bind at multiple sites to the 3' UTR of the mRNA and block translation by ribosomes.

The natural process begins in the cell when a long primary miRNA transcript (pri-miRNA) is transcribed by RNA polymerase II in the nucleus. Each miRNA comes from a gene that is dedicated to the production of a particular 22-nucleotide miRNA (Lim et al, 2003). At least 474 miRNAs have been identified and sequenced in the human genome as of November 2006 (Griffith-Jones, 2004) and probably 1500 exist which regulate 5300 genes (Zamore et al, 2005), although this figure could be as high as 7500 genes. [See the miRNA registry at www.microrna.sanger.ac.uk/Software/Rfam/mirna/index] The pri-miRNAs are substrates for nuclear RNase III enzymes called Drosha, which is linked with another protein DGCR8 and together form a microprocessor (Figs. 2 and 3).

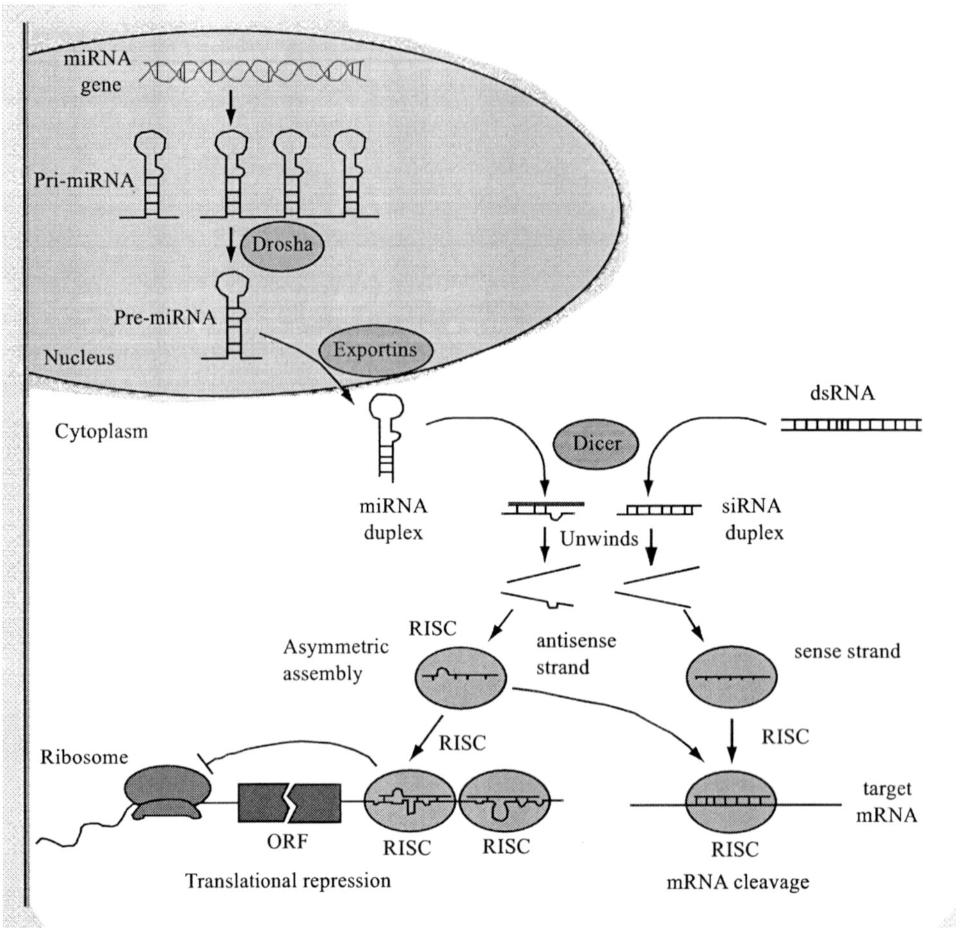

Figure 2. Production of miRNAs and siRNAs. miRNA is a non-coding RNA and the pri-RNA and pre-miRNA is partly processed by RNases in the nucleus, such as DROSHA, complexed with another protein DGCR8. It is exported out of the nucleus by EXPORTIN5 to the cytoplasm, where it is further processed to a 22 nt miRNA with 5' phosphate overhang. A short double stranded RNA, siRNA, can be introduced experimentally, which also binds to the RISC complex. The sense strand is degraded and the guide strand binds to the mRNA to carry out cleavage or translational repression.

Following this initial processing, these molecules fold to produce hairpin structures of about 70 nucleotides, labeled precursor miRNAs (pre-miRNA), which are recognised by the two nucleotide 3′overhang (for reviews see Ambros, 2004; Karagiannis et al, 2005; Mattick et al, 2005; He et al, 2004). They are escorted out of the nucleus through the nuclear pore by a transport receptor, EXPORTIN5, and a Ran GTPase (Kim, 2004). The pre-miRNAs are handed over to another RNase complex called DICER in the cytoplasm for a second processing to produce a miRNA or a siRNA duplex (Fig. 3). The only difference between a miRNA and a siRNA is that the former are normal products of the nucleus and the latter are synthetic. Generally, miRNA derive from RNA transcripts containing a stem loop and load onto the RISC as single-stranded RNA. siRNAs begin as double stranded RNA, usually depend on an artificial delivery system to the cell and require selection of the antisense strand for incorporation into the RISC (Valencia-Sanchez et al, 2006).

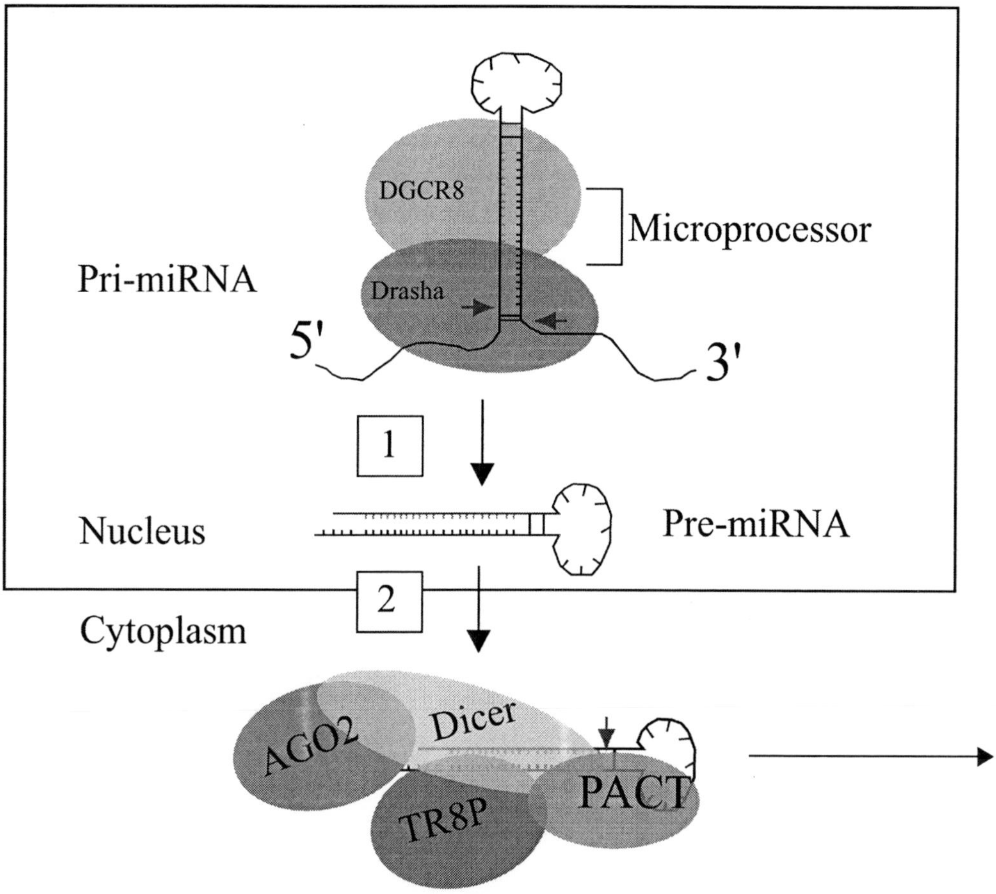

Figure 3. Processing of the miRNA. A microprocessor composed of DROSHA and DGCR8 cleave the pri-miRNA to a 70nt dsRNA with a stem loop. After export out of the nucleus into the cytoplasm, it is loaded onto the DICER complex, together with cofactors TAR RNA binding protein (TRBP), PACT and AGO2. The sense strand is cleaved by AGO2.

DICER is part of the larger protein complex that has several cofactors, a TAR RNA binding protein (TRBP), PACT and AGO2. (Chendrimada et al, 2005; Haase et al, 2005; Lee et al, 2006). TRBP recruits the DICER complex to AGO2 for miRNA processing. AGO2, a member of the family of Argonaute proteins, which are highly basic proteins of about 100 kD, contains two RNA binding domains. The PIWI domain binds the small RNA guide at its 5′end and the PAZ domain which binds the single stranded 3′ end of the small RNA (Liu et al, 2005). In humans, AGO2 is the only one of four Argonaute proteins that can cleave targeted mRNA. It was originally thought that the rejected strand or sense strand of the dsRNA was separated by a helicase, but it has been shown that the destruction of this strand is carried out by AGO2 (Matranga et al, 2005).

Both the miRNAs and siRNAs are in a staggered duplex with two unpaired nucleotides at either end. The siRNA must be 5′phosphorylated to enter the RISC complex. The antisense strand is selected by the thermodynamic stability of its 5′end, with weaker pairing favouring selection. Cleavage of the target mRNA is directed by the middle of the siRNA, measured 10 nucleotides from the 5′end of the siRNA. Mismatches at the site of cleavage inactivate siRNA while others are tolerated to various extents (Hutvagner et al, 2002). Up to nine mismatches at the 3' end of siRNA can still result in mRNA cleavage, but this takes place very slowly. Finally, the major factor in determining the efficacy of the siRNA is the profile of its thermodynamic stability of the antisense strand and the target mRNA, rather than the nucleotide match of the antisense strand to the mRNA target (Aronin, 2006).

Cleaved mRNA cannot be translated and its encoded protein is not produced. New studies have shown that siRNAs, bound to AGO2, can move the target mRNAs from the cytosol to sites of mRNA destruction called P-bodies (Liu et al, 2005). AGO2 associates with the enzymes that remove the 7meG cap, a prerequisite for their destruction in the P-body. The sequence of events may be that siRNAs carry out translational repression, which initiates a relocalisation to the P-body (Pillai et al, 2005). Once in the P-body the target mRNA is degraded. The siRNA is released from the complex so that it can return to the cytosol to carry out further repression of the same mRNA molecules.

Protein Checking System

What is the relevance of RNAi to methylation and cancer? Several dogmas abound that prevent the above mechanism from being fully utilised in our understanding of carcinogenesis. The above experiments were largely carried out in non-mammalian organisms such as plants, unicellular flagellates, fungi and yeast, particularly *Arabidopsis*, *Caernorhabditis elegans, Drosophila melanogaster, Tetrahymena thermophila* and *Schizosaccharomyces pombe*, all with smaller genomes subject to easier genetic manipulation than mammalian cells.

The first dogma that should be re-examined is that the mammalian cell does not possess a protein checking system, that is, a system which can check the functioning or tertiary structure of newly synthesized protein. The cell has a highly regulated system for degradation of proteins through ubiquitylation (for review see Kerscher et al, 2006). The highly conserved ubiquitin protein of 76 amino acids is covalently conjugated in a regulated multistep process by a ligase to the protein to be degraded, by as many as four ubiquitin moieties. The attachment of a chain of ubiquitin molecules or even just one ubiquitin marks the protein for degradation. Many short lived proteins, such as p53, with ubiquitin enzymatically attached through a substrate lysine to a COOH group of a terminal glycine, are broken down by the 26S proteasome. This is composed of a 20S catalytic chamber, capped at both ends with19S regulatory units. Ubiquitin status plays a major role in protein trafficking, and acts as a sorting signal at the multivescicular body and at the plasma membrane. Ubiquitylation can also induce endocytosis. The ubiquitin pathway is a major regulator of the protein quality of misfolded proteins entering the Endoplasmic Reticulum. These misshapen proteins are marked by ubiquitin for rapid degradation in the cytoplasm by the 26S proteasomes (Meusser et al, 2005).

However ubiquitin status is also regulated by ubiquitin cleavage, which is carried out by deubiquitylating enzymes called DUB (for review see Millard et al, 2006). The human genome codes for approximately 80 DUBs, which can be divided into five classes on the basis of their catalytic domains. A specific DUB, called USP4, regulates degradation of misfolded forms of G-coupled receptors at the Endoplasmic Reticulum. It is conceivable that a combination of ubiquitin ligases and DUBs are responsible for surveying inactive or dysfunctional proteins, which have resulted from just a single incorporation of the wrong amino acid at the active site. The mRNA, the source of the malfunctioning protein, could then be targeted by the RNAi and siRNA system for removal.

To initiate a protein check and subsequent degradation, it would be expected that the errant protein must be partially inactive. Not every mutant protein or mutated mRNA elicits a carcinogenic response. The DNA repair protein, O-(6)-methylguanine-DNA methyltransferase (MGMT), was engineered to contain eight mutations within the active site of the enzyme, although it still retained full activity (Loeb et al, 2003). This mutated mRNA and protein was tolerated with no ill effects, at least under cell culture conditions. It would be expected that the cell could operate a sophisticated system to check protein function, in much the same way the cell operates multiple DNA repair mechanisms that survey the newly transcribed DNA strand to ensure fidelity.

Recent evidence does support the existence of a RNA quality control system in budding yeast (Vanacova et al, 2005). It was found that the nuclear protein Trf4p, a polymerase β nucleotidyltransferase in conjunction with a RNA-binding protein, Air1p or Air2p, formed a complex with a RNA helicase Mtr4p. This complex was able to add a polyA tail (60-70A) to aberrant tRNAs, targeting them for destruction by the nuclear exosome. The recognition of aberrant tRNA was based on structural features. It was speculated there might also be a mRNA checking system based on polyadenylation in metazoans in the same way that bacterial mRNA is degraded (Zamore et al, 2005). However, no matter how rational this idea would appear, so far there is no hard evidence for a systematic protein checking mechanism for proper functioning in mammalian cells.

RNA-Dependent RNA Polymerase and sicRNA

The second dogma that should be re-examined is that there is no RNA-dependent RNA polymerase (RdRP) in mammalian cells that could synthesize a short antisense RNA. However, research has concentrated on finding a RNA-dependent RNA polymerase such as RNA polymerase II, that behaves like a nuclear RNA polymerase, synthesising molecules of a large size, many thousands of base pairs. A specific search for a cytoplasmic located RdRP or even a nuclear-located enzyme, that synthesizes short RNAs of a length as short as 20-25 bps and no longer than 200 bps, similar to a miRNA, is necessary. The cDNA of such an enzyme might be very different from the RdRP found in plants. Recent evidence for such a RdRP has come to light by investigations on the repetitive DNAs of the centromeres, which act as a substrate for kinetochores (Djupedal et al, 2006). Two subunits of RNApol II, Rpb1 and Rpb2, were shown to have specific roles in chromatin modification and siRNA production and may play a role in transcribing from RNA substrates (Kato et al, 2005; Schramke et al, 2005).

An important assumption of our hypothesis is that the cell, in self-defence against rogue mRNAs, is able to generate its own siRNA, which specifically attacks the mutated sequence to carry out degradation of the homologous mRNA. This may be initiated on a signal from the protein checking system or from a mRNA checking system, based on a misfolded protein structure or inadequate function. In the following discussion, sicRNA here refers to endogenously generated, Small Interfering Cellular RNA, made by the cell itself, using a 22 nucleotide sense strand of the mutant section of mRNA as a guide, in response to a dysfunctional protein. This mechanism would imply that the metazoan cell has a putative RdRP that is able to direct the synthesis of a short section of antisense RNA (maybe no more than 30 nucleotides) of the aberrant mRNA.

The resulting dsRNA, consisting of the mutant, short section of mRNA (the sicRNA) and the antisense strand is processed by DICER, binds to the RISC complex, the sense strand is degraded and the sicRNA marks the mRNA for degradation (Fig.4). An RNA Polymerase IV (RNApol IV) found in plants, fungi and *Caenorhabditis elegans*, can generate thousands of copies of an siRNA and is the basis of quelling (Mello et al, 2004). It is not suggested that the new sicRNA can move throughout the tissue, thus providing systemic resistance as in plants. So far no RNApol IV has been found in *Drosophila*, insects and mammals, but the evidence from the extensive data on cancer suggests that there is some sort of mechanism for generating such new sicRNAs. Perhaps in mammals, a sicRNA, using a different enzyme or a subunit of an RNA polymerase pol II or pol III, might be involved.

The sicRNA may be represented by the newly discovered class of 750 tiny non-coding RNAs (tncRNAs), of 20 nt, discovered in *C. elegans* (Ambros et al, 2003). They are precisely complementary to mRNA from more than 500 different genes and seem to be endogenous siRNAs. So far these tncRNAs have not been found in mammals. It is clear the defence system must be infinitely adaptable, as in any one oncogene or tumor suppressor gene, hundreds of different mutations have been identified. For example, there are 300 known mutations in the *APC* gene one cause of colon cancer; there are over 2,000 different mutations in the two breast cancer genes *BRCA1* and *BRCA2*; and 225 different mutations in the DNA mismatch repair gene, *MLH1,* the cause of a hereditary colon cancer Hereditary Non-Polyposis Colon Cancer (HNPCC). Usually only one organ of the body is targeted for malignancy in a familial cancer, although every cell in all tissues carries the mutation. Hence there must be a general, efficacious, adaptive mechanism producing sicRNAs for targeting thousands of differently mutated mRNAs that can arise. Otherwise, in an inherited cancer, such as Familial Adenomatous Polyposis (FAP), every cell in the body would be clogged with large amounts of aberrant APC mRNAs and dysfunctional APC protein, and every tissue would eventually produce a cancer.

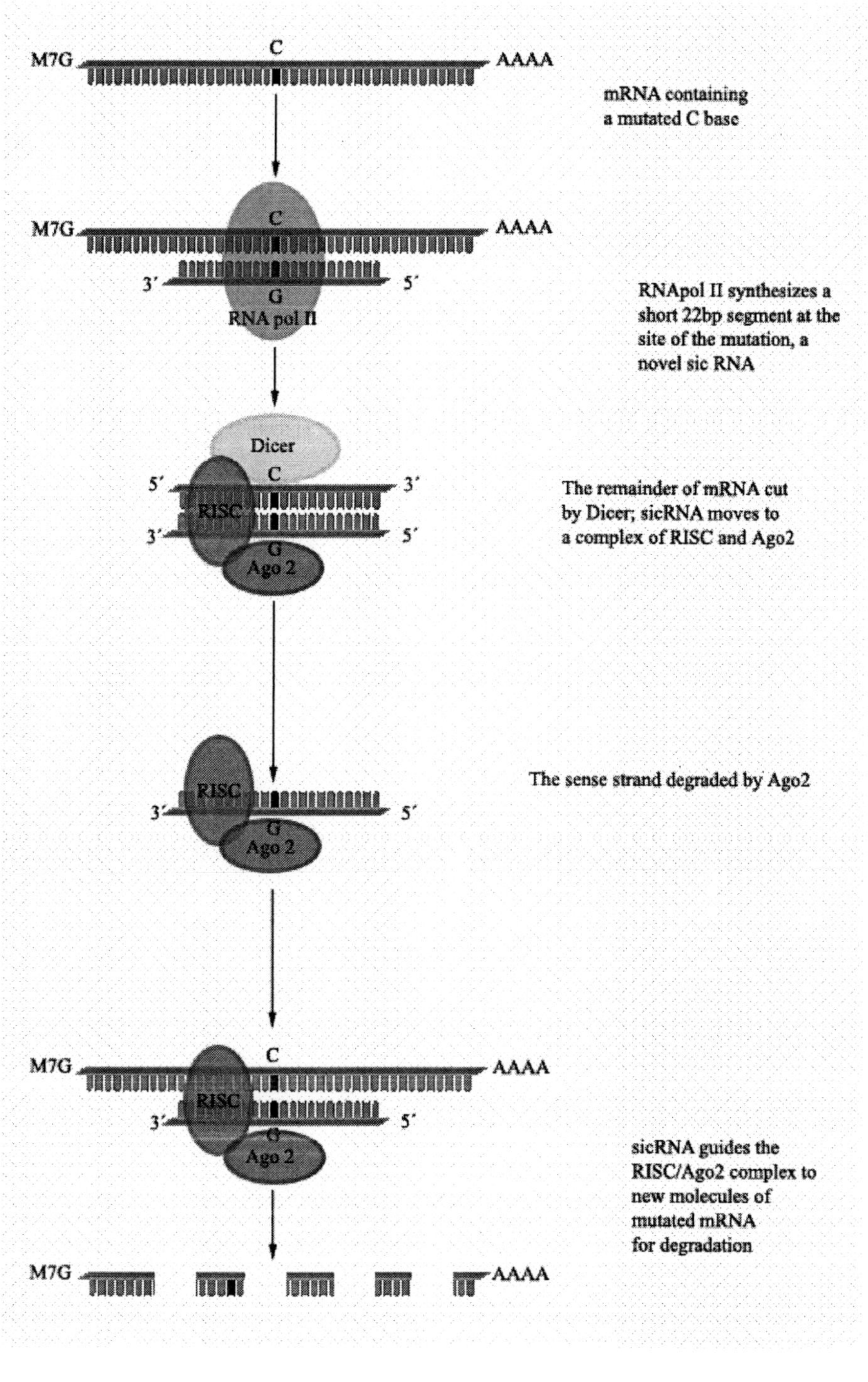

Figure 4. Putative synthesis of small interfering cellular RNA (sicRNAs). A mutated mRNA is detected by a protein checking system, perhaps based on a ubiquitin system or polyA tail. A new sicRNA, which is antisense to the mutation, is synthesised by a RNA-dependent RNA polymerase (RdRP), perhaps a subunit of RNA pol II. The double-stranded sicRNA binds to Argonaute 2 (Ago2), which degrades the sense strand of the sicRNA. The Ago2/RISC complex is guided to the mutated mRNA and removes it to the specialised P-body in the cytoplasm for degradation. The sicRNA is released along with Ago2, to seek out further molecules of mutated mRNA, targeting them for inhibition of translation and/or degradation.

Known Characteristics of siRNAs

Recent experimental data using siRNAs have delineated some of the parameters necessary for efficient elimination of any mutated mRNA. The sicRNAs would have to have exactly the same characteristics to function as a defence mechanism against mutated genes. SiRNAs, generated against different regions of an mRNA, were able to suppress translation to varying degrees. Examination of 36–40 different siRNAs designed to bind to three different regions of mRNA, namely the 5' untranslated region, the coding exon and 3' untranslated region (UTR), of CD54 and PTEN mRNAs found that 9–12 siRNAs inhibited a 200nt span in the 3' UTR expression by greater than 50% (Vickers et al, 2003). One or two mismatches placed in the central domain of the siRNA sequence were sufficient to disable effective targeting. It was concluded that the secondary structure of the mRNA strongly influenced the effectiveness of the antisense siRNA. Not all siRNAs designed to hybridise to a mRNA transcript were effective. The RNAi agent had to be a siRNA of <30 nt in length as use of larger conventional dsRNA (>500 nt) is limited by antiviral/interferon responses, which trigger activation of the Jak–Stat pathway and a generalised translational repression and apoptosis (Sledz et al, 2003; Paddison et al, 2002).

The specificity of the siRNA system is demonstrated by the fact that it can distinguish between alternatively spliced forms of a single protein. In *Drosophila* cultured cells, a gene containing a mRNA that is spliced into 12 different forms but one siRNA designed to target one isoform, did not degrade the other 11 isoforms (Celotto et al, 2002). Thus double-stranded siRNA can successfully discriminate between very similar mRNAs. In another experiment, three different siRNAs, antisense to the gene *p53*, were introduced into transgenic mice with Myc-induced lymphomagenesis by a retroviral vector (Hemann et al, 2003). The extent of inhibition of the p53 protein, by the three siRNAs, correlated with the severity and type of the resulting hyperplasia, with different siRNAs able to elicit different tumour phenotypes. One phenotype was a diploid tumour while another was highly metastatic, even though the same gene *p53* was targeted. This variation of clinical phenotype with differences in the mutated region of the same gene has been noted in FAP. Classical FAP with thousands of adenomas is usually associated with *APC* mutations in codons 169–1600. Extracolonic features of congenital hypertrophy of the retinal pigment epithelium is seen with patients with mutations in codons 463–1387 of *APC*. In Gardner's syndrome, characterised by osteomas, skin fibromas and epidermoid cysts in addition to polyposis, the mutations usually occur in the codons 1403–1578.

The RNAi system is so specific, that a siRNA is even able to sense one base change in a mutated mRNA. As an example of this, a 19nt siRNA sequence, ligated into a retroviral pSUPER vector, was designed to target the G-T mutation of the K-*ras* allele, generating a Gly to Val substitution at codon 12. This could specifically inhibit the mutant K-*ras* allele in human pancreatic carcinoma cell line CAPAN-1, while leaving the wild-type K-*ras* allele untouched (Brummelkamp et al, 2002a). A single base change mismatch in the sequence of the same siRNA completely abrogated this suppression of gene expression of the mutated K-*ras* (Brummelkamp et al, 2002b). The siRNA system can also distinguish between Single Nucleotide Polymorphisms (SNPs), as shown for the knock down of the gene responsible for Machado-Joseph Disease, spinocerebellar ataxia type 3 allele (Miller et al, 2003). Selective siRNA targeting of the disease-causing CAG repeats was not possible, but suppression was achieved through an siRNA of an associated SNP of the mutant allele. Thus, siRNAs have been shown to provide the basis of a mechanism to detect a single base mutation in a mutated mRNA. This is exactly what is needed in a defence system for the cell to protect against any mutation that may arise spontaneously or through the inheritance of one mutated allele of an oncogene or a tumour suppressor gene.

A range of candidate genes to target for elimination by siRNAs as a possible cancer treatment has been listed in a recent review (Izquierdo, 2005). A plethora of individual siRNAs against hundreds of genes has been generated, laying a foundation for new discoveries of oncogenes (Gartel et al, 2006). Clearly the siRNA and RISC system has sufficient variability and adaptive response to contain aberrant mRNAs in most tissues, except the tissue where a cancer results. Why does only one tissue develop a cancer?

Silencing of the Mutated Gene

Once the mutant protein has been identified, the mutant mRNA has been degraded or translation prevented by sicRNA machinery, the next line of defence must occur on the actual mutated gene in the nucleus, to conserve the energy of the cell and not waste valuable resources. Another dogma proven to be incorrect was that this type of gene silencing, called post-transcriptional gene silencing (PTGS), only occured in plants and *S. pombe*. However, two groups have shown that a siRNA targeted to the promoter site in the genome induced transcriptional silencing via DNA methylation. One siRNA targeted at the oncogene *erbB2* promoter induced specific CpG methylation (Morris et al, 2004), and similarly an siRNA directed at the promoter site of the gene Elongation Factor 1 alpha inhibited transcription by methylation of the promoter sequence in the nucleus of tissue culture cells (Kawaski et al, 2004). It is not necessary to have a siRNA directed against the promoter site, as a siRNA directed against the gene itself is sufficient to promote CpG methylation and hence transcriptional silencing. Thus, these experiments proved that siRNAs can play a role in mammalian systems in silencing genes by site-specific methylation of the promoter site (for a review see Morris, 2006).

Another key postulate of this theory is that in a cancer-prone tissue, the anti-sense strand sicRNA made against the mutant mRNA in the cytoplasm is able to be transported back into the nucleus during cell division and initiate one of three outcomes, hypermethylation or hypomethylation of the promoter site of the mutant gene or LOH of the gene itself. It is thought the sicRNA enters the nucleus during cell division when the nuclear membrane breaks down. The factors determining which one of these three results occurs are not known. The RISC complex is not found in the nucleus, so a different protein complex binds to the sicRNA to cause its translocation into the nucleus and activation of the methylating machinery. There is some evidence that a siRNA bound to the protein complex of an RNA-Induced Initiation of Transcription gene Silencing complex (RITS) can carry out methylation. The sicRNA binds to the homologous DNA being transcribed by RNA polymerase II and methylation of the promoter site ensues, switching off the mutant gene. I will not detail how the process of silencing by methylation occurs through modification of the lysine groups on

Histone3, phosphorylation and hypoacetylation (for reviews see Klose et al, 2006 and Jenuwein, 2006).

LOH and Methylation

LOH is a feature of carcinogenesis with many cancers showing LOH of at least 25% and sometimes as much as 50% of their alleles (Lengauer et al, 1998). LOH has been found to occur in preneoplastic tissue, before any other standard pathological signs of dysplasia are visible. LOH occurs in the kidney neoplasm of Wilm's tumor in chromosome 11p15.5 and LOH is a common event in colon cancer at 1p32 and 18q21–23 (Kambara et al, 2001). Allelic loss was found at three chromosomal sites as a first step in pancreatic cancer, at 9p21, the location of the *p14ARF* gene, 17p13, (*p53* gene) and 18q21, (*SMAD4* gene) (Luttges et al, 2001). LOH at *p14ARF* and *p16INK4A* is a common event in hepatocellular carcinomas (Herath et al, 2002). Similar examples of LOH can be shown for non-small cell lung cancer at chromosome 3p, endometrioid ovarian carcinomas and bladder cancers (for a review see Tycko, 2003). LOH is considered to be caused by chromosomal rearrangements at various locations, deletion, mitotic recombination or chromosomal loss.

Few investigations have examined the genetic basis of LOH but siRNAs could play a key mechanistic role in LOH in human tumours. However siRNAs have been found to carry out the directed elimination of DNA sequences, which could be viewed as another type of LOH in the ciliate protozoan, *Tetrahymena thermophila*, a unicellular eukaryote (Mochizuki et al, 2002). Tetrahymena has two structurally and functionally different nuclei in a single cell. The diploid, germline micronucleus, transcriptionally inactive, called the MIC-DNA and the polyploid, somatic macronucleus, called MAC-DNA, which is transcriptionally active. The MIC undergoes meiosis during the sexual process of conjugation to form two haploid pronuclei, one of which is reciprocally exchanged between the two conjugating cells. The migratory and stationary pronuclei then fuse to create a zygotic nucleus that divides to produce the next generation of new MACS and MICS. After the new MAC forms, the old, parental MAC is destroyed. During development of the new MAC, massive rearrangement of the genome occurs (for review see Mochizuki and Gorovsky, 2004).

This programmed DNA degradation occurs at some 6000 sites of genomic MIC-DNA or 15% of the genome. These Internal Eliminated Sequences (IES), range in size from 0.5 to 20 kb, and are mostly moderately repeated DNA sequences in the MIC. IES elimination sites are healed by religation of the remaining strands. The sequences of these repeats vary remarkably in different IESs due to conjugation, and no consensus sequences have been found in or around IESs that could explain how they are targeted for elimination. Recent data can now provide a mechanism for this programmed degradation based on RNAi, using a large number of wild-type siRNAs, 26–31 nt long which preferentially hybridise to these different DNA sequences. These siRNAs, called scanRNAs (scnRNAs) are expressed during conjugation and bind to Twi1p, a member of the PPD (PAZ-Piwi domain)/Argonaute protein family. In the Mochizuki and Gorovsky model, the MIC genome is transcribed bidirectionally to make double-stranded RNAs (dsRNAs) that are processed to scnRNAs in the cytoplasm. There, scnRNAs complex with Twi1p and are transferred to the old MAC, where the scnRNAs homologous to MAC-DNA sequences are degraded. The remaining MIC-specific scnRNAs, still complexed with Twi1p, are then transferred to the new MAC, where they promote elimination of the IES sequences to which they are homologous. This is a method of destroying foreign DNA, viruses and transposons in the new MAC. IES elimination can be seen as the ultimate form of gene silencing.

Recently it was found that *T. thermophila* contains two distinct groups of siRNAs, and is the first unicellular organism known to express more than one class of endogenous siRNAs. A new class of 23-24 nt siRNAs, called sRNAs, may serve as guides for RNA cleavage, targeting transcripts from the antisense strand of sRNA clusters or related RNAs from other loci, playing a similar role as in mammalian cells. The 23–24-nt sRNAs are distinct from the conjugation-specific 27–30-nt siRNAs in size, developmental expression, genomic origin, and putative function. The unique features of the 23–24-nt sRNAs reveal the existence of a greater diversity in the biogenesis, function, and regulation of sRNAs than previously known (Lee et al, 2006). Different classes of siRNAs may also be functioning in mammals.

The IES elimination of DNA also involves a histone methyltransferase (HMT) which methylates a lysine on histone H3 to produce a H3K9Me and hypoacetylation of the histone. Chromodomain proteins can bind to the methylated H3K9 and/or to RNA. The *Tetrahymena* sequences are marked for deletion by the presence of H3K9Me in the MIC-DNA chromatin. The H3K9-specific HMT may itself be a chromodomain protein or may bind to the chromodomain proteins that accumulate on the sequences recognized by the small scnRNAs. Thus the H3K9 HMT and chromodomain proteins can interact in a positive loop to spread and maintain methylation and replicate heterochromatin. A key role was also played by another gene coding for the protein Programmed DNA Degradation Signal (PDD1p and PDD3p) which binds to H3K9Me as did HP1 (Taverna et al, 2002). PDD1 contains multiple chromo-domains to connect the targeted DNA sequence though pairing with a siRNA to H3K9Me (Jenuwein, 2002)

So it appears that targeted CpG methylated DNAs can either be stably silenced by a chromodomain protein C complex or going one step further, the CpG methylated DNA can be attacked by DNA nucleases, leading to DNA elimination altogether. We have detailed these important *Tetrahymena* experiments, because although the organisation of the unicellular eukayote DNA into micro and macro nuclei is distinctly different from the mammalian nucleus, these experiments show that hypermethylation and LOH could be two sequential steps of the one process, mediated by siRNAs. So far this process has not been described in metazoans but there are many features similar to that of the formation of heterochromatin. This process could be evolutionary conserved in eukaryotes, one expression being that of the phenomena of LOH in cancers.

A mechanism to explain LOH in cancers, based on the *Tetrahymena* sequence of events is proposed (Fig. 5). Firstly, the sicRNA synthesised against the mutated mRNA would have to gain access to the nucleus. It is proposed a PPD protein, with a Piwi and PAZ domains, possibly AGO2 allows the sicRNA to bind to a small region of homologous DNA on a chromosome, being transcribed. The presence of the PPD protein attracts hypoacetylation and a histone methyl transferase to produce a H3K9Me. A feature of this mechanism is the possibility of spreading the methylation across a DNA region with no homology to the original sicRNA. The accumulation of proteins and heterochromatin state activates a DNA nuclease. Even in *Tetrahymena*, very little is known of the recruitment of nucleases. Nothing is known of the mechanism that determines which event takes place, silencing by physical degradation of DNA or hypermethylation. However the similarities with the LOH phenomena in cancer is unmistakable.

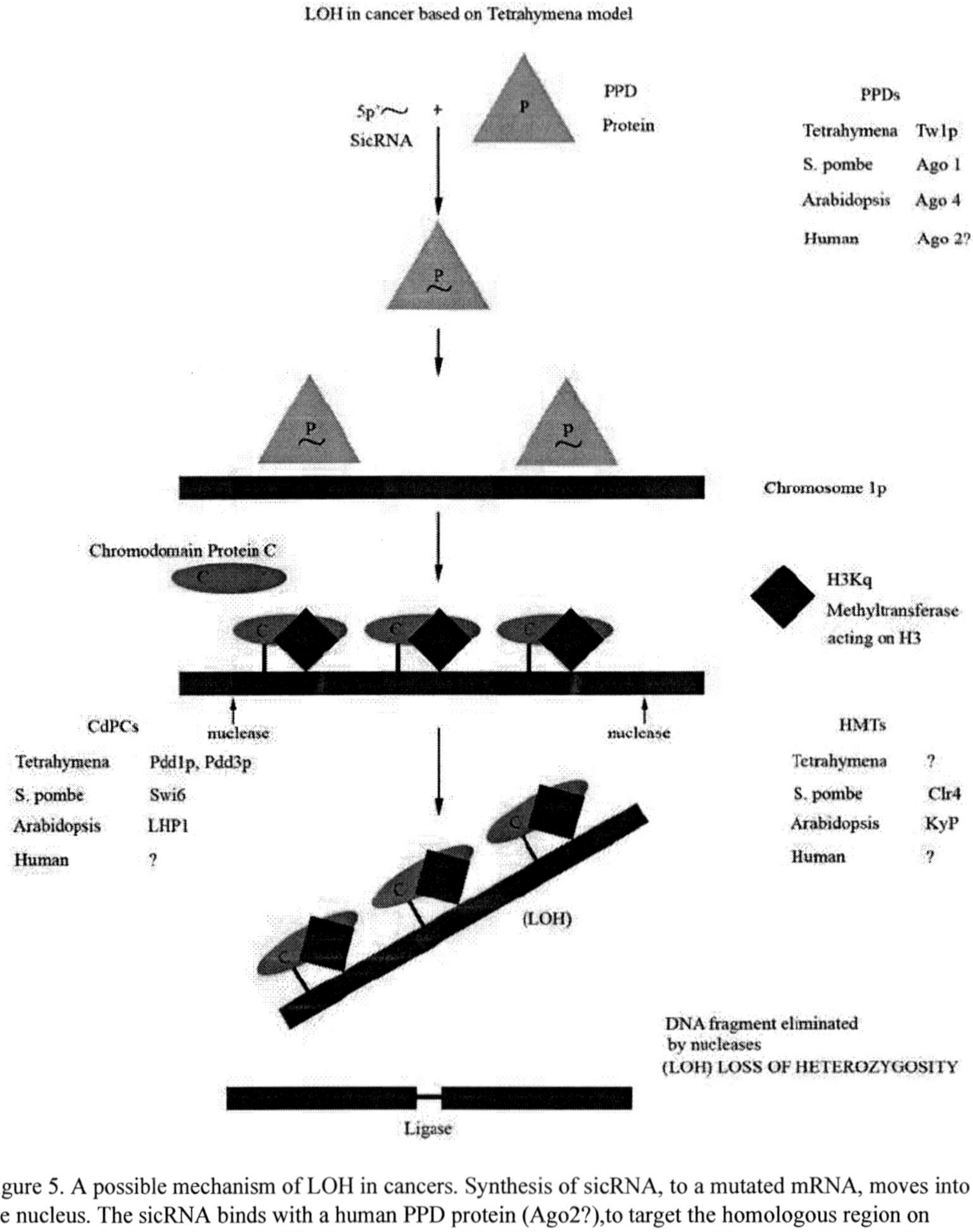

Figure 5. A possible mechanism of LOH in cancers. Synthesis of sicRNA, to a mutated mRNA, moves into the nucleus. The sicRNA binds with a human PPD protein (Ago2?),to target the homologous region on chromosomal DNA to initiate LOH. The presence of the PPD protein attracts a histone methyl transferase and a chromodomain protein C which together allows the formation of heterochromatin. The H3K9 and chromodomain structure also encourage a spreading of methylation to sites not initially covered by the sicRNA target region. The two methylated ends are cut by DNA nucleases to achieve LOH. The remaining DNA is ligated. This is based on the Mochizuki and Gorovsky model described for Tetrahymena.

Contradictory results have been obtained where RNAi-mediated silencing has occurred in the absence of DNA methylation. However, an explanation could be that LOH of the targeted gene has taken place instead (Park et al, 2004). The LOH mechanism could also be applied to 30% of genes not controlled by CpG methylation of promoter sites. It would be interesting to examine how the cell silences a germline mutant gene in tissues, which do not give rise to a cancer. Is it mediated by RNAi and are there any consequences that impinge on the nucleus containing the mutant gene? Thus either LOH and/or methylation of CpG islands in the promoter site may be the first two initial steps that occur in the pretumor stage in stem cells.

For example, familial colon cancer patients (FAP) with a truncating mutation close to codon 1317 of the APC gene, almost invariably show LOH of the remaining allele, not another mutation (Lamlum et al, 1999). In addition, the specific sequence around the mutation in a gene may determine the phenotype that subsequently develops. Hereditary desmoid disease as well as multiple, colorectal adenomas occurred in a family which had a frameshift mutation at codon 1924 of the APC gene, resulting in a truncated protein of 1947 amino acids (Lamlum et al, 2000). It was thought this specific mutation led to desmoid cancers. The tumour suppressor genes *p14ARF* and *p16INK4A*, both located at 9p21 under different promoters, are common targets in cancers. In lung and breast carcinomas homozygous deletion of the *INK14/ARF* locus is common but in colon cancer promoter hypermethylation is more frequent (Esteller et al, 2000). Several human miRNAs genes have been found to occupy a position in the genome close to LOH loci in many cancers (Calin et al, 2004). Further, the LOH site found in tumours may not be indicative of the position of a tumour suppressor gene but simply the site where a sicRNA binds to the homologous DNA to carry out a deletion.

Cell Mosaicism Due to LOH

LOH could be another mechanism by which the cell maintains a dialectical equilibrium. It has been shown that human tissues contain a large number of cells carrying inactivating mutations, giving rise to what has been called a genetic mosaic of the tissue. The mechanism in this case was thought to be due to unequal crossing over during mitotic recombination or due to repair of a stalled replication fork after double-strand breaks (Tycho, 2003).

As an example of tissue mosaicism, the staining for expression of the non-essential enzyme O-acetyl transferase in the colonic mucosa showed that the number of crypts that had undergone spontaneous LOH to become the negative homozygote was calculated at 6/10,000 crypts (Jass et al, 1994). This figure extrapolates to thousands of crypts in a whole colon that has lost the expression of O-acetyl transferase. Again hypermethylation could be the cause of the loss of expression as this work was carried out before the gene silencing by hypermethylation of promoters was understood. Since then it has been shown that an apparent interplay between *de novo* methylation and demethylation at each cell division gives rise to a heterogeneous pattern of methylation for any one section of DNA (Clark et al, 2002).

However there could be another reason for this LOH which reflects a continual loss of expression of genes due to a mutation, producing a sicRNA which then degrades the homologous region in the genome. LOH of small regions may be a very common event in all tissues due to spontaneous mutations and play a bigger role than previously thought in initiating a pretumor. As evidence of this possibility, in normal tissue, adjacent to a breast cancer, a high frequency of LOH was found for markers on chromosome 3p24 (Deng et al, 1996). There is no reason to doubt that other tissues contain clones of cells with LOH or hypermethylation encompassing other chromosomal regions and may contribute to the earliest formation of a cancer.

Levels of miRNAs in Cancers

One key piece of evidence that miRNAs have a role in carcinogenesis is that cancers have greatly disturbed levels of miRNAs, generally much reduced compared to normal tissue. The first indication of this was the discovery that the miR-143/miR-145 cluster is down regulated in colon cancers (Michael et al, 2003). In addition, recent work found another possible 133 novel candidates of miRNA from human colorectal cancer cell lines, 200 known mature miRNA forms and 112 previously uncharacterised miRNA forms (Cummins et al, 2006). Further evidence came from a study of the human *BIC* gene, which encodes the *miR-155* at nucleotides 241-262 on chromosome 21. The hairpin precursor *miR-155/ḆIC* is 100 times upregulated, not down regulated, in children with Burkitt lymphoma but not in other hematologic malignancies such as pre-B/common or T-cell leukaemia (Metzler et al, 2004). Their results suggested that *miR-155* could function in cooperation with *MYC*, an oncogene which acts as a transcription factor for cell proliferation, in the transformation of B cells and may play a role in the late stages of tumour progression.

In colon and pancreatic cancers (Hilgers et al, 1999), LOH is frequently found at 1p34-35, a region where three closely related Argonaute family members of the RISC complex are located, namely hAgo3, EIF2C1/hAgo1 and Ago4 (Gong et al, 2005). Another member of the Argonaute family, HIWI, is located on chromosome 12q24.33, which is linked to testicular germ cell tumours of adolescents and adults. HIWI is overexpressed in seminomas, again implicating RISC/miRNA in some aspects of carcinogenesis (Qiao et al, 2002). Two other miRNAs, *miR-15a* and *miR-16a*, are located at chromosome 13q14, a region deleted in more than 68% of B cell Chronic Lymphocytic Leukemias (CLL). 13q14 lies between exons 2 and 5 of the Leu-2 gene. Deletions at this region also occur in approximately 50% of mantle cell lymphoma, in 16–40% of multiple myeloma and in 60% of prostate cancers (Gong et al, 2005). The 13q14 region contains at least four non-coding genes, including *miR-15a, miR-16a* as well as *Leu-1* and *Leu-2*. The same miRNAs, *miR-15a* and *miR-16-1* were also expressed at lower levels in pituitary adenomas (Bottoni et al, 2005). These are just a few examples of the possible link between disruption of an important miRNA gene that may be crucial in maintaining regulatory control, by a oncogene or tumor suppressor gene in carcinogenesis.

Another finding is the strong correlation between the location of miRNAs and Homeobox genes (*HOX*). *HOX*-containing genes are a family of encoding transcription factor genes that play crucial roles during normal development and in oncogenesis. The HOX genes, such as *HOXB4, HOXB5, HOXC9, HOXC10, HOXD4* and *HOXD8* are deregulated in a number of solid and hematopoietic cancers (Owens et al, 2002). The *HOXB* cluster on 17q21 includes two miRNAs, *miR-10a and miR-196*, located upstream of *HOXB*. The *HOXC* cluster at 12q13 contains *miR-196-2* and *miR-10b* maps to the HOXD cluster at 2q31. This data suggests that miRNAs in close proximity to the HOX clusters may be altered along with these HOX genes in human cancers (Calin et al, 2004). Oncogenic *miR-21* was also strongly overexpressed in human brain tumour and glioblastomas (Chan et al, 2005).

A recent finding indicated that the change of miRNA levels in cancer tissues is not due to a mutation but some disturbance in their expression. Screening 15 miRNAs linked to tumorigenesis by virtue of their mRNA targets or their chromosomal localization in 91 cancer-derived cell lines, no evidence of mutations that altered the sequence of the mature miRNA could be found (Diederichs et al, 2006). The let-7 family of miRNAs which targets oncogenic *Ras* and miRNAs which are located to sites of frequent chromosomal instability (*miR-143, miR-145, miR-26a-1, and miR-21*) were examined in detail. Some 16 sequence aberrations in miRNA precursors was identified, including some with effect on their secondary structure, but none altered the ability of precursors to be processed to the mature form *in vivo*.

Further miRNA genes are frequently located at fragile sites as well as in minimal regions of LOH, minimal regions of amplification or common breakpoint regions. Some common chromosomal fragile sites (FRAs) predispose to DNA instability in cancer cells. Indeed, FRAs are preferential sites of sister chromatid exchange, translocation, deletion and amplification or integration of plasmid DNA and tumour associated viruses such as human papilloma virus (HPV16). Overall, Calin et al (2005) found that 52% (98/186) of miRNA genes are in cancer-associated genomic regions or in fragile sites. Hence, there seems to be a strong link between the position of miRNAs and the location of cancer-associated genomic regions. A new bead-based flow cytometric miRNA expression profiling method was developed to analyse 217 mammalian miRNAs from a panel of more than 200 human cancers (Li et al, 2005). A general down regulation of miRNAs was observed compared to normal tissues, indicating that the majority of miRNAs may be tumor suppressors. In addition, cancers from similar developmental origins were accurately grouped together, whereas mRNA microarrays did not offer a correct classification. This data also suggested that miRNA profiling may be useful in cancer diagnosis.

Cancer Is a Result of Interference of miRNA by a sicRNA

Another key postulate of my theory is that the methylation or LOH of an oncogene or tumour suppressor due to the action of a sicRNA interferes with the expression of an miRNA, fundamental to that particular tissue. The sicRNA is generated to cope with any mutation, and once linked up with the RITS machinery, it can affect CpG methylation of the targeted gene. It should be noted that CpG methylation can also switch on gene expression. Thus the sicRNA of a particular oncogene could theoretically bind to sites all over the genome by annealing to 6-8 nucleotides on an homologous DNA strand, but this would produce a cancer in many different tissues, depending on the sequence of the mutated site. As this doesn't happen and a mutation of an oncogene always produces a cancer in one specific tissue, the answer must be linked to the production of specific miRNAs for that particular tissue. For example, an inherited mutation in the *MLH1* gene always initiates a cancer in the large colon, and sometimes in ovarian tissue, but these are derived from the same embryonic tissue, but not in the lung nor liver nor prostate gland, even though the germline mutation is present on one allele in every cell in the body. An examination of what is currently known of miRNAs supports this argument.

An important property of miRNAs is their ability to target clusters of functionally related genes. In *Drosophila*, miR-7 targeted the Notch gene and miR-277 played a role in the metabolic switch regulating amino acid catabolism (Stark et al, 2003; Ambros, 2004). The cell death regulator Bantam, was originally thought to be a protein-coding gene for cell proliferation in *Drosophila*, but has now been found to encode a hairpin structure and is in fact an miRNA, although its targets are unknown (Xu et al, 2004). Thus miRNAs are a perfect tool for disturbing the precise phalange of proteins, which define any tissue, a feature grossly distorted in most cancers.

MiRNAs vary from more than 1000 molecules per cell to some exceeding 50,000 molecules per cell (Novina et al, 2004). It is therefore plausible that a single miRNA species could modulate the expression of a 100 mRNA targets (Bartel et al, 2004). The other possibility is that five different miRNAs are necessary to target and destroy an mRNA. Recent computation has determined that miRNAs may target at least 30% of human genes by the 3' untranslated region (UTR) of the mRNA, where the 'seed sequence' consists of nucleotides 2–7 of the miRNA flanked on either side by an adenosine (Lewis et al, 2005). Thus it's possible that a miRNA exists for all tissues, regulating the protein profile.

The exact consequences of such interference with a miRNA is not known. It is proposed a sicRNA synthesised by the cell to stop translation of a rogue mRNA, enters the nucleus during mitosis. The sicRNA attached to a protein complex, binds to a homologous region of the gene, methylates the promoter site and inadvertently prevents the production of a miRNA close to that site or within the site. In a slightly different scenario, some molecules of the sicRNA bind to a miRNA undergoing transcription and abrogates its action. In an attempt to silence the production of the wrong genes for that tissue, other genes are silenced by the same sicRNA mechanism. This could result in hundreds of genes being methylated and silenced and more miRNAs being disrupted.

Most cells with a disabled miRNA will be eliminated by apoptosis. Thus for the stem cell to survive, two processes must be deactivated, apoptosis and normal control of cell proliferation. In another scenario, the sicRNAs may compete with a small modulatory dsRNA (smRNA), such as found in the neurone, described in detail below (Kuwabara et al, 2004), and switch off synthesis of a large group of tissue specific mRNAs. The type of miRNA that is disabled may depend on it being actively transcribed at the same time the sicRNA enters the nucleus. In the next round of cell division, a different miRNA could be inactivated, producing a cancer cell with a number of different characteristics. It may need several rounds of cell division before a second miRNA is disabled. This would add enormous variation to the resulting phenotype, as described above with different siRNAs against *p53*. Hence the cell undergoes a disintegration and a dedifferentiation of the particular architecture of that tissue. The disruption of a specific miRNA marks a qualitative step that signals the progress from a normal stem cell to a harmless Aberrant Crypt Foci (ACF), to an adenoma or a hyperplastic polyp, to a serrated adenoma, at least in colon cancer (Wynter et al, 2004).

In a familial cancer, in tissues that do not develop a cancer but also contain the mutant germline allele, it would be expected that the mutant allele is silenced by methylation. End of story. The sicRNA produced to deal with that specific mutation carries out CpG methylation of the gene or LOH. This process does not interfere with the production of any miRNA relevant to that tissue. The cell survives due to the normal functioning of the single wild type allele.

Thus cancer is not a linear development of clonal expansion of cells with an increasing number of mutations, but a series of discrete steps with a qualitative change at particular points during the development of tumorigenesis, because specific miRNAs or smRNAs are disrupted. The majority of aberrant cells will undergo apoptosis. For example, many dozens of ACFs can be visualised by methylene blue in a length of colon but only a miniscule number, if any, of these will form an adenoma, and only 5% of the adenomas will develop into a colon cancer. The conversion of an ACF to an adenoma or a hyperplastic polyp may depend on the sicRNA eventually blocking production of a miRNA specific for colon mucosal cells.

It has been suggested that each tissue type might have a unique miRNA profile to modulate tissue-specific patterning, as has been recently found in neurons. The key transcriptional regulator that restricts neuron-specific genes to neuronal cells is a zinc finger protein, known as NRSF/REST (Kuwabara et al, 2004). NRSF/REST binds specifically to a DNA response element (NRSE/RE1), which in turn is encoded in the promoter regions of about 60 genes controlling all aspects of neuronal development such as neurotransmitter receptors, ion channels, synaptic vessels, etc. NRSF/REST mediates transcriptional repression through the association of the N-terminal repressor domain with the histone deacetylase HDAC1/2 and also through a complex with methyl-CpG binding proteins (MeCP2). It was found that a small dsRNA of 20nt could modulate the NRSE/RE1 and NRSF/REST protein machinery to switch on neuronal gene expression from a repressed state to an active state. This new element was called a small modulatory dsRNA (smRNA). This discovery challenges the previous findings that miRNAs could only decrease gene expression. This 20nt smRNA, which should diffuse out of the nucleus because of it small size, was somehow restricted to the nucleus, perhaps by particular protein complexes.

What is currently known about the role of miRNAs could explain many of the features of cancer cells, such as dedifferentiation and loss of all special features of the cells belonging to that tissue. Once the dysregulation of both cell proliferation and apoptosis pathways have been initiated, the driving force of carcinogenesis is increasing loss of miRNA function. The mature miRNAs control a range of developmental events, such as timing of cell-fate decision, stem cell maintenance, apoptosis and organ morphogenesis. MiRNAs have been found to downregulate a great number of mRNA transcripts, thereby helping to define tissue-specific gene expression in humans (Lim et al, 2005). MiRNAs can also target some of the 1850 known transcription factors, giving this class of RNAs a role in integration of genetic regulation (Herr et al, 2004).

siRNA as a Cure for Cancer and Genetic Diseases-the Dangers

Thousands of siRNAs have now been experimentally shown to reduce the growth rate of tumours (Gong et al, 2005). Many genetic diseases such as lysosomal storage diseases, those of the central nervous system and hemophilia are being treated by adeno-associated viral vectors containing a siRNA (Warrington et al, 2006). However, the problem of specific targeting of an siRNA to an oncogenic mRNA is commonly recognised, as many off-target effects can occur and it is difficult to construct sufficient controls. This was highlighted by an experiment in which the introduction of a siRNA caused changes in over 1000 genes, detected by microarray hybridization. (Persengiev et al, 2004). The target of this siRNA was not even expressed in the cell line under examination. Another artifact affecting molecular oncology is that dsRNA can suppress *p53,* and therefore lead to false conclusions on growth reduction (Marques et al, 2005). The siRNA must be designed in such a way as not to silence the wild-type allele, but be able to distinguish the mutant allele of an oncogene. Similarly, each SNP site of a wild type allele might need to be tested to ensure it is not eliminated along with the mutated allele.

Another factor, which is often overlooked, is that high levels of the introduced siRNA will be processed by the same machinery that deals with processing endogenous miRNA. The stoichiometry of proteins involved in the RISC complex such as AGO2 and R2D2 in relation to the miRNA are not known (Aronin, 2006). This could compete with the miRNA and decrease the efficiency of the endogenous miRNA (Gartel et al, 2006). Another off-target effect is that the siRNA changes the expression profile of unintended mRNAs, further confusing interpretation of the resulting data (Scacheri et al, 2004). However, another very serious off-target effect is that sequences designed as siRNA can act in a manner similar to a miRNA. Some miRNAs prevent target translation rather than degradation. This means the siRNA could act as a pleiotropic regulator of a large number of mRNAs. Relaxed target specificity has been evoked as the reason for the off-target effects of some siRNAs (Doench et al, 2003).

However, there are even more difficult problems with the therapy of siRNAs in that now it is known that siRNAs play a role in segregation of the chromatids during mitosis. They are transported into the nucleus in an association with the RITS complex to silence centromeric heterochromatin. Therefore, the synthetic siRNAs could be the initiator of the process in which the tissue develops a cancer, with chromosomal abnormalities. At present, the new exciting discoveries of miRNAs and their ability to knockout specific genes is generating much enthusiasm for the development of therapeutic gene silencing, especially for patients with an incurable familial genetic disorder. But this method needs very careful evaluation. The application of the therapy can go badly wrong, as indicated by an experiment on eight young patients suffering from X-linked severe combined immunodeficiency (SCID) (Hacein-Bey-Abina et al, 2003). The patients, after injection with a retroviral-based therapy, the lentivirus vector containing the corrected gene, initially improved. However, 30–34 months later, two patients developed leukaemia, indicating a serious problem with the delivery system as the proviral integration site disrupted a key gene, the LMO-2 gene (Dykxhoorn et al, 2003). The treatment had to be stopped. This is a clear indication that there are many problems with the cell delivery system and additional problems with tissue selectivity.

Thorough evaluation is even more urgent now that human clinical trials based on siRNAs for AIDS, neurological diseases, Hepatitis C, Adult Macular Degeneration and Respiratory Syncytical Virus are already under way. All clinical trials of synthetic siRNAs should be put on hold until far more information is available as to the off-target effects of the specific siRNA, especially effects in the cell nucleus. siRNA may affect gene expression at the transcriptional level, making the potential sites of additional interaction infinitely greater with the attendant risks (Kawasaki et al, 2004; Morris et al, 2004). The resulting action of a specific siRNA may depend on the genetic make-up of the individual and this would have to be tested before any clinical trial took place. At the present time, it is not clear how these sorts of interactions could be properly evaluated.

We should learn from the past scientific errors; the mad rush to patent discoveries that are not fully understood. The genetic modification of plants without thought of the ramifications is one such example. To prevent a further erosion of general confidence in scientific endeavor, all clinical trials of synthetic siRNAs should be halted until the RNAi system is more clearly defined and comprehensively understood. The only case where a clinical trial might be authorised would be that of saving a life.

Conclusion

In summary, a key, novel idea, proposed here, is that the cell has the ability to recognise a mutated protein and/or mRNA. Secondly, the cell can generate its own short interfering RNA (siRNA) using an RNA polymerase to destroy mutated mRNA, even when only a single base pair in the gene has mutated. The anti-sense strand of the short RNA molecule (called sicRNA), targets the mutated mRNA of an oncogene or a tumour suppressor. The resulting double stranded RNA, using the RNA-induced silencing complex (RISC) in the cytoplasm, dices the mutated mRNA. During cell mitosis, the sicRNA complex can move into the nucleus to target the mutated gene. The sicRNA triggers the assembly of protein complexes leading to epigenetic modification of the promoter site of the mutant gene, specifically methylation of cytosines. In some instances, instead of methylation, the homologous DNA is degraded, leading to loss of heterozygosity (LOH). The factors controlling these two actions are unknown, but the result is gene silencing or physical destruction of the mutant gene. The cell survives dependent on the functioning of the single, wild-type allele.

An error in RNAi defence occurs when the sicRNA enters the nucleus, and by methylation of the target gene, inadvertently interferes with the production of a miRNA specific for that tissue. This produces a change in the profile of correct proteins for that tissue. The variability of the cancer response may depend on which specific miRNA is being transcribed at the time the sicRNA enters the nucleus. In the vast majority of these cases, the aberrant miRNA production will have no effect on cell function as apoptosis eliminates nonviable cells. On a rare occasion, a preneoplastic stem cell is initiated when miRNA interference switches on/off a gene involved in apoptosis, as well as a gene involved in cell proliferation and DNA damage surveillance.

A malignancy develops when the sicRNAs fortuitously targets a pleiotropic microRNA (miRNA), or a smRNA (as described for neurons) or activates a transcription factor, resulting in the translation of a large number of new genes, alien to that tissue. This leads to dedifferentiation of the tissue, CpG methylation changes, a resculpting of the histone code, chromosomal rearrangements, along a number of specific pathways, the gain of immortality and the dissemination of a metastatic cancer. The driving force of carcinogenesis is the sequential loss of active miRNAs. This model is slightly adjusted from a previous hypothesis, where it was postulated that the sicRNA could attack any homologous 6-8 nt DNA sequence in the nucleus (Wynter, 2006). The interference of sicRNA on tissue specific miRNAs confines tumorigenesis to one tissue only.

The hypotheses of this model that need to be proven experimentally in humans are:

1. There is a protein/mRNA checking system for aberrations, perhaps based on poly A, similar to tRNA in yeast or a system, assessed by the number of ubiquitylation and deubiquitylation sites which can correctly diagnose a misfolded protein.
2. The cell can generate its own siRNAs by use of a mammalian RNA-dependant RNA polymerase with a sequence that is the anti-sense strand to bind to any short section (20–25 nt) of mutated mRNA that may arise.
3. The newly synthesised siRNA, called sicRNA, can hinder translation or degrade the mutated mRNA through the RISC complex.
4. In all tissues, the sicRNA/RITS complex can move into the nucleus during cell division and silence the specific mutated gene by CpG methylation or LOH. In non-cancerous tissue, which also carries the mutant allele, this is the end of the story.
5. A pretumor begins when CpG methylation and/or LOH disrupt a key miRNA specific for that tissue. For the stem cell to survive, at least two processes, one involved in cell proliferation and the other in apoptosis, must be disabled.
6. A qualitative leap in pathogenesis occurs when a sicRNA abrogates a pleiotropic miRNA, essential for that tissue, leading to cell dedifferentiation.
7. Cancers can be initiated by many different pathways, utilising sicRNAs, either elicited by a familial germline mutation or by a spontaneous mutation which is deleterious to that tissue, because of the disruption of a specific miRNA. The exact phenotype even in the same individual will depend on the initial error and the order of subsequent random events.

MiRNAs and siRNAs can theoretically explain all the diverse phenomena related to carcinogenesis. Indeed, some miRNAs should now be classified as classical oncogenes or tumour suppressor genes (Gong et al, 2005). The above is only a sketchy outline of the phenomena of carcinogenesis, RNAi and all the methylation phenomena, including LOH. Detailed experimental work will fill in the structural features of RNA/DNA hybrids and conformational changes of protein complexes with RNA, affecting function. Obviously there exists a wealth of complex interactions, many unexpected connections, protein acrobatics involving conformational changes and many complicated feedback loops, which control all aspects of cell proliferation and homeostatic maintenance. At the moment, all we can do is only guess at some of these features. The processes of miRNA and RNAi are so fundamental to the functioning of the cell that RNAi errors may also provide the answers to some diseases that have proven refractory to classical genetic analysis, such as schizophrenia and autism. Even emotional memory may be encoded in miRNA (Vendramini, 2005).

Cancer has always afflicted humans, although the numbers are dramatically increasing, with at least 7 million people world-wide expected to die each year from a cancer. It is not just a modern disease; the first description of cancer can be found in an Egyptian papyrus, dating back to approximately 1600BC. Clearly, many important pieces of the RNAi-cancer puzzle are missing, but sufficient data is available to begin generating hypotheses to try and identify experimentally the blank pieces of the jigsaw.

Acknowledgements

The authors would like to thank Dr Marcelo A. Alfonzo-Gonzalez for help with figures and the CDCH of the University Central de Venezuela for a grant PI 09.6030.2005 to Prof. Itala Lippo de Becemberg. References were finalized in Nov, 2006.

References

Ambros V. (2004) The functions of animal microRNAs. *Nature.* 431, 350-5.

Ambros V, Lee RC, Lavanway A, Williams PT, Jewell D. (2003) MicroRNAs and other tiny endogenous RNAs in *C. elegans. Curr. Biol.* 13, 807-18.

Aronin N. (2006) Target selectivity in mRNA silencing. *Gene Ther.* 13, 509-16.

Bartel DP, Chen C-Z. (2004) Micromanagers of gene expression: the potentially widespread influence of metazoan microRNAs. *Nat. Rev. Genet.* 5, 396-400.

Bottoni A, Piccin D, Tagliati F, Luchin A, Zaatelli MC, degli Uberti EC. (2005) miR-15a and miR-16-1 down-regulation in pituitary adenomas. *J. Cell Physiol.* 204, 280-285.

Brummelkamp TR, Bernards R, Agami R. (2002a) Stable suppression of tumorigenicity by virus-mediated RNA interference. *Cancer Cell.* 2, 243-7.

Brummelkamp TR, Bernards R, Agami R. (2002b) A system for stable expression of short interfering RNAs in mammalian cells. *Science.* 296, 550-3.

Calin GA, Sevignani C, Dumitru CD, Hyslop T, Noch E, Yendamuri S et al (2004) Human microRNA genes are frequently located at fragile sites and genomic regions involved in cancers. *Proc. Natl. Acad. Sci. USA.* 101, 2999-3004.

Caplen NJ, Parrish S, Imani F, Fire A, Morgan RA. (2001) Specific inhibition of gene expression by small double-stranded RNAs in invertebrate and vertebrate systems. *Proc. Natl. Acad. Sci. USA.* 98, 9742– 7.

Celotto AM, Graveley BR. (2002) Exon-specific RNAi: a tool for dissecting the functional relevance of alternative splicing. *RNA.* 8, 718-24.

Chan JA, Krichevsky AM, Kosik KS (2005) MicroRNA-21 is an antiapoptotic factor in human glioblastoma cells. *Cancer Res.* 65, 6029-33.

Chendrimada TP, Gregory RI, Kumaraswamy E, Norman J, Cooch N, Nishikura K et al. (2005) TRBP recruits the Dicer complex to Ago2 for microRNA processing and gene silencing. *Nature.* 436, 740-4.

Clark SJ, Melki J. (2002) DNA methylation and gene silencing in cancer: which is the guilty party? *Oncogene.* 21, 5380-7.

Cummins JM, He Y, Leaery RJ, Pagliarini R, Diaz LA, Sjoblom T. et al (2006) The colorectal microRNAome. *Proc. Natl. Acad. Sciences USA.* 103, 3687-92.

Deng G, Lu Y, Zlotnikov G, Thor AD, Smith HS. (1996) Loss of heterozygosity in normal tissue adjacent to breast carcinomas. *Science.* 274, 2057-9.

Denli AM, Tops BB, Plasterk RH, Ketting RF, Hannon GJ. (2004) Processing of primary microRNAs by the Microprocessor complex. *Nature.* 432, 231-5.

Diederichs S. Haber DA. (2006) Sequence variations of microRNAs in human cancers: Alterations in predicted secondary structure do not affect processing.

Djupedal I, Portoso M, Spahr H, Bonilla C, Gustafsson CM, Allsshire RC et al.(2006) RNA Pol II subunit Rpb7 promotes centromeric transcription and RNAi-directed chromatin silencing. *Genes Dev.* 19, 2301-6.

Doench JG, Petersen CP, Sharp PA. (2003) siRNAs can function as miRNAs. *Genes Dev.* 17, 438-442.

Dykxhoorn, DM, Novina, CD, Sharp, PA. (2003) Killing the messenger: short RNAs that silence gene expression. *Nature Rev.* 4, 457-67.

Elbashir SH, Harborth J, Lendeckal W, Yalcin A, Weber K, Tuschl T. (2001) Duplexes of 21-nucleotide RNAs mediate RNA interference in cultured mammalian cells. *Nature.* 411 494-8.

Esteller M, Sparks A, Toyota M, Sanchez-Cespeded M, Capella G, Peinado M et al. Analysis of Adenomatous Polyposis Coil promoter hypermethylation in human cancer. *Cancer Res.* 60, 4366-71.

Fire A, Xu S, Montgomery MK, Kostas SA, Driver SE, Mello CC. (1998) Potent and specific genetic interference by double-stranded RNA in *Caenorhabditis elegans. Nature.* 391, 806-11.

Filipowicz W. (2005) RNAi: the nuts and bolts of the RISC machine. *Cell.* 122, 17-20.

Gartel AL, Kandel ES. (2006) RNA interference in cancer. *Biomol. Engineer.* 23, 17-34.

Gong H, Liu C-M, Liu D-P, Liang C-C. (2005) The role of small RNAs in human diseases: potential troublemaker and therapeutic tools. *Med. Res. Rev.* 25, 361-81.

Griffith-Jones S. (2004) The Micro RNA Registry, NAR 32, *Database Issue.* D109-D111.

Haase AD, Jaskiewicz L, Zhang H, Laine S, Sack R, Gatignol A, Filipowicz W (2005) TRBP, a regulator of cellular PKR and HIV-1 virus expression, interacts with Dicer and functions in RNA silencing. *EMBO Reports.* 6, 961-7.

Hacein-Bey-Abina, S, von Kalle, C, Schmidt, M, Le Deist, F. (2003) A serious adverse event after successful gene therapy for X-linked severe combined immunodeficiency. *New Engl. J. Med.* 348, 255-6.

Haley B, Zamore PD. (2004) Kinetic analysis of the RNAi enzyme complex. *Nat. Struct. Mol. Biol.* 11, 599-606.

Hall IM, Shankaranarayana GD, Noma KI, Ayoub N, Cohen A, Grewal, SI.(2002) Establishment and maintenance of a heterochromatin domain. *Science.* 297, 2215-8.

Hall PA, Russell SH. (2005) New perspectives on neoplasia and the RNA world. *Hematol. Oncol.* 23, 49-53.

He L, Hannon GJ. (2004) MicroRNAs: small RNAs with a big role in gene regulation. *Nat. Rev. Genet.* 5, 522-31.

Hemann MT, Fridman JS, Zilfou JT, Hernando E, Paddison PJ, Cordon-Cardo C et al (2003) An epi-allelic series of p53 hypomorphs created by stable RNAi produces distinct tumor phenotypes *in vivo*. *Nat. Genet.* 33, 396-400.

Herath N, Kew M, Walsh M, Young J, Powell L, Leggett BA et al (2002) *J. Gastroenterol. Hepatol.* 17, 301-7.

Herr AJ, Baulcombe DC. (2004) RNA silencing in plants. *Cold Spring Harb. Symp. Quant. Biol.* 69, 363-70.

Hilgers W, Tang DJ, Sugar AY, Shekher MC, Hruban RH, Kern SE. (1999) High resolution deletion mapping of chromosome arm 1p in pancreatic cancer identifies a major consensus region at 1p35. *Genes Chromosomes Can.* 24, 351-5.

Hutvagner G, Zamore PD. (2002) A microRNA in a multiple-turnover RNAi enzyme complex. *Science.* 297, 2056-60.

Izquierdo M. (2005) Short interfering RNAs as a tool for cancer gene therapy. *Cancer Gene Ther.* 12, 217-27.

Jass JR, Edgar S. Unicryptal loss of heterozygosity in Hereditary Non-Polyposis Colorectal Cancer. *Pathology.* 26, 414-7.

Jenuwein T. (2002) An RNA-guided pathway for the epigenome. *Science.* 297, 2215-8.

Jenuwein T. (2006) The epigenetic magic of histone lysine methylation. *FEBS J.* 273, 3121-35.

Kambara T, Matsubara N, Nakagawa H, Notohara K, Nagasaka T, Yoshino T et al. (2001) High frequency of low-level microsatellite instability in early colorectal cancer. *Cancer Res.* 61, 7743-6.

Karagiannis T, EL-Osta A. (2005) RNA interference and potential therapeutic applications of short interfering RNAs. *Cancer Gene Ther.* 12, 787-95.

Kato H, Goto D, Martienssen R, Urano T, Furukawa K, Murakemi Y. (2005) RNA polymerase II is required for RNAi-dependent heterochromatin assembly. *Science.* 309, 467-9.

Kawasaki H, Taira K. (2004) Induction of DNA methylation and gene silencing by short interfering RNAs in human cells. *Nature.* 431, 211-7.

Kerscher O, Felberbaum R, Hochstrasser M. (2006) Modification of proteins by ubiquitin and ubiquitin-like proteins. *Annu. Rev. Cell Develop. Biol.* 22, 159-80

Klose RJ, Bird AP (2006) Genomic DNA methylation: the mark and its mediators. *Trends Biochem. Sci.* 31, 89-97.

Kuwabara T, Hsieh J, Nakashima K, Taira K, Gage FH. (2004) A small modulatory dsRNA specifies the fate of adult neural stem cells. *Cell.* 116, 779-93.

Lamlum H, Ilyas M, Rowan A, Clark S, Johnson V, Bell J et al (1999) The type of somatic mutation at APC in FAP is determined by the site of the germline mutation: A new facet to Knudson's 'two-hit' hypothesis. *Nat. Med.* 5, 1071-5.

Lamlum H, Tassan NA, Jaeger E, Frayling I, Sieber O, Reza FB et al. (2000) Germline APC variants in patients with multiple colorectal adenomas, with evidence foe the particular importance of E1317Q. *Hum. Mol. Genet.* 9, 2215-21.

Lee SR, Collins K. (2006) Two classes of endogenous small RNAs in *Tetrahymena thermophila. Genes and Dev.* 20, 28-33.

Lee Y, Hur I, Park SY, Kim YK, Suh MR, Kim VN. (2006) The role of PACT in the RNA silencing pathway. *EMBO Journal.* 25, 522-32.

Lengauer C, Kinzler KW, Vogelstein B. (1998) Genetic instabilities in human cancers. *Nature.* 396, 643-9.

Lewis BP, Burge CB, Bartel DP. (2004) The role of PACT in the RNA silencing pathway. *Cell.* 120, 15-20.

Li S, Crothers J, Haqq CM, Blackburn EH. (2005) Cellular and gene expression responses involved in the rapid growth inhibition of human cancer cells by RNA interference-mediated depletion of telomerase RNA. *J. Biol. Chem.* 280, 23709-17.

Lim LP, Glasner ME, Yekta S, Burge CB, Bartel DP. (2003) Vertebrate microRNA genes. *Science.* 299, 1540.

Lim LP, Lau NC, Garrett-Engele P, Grimson A, Schelter JM, Castle J et al. (2005) Microarray analysis shows that some microRNAs down regulate large numbers of target mRNAs. *Nature.* 433, 769-73.

Liu J, Valencia-Sanchez MA, Hannon GJ, Parker R. (2005) MicroRNA-dependent localization of targeted mRNAs to mammalian P-bodies. *Nature Cell Biol.* 7, 719-23.

Loeb LA, Loeb KR, Anderson JP. (2003) Multiple mutations and cancer. *Proc. Natl. Acad. Sci. USA.* 100, 766-81.

Luttges J, Galehdari H, Brocker V, Schwarte-Waldhoff I, Henne-Bruns D, Kloppel G et al. (2001) Allelic loss is often the first hit in the biallelic in activation of the P53 and DCP4 genes during pancreatic carcinogenesis. *Am. J. Pathol.* 158, 1677-83.

Marques JT, Rebouillat D, Ramana CV, Murakami J, Hill JE, Gudkov A et al. (2005) Down-regulation of p53 by double-stranded RNA modulates the antiviral response. *J. Virol.* 79, 11105-14.

Matranga C, Tomari Y, Shin C, Bartel DP, Zamore PD. (2005) Passenger-strand cleavage facilitates assembly of siRNA into Ago2-containing RNAi enzyme complexes. *Cell.* 123, 607-20.

Mattick J, Makunin IV. (2005) Small regulatory RNAs in mammals. *Hum. Mol. Genet.* Apr15; 14, R121-32.

Mello CC, Conte D. (2004) Revealing the world of RNA interference. *Nature.* 431, 338-42.

Metzler M, Wilda M, Busch K, Viehmann S, Borkhardt A. (2004) High expression of prescursor microRNA-155/BIC RNA in children with Burkitt lymphoma. *Genes Chromosome Cancer.* 39, 167-9.

Meusser B, Hirsch C, Jaroosch E, Sommer T. (2005) ERAD: the long road to destruction. *Nature Cell Biol.* 7, 766-72.

Michael MZ, O'Connor SM, van Holst , Pellekaan NG, Young GP, James RJ. (2003) Reduced accumulation of specific microRNAs in colorectal neoplasia. *Mol. Cancer Res.* 1, 882-91.

Millard SM, Wood SA. (2006) Riding the DUBway: regulation of protein trafficking by deubiquitylating enzymes. *J. Cell Biol.* 173, 463-8.

Miller VM, Xia H, Marrs GL, Gouvion CM, Lee G, Davidson BL, Paulson HL. (2003) Allele-specific silencing of dominant disease genes. *Proc. Natl. Acad. Sci. USA.* 100, 7195-200.

Mochizuki K, Fine NA, Fujisawa T, Gorovsky MA. (2002) Analysis of a PIWI-related gene implicates small RNAs in genome rearrangement in Tetrahymena. *Cell.* 110, 689-99.

Mochizuki K, Gorovsky MA. (2004) Small RNAs in genome rearrangement in Tetrahymena. *Curr. Opinion Genetics and Dev.* 14, 181-7.

Mochizuki K, Gorovsky MA. (2005) A Dicer-like protein in Tetrahymena has distinct functions in rearrangement, chromosome segregation, and meiotic prophase. *Genes and Dev.* 19, 77-89.

Morris KV, Chan SW, Jacobsen SE, Looney DJ. (2004) Small interfering RNA-induced transcriptional gene silencing in human cells. *Science.* 305, 1289-92.

Morris KV. (2006) Therapeutic potential of siRNA-mediated transcriptional gene silencing. *Biotechniques.* Apr; Suppl: 7-13.

Napoli C, Lemieux C, Jorgensen R. (1990), *Plant Cell.* 2, 279-89.

Novina CD, Sharp PA. (2004) The RNAi revolution. *Nature.* 430, 161-4.

Owens BM, Hawley RG. (2002) HOX and non-HOX homeobox genes in leukemic hematopoiesis. *Stem Cell.* 20, 364-79.

Park CW, Chen Z, Kren BT, Steer CJ (2004) Double-stranded siRNA targeted to the huntingtin gene does not induce DNA methylation. *Biochem. Biophysic. Res. Commun.* 323, 275-80.

Paddison PJ, Caudy AA, Hannon GJ (2002) Stable suppression of gene expression by RNAi in mammalian cells. *Proc. Natl. Acad. Sci. USA.* 99, 1443-8.

Persengiev SP, Zhu X, Green MR. (2004) Nonspecific, concentration-dependent stimulation and repression of mammalian gene expression by small interfering RNAs (siRNAs). *RNA.* 10, 12-18.

Pillai RS, Bhattaachayva SN, Artus CG, Zoller T, Cougot N, Basyuk E et al. (2005) Inhibition of translational initiation by Let-7 MicroRNA in human cells. *Science.* 309, 1573-6.

Qiao D, Zeeman AM, Deng W, Looijenga LH, Lin H. (2002), *Oncogene.* 21, 3988-99.

Scacheri P, Rozenblatt-Rosen O, Caplen NJ, Wolfsberg TG et al (2004), *Proceed. Natl. Acad. Sciences USA.* 101, 1892-7.

Schramke V, Sheedy DM, Denli AM, Bonila C, Ekwal K, Hanon GJ et al (2005) RNA-interference-directed chromatin modification coupled to RNA polymerase II transcription. *Nature.* 435, 1275-9.

Sledz CA, Holko M, de Veer MJ, Silverman RH, Williams BRG. (2003) *Nat. Cell Biol.* 5, 834-9.

Stark A, Brennecke J, Russell RB, Cohen SM. (2003) *PLoS Biol.* 1, E60.

Taverna SD, Coyne RS Allis CD. (2002) Methylation of histone h3 at lysine 9 targets programmed DNA elimination in *Tetrahymena. Cell.* 110, 701-11

Tycko B. (2003) Genetic and epigenetic mosaicism in cancer precursor tissues. *Ann. NY Acad. Sci.* 983, 43-54.

Valencia-Sanchez MA, Liu J, Hannon GJ, Parker R. (2006) Control of translation and mRNA degradation by miRNAs and siRNAs. *Genes Dev.* 20, 512-24.

Vanacova S, Wolf J, Martin G, Blank D, Dettwiler S, Friedlein A et al (2005), *PLoS Biol.* 3, e189.

Vendramini D. (2005) *Medical Hypotheses.* 64, 512-519.

Vickers TA, Koo S, Bennett CF, Crooke ST, Dean NM, Baker BF. (2003), *J. Biol. Chem.* 278, 7108-18.

Warrington KH, Herzog RW. (2006), *Human Genetics.* 119, 571-603.

Wynter CVA, Walsh MD, Higuchi T, Leggett BA, Young J, Jass JR (2004) Methylation patterns define two types of hyperplastic polyp associated with colorectal cancer. *Gut.* 53, 573-80.

Wynter CVA (2006) The dialectics of cancer: A theory of the initiation and development of cancer through errors in RNAi. *Medical Hypotheses.* 66, 612-635.

Xu P, Guo M, Hay BA. (2004), *Trends Genet.* 20, 617-24.

Zamore PD, Haley B. (2005) Ribo-gnome: the big world of small RNAs. *Science.* 309, 1519-24.

In: DNA Methylation Research Trends
Editor: Taku B. Kobayashi, pp. 237-259

ISBN: 978-1-60021-721-0

Chapter IX

Methylation in Cancer Research

Juan Carlos Roa*[*] *and Patricia García
Department of Pathology. Universidad de La Frontera, Temuco, Chile.

Abstract

DNA methylation is an epigenetic modification characterized by the incorporation of methyl group at the 5-position of cytosine within of CpG dinucleotides. Epigenetic regulation plays a relevant role in development and DNA methylation is the mechanism that ensures cell-type specific expression of a limited range of genes. Different diseases, among them cancer, present altered methylation patterns that change the perfect epigenetic balance of normal cells. In cancer cells, the aberrant pattern is characterized by a global genomic hypomethylation and promoter hypermethylation. Such processes lead to genomic instability and gene silencing, respectively. Transcriptional repression associated with aberrant methylation of CpG islands on promoter areas is a consequence of several chromatin structural changes produced by the interaction of methylcytosines with different protein complexes that recruit histone-modifying enzymes, such as histone acetylases and methylases. Genes involved in different metabolic pathways may undergo aberrant methylation, including DNA repair genes, cellular cycle regulators, genes associated with apoptosis, cell invasion and cellular detoxification among others. Analyses of DNA methylation status has enabled methylator phenotypes associated with specific tumor groups to be established, advances that were partly possible due to the surge of new methodologies in this field of research. The main areas of clinical application that can benefit from biomarkers based on aberrant DNA methylation patterns are those related to diagnosis, tumor behavior and cancer treatment. Since methylation changes often precede apparent malignant changes in carcinogenesis, they can be used in the early diagnosis of cancer detecting tumor cells from biological samples. In addition, methylated genes that alter different metabolic pathways may be used as prognostic indicators, for example those related to angiogenesis (TIMP family) or cellular invasion (cadherins). Furthermore, the nature of the affected genes may help

[*] Correspondence to: Dr. Juan Carlos Roa. Department of Pathology. Faculty of Medicine. Universidad de la Frontera. Manuel Montt 112 Zip code 478-1176 Temuco, Chile; 56-45-216273 56-45-264909; Jcroa@ufro.cl

to predict the chemotherapy and/or hormonotherapy response. Finally, the potential to reverse DNA methylation and reactivate those genes affected is an attractive option for its use as a new therapeutic target in cancer treatment. In oncology research, the study of factors that determine the pathogenesis has enabled different neoplasias to be characterized and thus advances in the management of cancer patients. In this respect, understanding epigenetic mechanisms and the processes that they involve can bring new and promising expectations for cancer treatment, not forgetting that cancer is also an epigenetic disease.

Introduction

Epigenetics is a fascinating field of research that has evolved rapidly, attracting the attention of many investigators around the world. This discipline has been studied using various approaches in order to understand its importance in organism development and its association with such biological processes as control of gene expression, DNA replication and cell cycle, DNA repair, imprinting, inactivation of the X-chromosome in females and much more. The scientific interest has focused particularly on the study of the strong association between alterations in the epigenetic patterns and different diseases, cancer among them. Many years of continuous and extensive research has generated a huge volume of scientific papers, reviews and some interesting books. As we mentioned above, the epigenetic studies include a wide variety of subjects and its association with cancer is only one important part. In this chapter, we first summarize the basics of DNA methylation in the genome, its role in development and its importance as a transcriptional repression mechanism. We also present a description of DNA methylation changes in human cancers related with the loss and gain of methylcytosines in carcinogenesis, with an emphasis on recent findings. Finally, we discuss the possible clinical applications of aberrant hypermethylation profiles in the detection and diagnosis of cancer, as well as in prognosis and therapeutic intervention. To limit the number of references throughout this chapter, we will usually cite reviews rather than original articles when we deal with matters of a more general or well established nature.

So what does "epigenetics" actually mean? This term refers to stable patterns of gene expression that occur without changes in the DNA sequence. To be more precise, epigenetics has been defined as: "The study of mitotically and/or meiotically heritable changes in gene function that cannot be explained by changes in DNA sequence" [1]. Epigenetic regulation plays an important role in animal and plant development as well as throughout adult life and is required to achieve stable expression or repression of genes in specific cell types or at defined developmental stages. Epigenetic modifications that influence gene expression include DNA methylation, histone modification (acetylation, methylation and phosphorylation) and chromatin remodeling. These epigenetic mechanisms are interrelated and alterations in the established epigenetic patterns may result in changes in gene expression causing transcriptional repression [2].

Possibly the most studied epigenetic modification is DNA methylation. This is an epigenetic marker involved in control mechanisms of a variety of biological processes and it exists in both prokaryote and eukaryote organisms. In mammalian cells, DNA methylation occurs through the covalent bonding of a methyl group at the 5-position of cytosine in the dinucleotide doublet cytosine guanine, CpG. In general, the CpG dinucleotide is severely under-represented in the human genome, making up approximately only 1% of all nucleotide bases. The majority of CpG dinucleotides in the human genome are dispersed non-randomly across retrotransposons or throughout the coding regions and introns of genes whereas up to 15% occur in CpG rich regions called CpG islands [3, 4]. These dense CpG regions mainly reside in the 5' region, including the promoter, untranslated region (UTR) and first exon of approximately 60% of all human genes. In normal cells, most CpG dinucleotides located in repetitive genomic regions outside CpG islands are methylated, including satellite DNA and parasitic elements [5]. CpG islands are generally unmethylated in normal cells at all stages of development and in all tissue types, with the exception of CpG islands on the inactive X chromosome in females, imprinted genes, germline-specific genes and tissue-specific genes [6].

In mammals, DNA methylation is essential for normal development, being required for the imprinting process (inactivation of one allele of a gene) and X-chromosome inactivation in women. The methylation patterns are established during early embryogenesis and maintained for many cell generations by a complex interplay of at least three independent DNA methyltransferase (DNMT) enzymes that transfer a methyl group to cytosine residues within the CpG dinucleotide recognition sequence [7].

To date, three members of the DNMT family have been described in mammalian cells. The first DNMT identified by biochemical assay was DNMT1 [8], which exhibits a 10–40-fold preference for hemimethylated DNA [9]. This enzyme is required for proper embryonic development, imprinting and X-chromosome inactivation. At each cell cycle, the methyltransferase DNMT1 maintains and perpetuates the DNA methylation pattern when putting a methyl group onto the newly replicated strand at each replication cycle [10]. The remaining enzymes with DNA methyltransferase activity (DNMT3a and DNMT3b) were identified by searches of expressed sequence tag (EST) databases [11]. Experimental data indicated that these DNA methyltransferases did not display preference by hemimethylated substrates [12]. Due to this property and to additional experiments that evaluated the effect of inactivation of the genes encoding DNMT3a and 3b in mice, it was suggested that these enzymes were those mainly responsible for *de novo* methylation [13]. However, there is sufficient evidence to indicate that all three DNMTs cooperate to maintain DNA-methylation patterns and possess both *de novo* and maintenance functions *in vivo* [14-16].

DNA Methylation and Transcriptional Repression

The role of DNA methylation in the differential regulation of gene expression was hypothesized many years ago. The potential mechanism was suggested by a number of early observations which revealed that specific methylation patterns within or adjacent to genes had an inverse correlation to gene expression causing transcriptional repression [17].

The efforts made over the last decade in order to understand the molecular mechanisms that underlie methylation-dependent repression has permitted clarification of the probable mechanisms by which DNA methylation can lead to transcriptional repression [18, 19].

Initially, it was proposed that local cytosine methylation of a particular sequence could directly interfere with transcription-factor binding to their recognition sites, thus leading to suppression of gene transcription [20]. A second mode of repression involves attraction of methyl-cytosine binding proteins that interact with histone deacetylases (HDACs) and chromatin remodeling factors leading to local condensation of chromatin with inaccessible promoters to transcription factors. The critical discovery was made in 1992 when Bird et al. characterized a methyl-CpG binding protein (MeCP2) that selectively recognized methylated DNA and act as transcriptional repressor [21, 22]. Later, other members of this protein family, which contains specific DNA-methyl binding domains (MBD), were identified by database searches [23]. They include MBD1, MBD2, MBD3 and MBD4. Whereas MDB4 is associated with DNA repair, the other three and MeCP2 have been shown to repress transcription by interaction with histone deacetylase complexes [24].

Post-translational modifications (PTMs) of histones play an important role in many cellular processes, notably gene regulation. The recruitment of histone-modifying enzymes by methyl-binding proteins may change the established histone code and modify the chromatin state from a transcriptionally competent (euchromatin) to a transcriptionally silent state (heterochromatin). For example, in the silenced allele of imprinted genes, it is possible to find specific patterns of post-translational histone modifications that are associated with transcriptional repression: deacetylation of histone H3 and methylation of H3 at Lys 9 or H4 at Lys 20 [25-27].

DNA-methyltransferases can also act as direct transcriptional repressors when interacting with crucial chromatin-remodeling factors. These enzymes cooperate to establish and maintain DNA-cytosine-methylation patterns and then begin to be considerate multifunctional proteins which play a key role in connecting DNA methylation and histone modification. A direct association *in vivo* between DNMT1 and two histone deacetylases (HDAC1 and HDAC2) has been already reported. The transcriptional repression domain identified in DNMT1 operates by recruiting histone deacetylase activity and generating an altered chromatin state [28, 29].

In summary, DNA methylation can contribute to transcriptional silencing through establishment of several repressive transcription complexes, which include methyl-CpG binding domain proteins (MBDs), histone deacetylases (HDACs), histone methylases (HMTs), DNA methyltransferases as well as other chromatin remodeling complexes. Thus, the maintenance of differential DNA methylation, essential for mammalian development and for the normal functioning of the adult organism, is a highly complex process that requires many different enzymatic activities. When this complex epigenetic machinery fails in the properly established and/or maintained of epigenetic information transcriptional disorders appears that can cause several major pathologies, including cancer, syndromes involving chromosomal instabilities and mental retardation, [30, 31].

DNA Methylation and Carcinogenesis

For many years, cancer was mainly considered a genetic disease but in the early 1980s a link between DNA methylation and cancer was demonstrated for the first time [32]. Since then, interest in understanding the role of DNA methylation in carcinogenesis and in explaining the mechanisms involved in the development of altered epigenetic patterns in cancer cells has grown. By studying those mechanisms in detail, knowledge has been gained about how DNA methylation patterns become disrupted in tumor cells and how these defects may be corrected [2].

As we mentioned earlier, DNA methylation patterns established during the process of early human embryogenesis remain relatively stable during adult life. The presence of altered patterns can sometimes become abnormal during fetal development, thereby creating a predisposition to pediatric cancers, and they can change during normal ageing and contribute to common cancer risk in adults. They can also support clonal evolution in human cancers, contributing to tumor progression. Indeed, it has been corroborated that methylation of some CpG islands in non-malignant tissues increase with age, although the total genomic content of 5-methylcytosine declines [2, 33, 34].

These two events are also present, but are much more pronounced, in cancer cells which experience an imbalance of DNA methylation through both gain and loss of 5-methylcytosine content. The loss of methylation in tumor cells occurs on methylated CpG from repetitive sequences and is known as "global" or "genome-wide" HYPOMETHYLATION. The second alteration of DNA methylation common in human cancers, characterized by increased methylation at some CpG islands located in gene regulatory regions, has been designated as HYPERMETHYLATION. Both events in general provide great advantage to the cancer cell, inducing genomic instability, loss of imprinting, activation of retrotransposon sequences and transcriptional silencing of tumor suppressor genes [35].

Hypomethylation and Cancer

We could say that the history of cancer epigenetics began when Feinberg and Vogelstein found DNA methylation aberrations in human cancer cells which corresponded to global hypomethylation. Using Southern blotting, they analyzed DNA that had been digested with methylation-sensitive restriction enzymes in search of differences between normal and cancer tissues, finding that a substantial proportion of CpGs were unmethylated in cancer cells in comparison with their corresponding normal tissue counterparts [32]. In the same year, Ehrlich et al. published that content of the 5-methylcytosine was globally reduced in human tumors. These studies revealed an overall decrease in methylcytosine content of up to 70% in human cancers compared with the level in normal somatic cells [36].

Loss of genomic methylation is a frequent event in some cancer types and may occur at different stages, being detected in the early stages of some cancers or even in their benign precursors as occurs with colon adenomas. In addition, hypomethylation is a common event in metastatic carcinomas as well as in their metastases and global hypomethylation seen in a number of cancers, such as breast, cervical and brain, and shows a progressive increase with the degree of malignancy correlating to the severity and metastatic potential of many tumor types [37, 38].

Hypomethylation in tumor cells is primarily due to the loss of methylation from non-coding repetitive regions of the genome such as satellites and retrotransposons, although the loss of methylcytosine in further repetitive sequences such as rRNA genes and single-copy sequences also contribute to the overall decrease [39].

Satellite sequences are normally methylated and this contributes to the establishment of heterochromatin. In many cancers, the pericentric heterochromatin regions on chromosomes 1 and 16 are heavily hypomethylated [40]. It has been reported that mutations in the DNA methyltransferase DNMT3B affect particularly the methylation of these sequences, which causes the instability of the chromatin [41, 42].

Demethylation of mobile elements like retrotransposons also contributes to genomic instability in human cancer. Retrotransposons, known as class I transposable elements, consist of two sub-types, the long terminal repeat (LTR) and the non-LTR retrotransposons. LTR retrotransposons make up approximately 8% of the human genome [43]. The non-LTR retrotransposons consists of two sub-types: long interspersed nuclear elements (LINEs) and short interspersed nuclear elements (SINEs). Among them, the L1 (LINE 1) transposon represents ~18% of the human genome and its association with human cancers has been studied [44-46]. In general, suppression of retroelement activity prevents not only retrotransposition, but also various disturbances of transcription by retroelement promoters, interference by retroelement enhancers, the activity of retroelement-encoded enzymes and illegitimate recombination between homologous elements. Failure to silence retroelements in cancer cells could therefore permit adverse activities of retroelements as well as impede any beneficial effects [47]. In somatic cells, L1 transcription and retrotransposition are prevented by a variety of control mechanisms, including methylation of L1 DNA and specifically L1 promoters. The L1 mutational insertions have been found to disrupt the APC gene and the CMYC gene in colon and breast cancer, respectively [48].

Gene-specific demethylation also occurs although does not contribute significantly to the decrease in the global content of methylcytosines. Originally it was thought that genomic hypomethylation could be associated with an overexpression of oncogenes such as MYC and RAS. In most such instances, further analyses revealed that the sites demethylated in cancers were not crucial to gene regulation [49, 50]; nevertheless, for some single-copy genes that are hypomethylated in human cancers, a significant correlation between demethylation and gene expression has been demonstrated [39]. For example, a subset of male germline-specific genes (the MAGE-type genes) has been particularly well investigated as they are activated in many human tumors. These genes are silent in all normal tissues except for the testis and placenta, but in a variety of human cancers a close correlation has been shown between demethylation of their promoter CpG islands and re-expression [51-54]. Analyses of the clinicopathological significance of demethylation of MAGE genes in gastric carcinoma have shown that demethylation occurs during progressive stages of carcinogenesis and may be associated with aggressive biological behavior in the tumor [53]. Similar studies in non-small cell lung cancers have suggested that MAGE genes are activated very early in lung carcinogenesis and may be considered as targets for lung cancer prevention [54].

Demethylation accompanied by increased expression has also been reported for the Maspin gene, a member of the serpin family, in ovarian cancer [55], gastric cancer [56] and pancreatic cancer [57]. Furthermore, overexpression of Maspin has been associated with morphological de-differentiation in thyroid cancers [58]. Other representative examples of hypomethylated genes in human cancers are the putative oncogene γ-synuclein (SNCG) in breast and ovarian cancers [59] and the S100 calcium binding protein A4 (S100A4) gene in colon cancer [60].

Hypermethylation and Cancer

Hypermethylation is the second kind of methylation defect that is observed in a wide variety of human cancers. Research on gene-specific hypermethylation events has been one of the main subjects of the last decade due to its importance in carcinogenesis as an effective transcriptional silencing mechanism. All available experimental evidence suggests that epigenetic silencing is a more frequent mechanism of gene inactivation than genetic change or mutational inactivation. CpG islands surround the transcriptional start sites of many human genes, among them several tumor suppressor genes. These sites are normally devoid of methylcytosine in somatic cells of normal tissue, but in some cancers most CpG islands in the human genome become methylated. The implication of this phenomenon in carcinogenesis has become quite clear and it is related to inactivation or silencing of the affected genes which may initiate or cause progression of common human cancers [35].

The first discovery that suggested a participation of aberrant hypermethylation in the tumor-suppressor gene inactivation was done in 1989. Greger *et al.* investigated the methylation status of the retinoblastoma suppressor gene (RB gene) in blood and tumor cells from a group of retinoblastoma patients. Based on their findings, they suggested that changes in the methylation pattern of the RB gene could play a role in the development and spontaneous regression of some retinoblastoma tumors [61]. Later, it was demonstrated that CpG island hypermethylation is also a common mechanism for inactivating the tumor suppressor gene p16INK4a in human cancer. Since then, numerous genes have been found to undergo aberrant methylation of their CpG islands (Table 1). This great advance in gene-specific methylation research was partly due to the introduction of valuable experimental techniques such as sodium bisulfite modification [62] and methylation-specific polymerase chain reaction [63] that enabled the detection of those CpG islands that are differentially methylated with greater efficiency and specificity.

The importance of CpG island hypermethylation in cancer can be appreciated by considering the nature of some genes involved and the frequency of the process. Genes susceptible to become aberrantly hypermethylated and silenced in nearly every tumor type are those involved in cell-cycle regulation ($p16^{INK4a}$, $p15^{INK4b}$, $p14^{ARF}$, *Rb*), tumor cell invasion (*CDH1, CDH13, TIMP3*), DNA repair (*MGMT, hMLH1*), chromatin remodeling, cell signaling (*APC, RASSF1*), detoxification (*RARb2, GSTP1*) and apoptosis (*DAPK1, TMS1*) [64, 65].

Certain genes are commonly methylated in a variety of cancers, while other genes present strong specificity with particular tumor types. This characteristic has allowed profiles of aberrant hypermethylation for distinct cancers to be defined [66]. The mechanisms involved in targeting gene-specific methylation in different tumor types are not yet known, but it is hypothesized that the preferential methylation of a certain gene in certain tumor types versus others is due to their inactivation conferring a selective advantage only on the first group [35]. Another option is that aberrant DNA methylation is directly targeted. It has been proposed that fusion proteins, such as PML-RAR, induces gene hypermethylation and silencing by recruiting DNA methyltransferases to target promoters and that hypermethylation contributes to its leukemogenic potential [67]. However, further studies are necessary to explain the molecular mechanisms involved in selecting and targeting CpG islands that become hypermethylated in specific tumor cells.

Depending on the type of cancer, the altered cellular pathways by abnormalities in DNA methylation may have a significant impact on development and progression of carcinogenesis. In addition, the density and spread of CpG dinucleotide methylation, the level of methylation within a tumor (methylated/unmethylated DNA copy number) and the frequency of methylation (percentage of tumors that are methylated at a given locus) are all important mediators of methylation effects in carcinogenesis [35]. Nevertheless, the general consequences of transcriptional repression of tumor suppressor genes are dramatic and confer a more malignant behavior on the tumor cell. For instance, the gene encoding the cyclin-dependent kinase inhibitor, $p16^{INK4a}$, whose expression is necessary for maintaining the retinoblastoma (Rb) protein in its active state, is frequently methylated in a variety of human tumors by affecting control of the cell cycle and therefore inducing proliferation [68].

The p53 network may also be altered via the methylation-mediated silencing of $p14^{ARF}$ gene, whose function is to block p53 degradation by inhibiting the E3 ubiquitin-ligase activity associated with MDM2. Aberrant hypermethylation of $p14^{ARF}$ is a relatively common epigenetic event in several primary tumors such as colorectal, gastric, renal, esophageal and endometrial neoplasms and gliomas and its methylated state is associated with abnormal cellular localization of MDM2 and therefore affects p53 protein activity [69].

Other important cell pathways that can be affected by the aberrant methylation of suppressor genes are the APC/b-catenin/E-cadherin network, DNA repair and apoptotic pathways. For instance, APC commonly undergoes aberrant methylation in tumors originating in the gastrointestinal tract including stomach, pancreas, liver and esophagus. Lack of APC protein that result in increased ß-catenin levels lead to activation of growth-promoting genes such as *c-myc*. Aberrant APC methylation occurs early in colorectal carcinogenesis and affect mainly wild-type APC, although *APC* gene mutation is more common in this neoplasia. Contrary, promoter hypermethylation seems to be the principle mechanism for *APC* inactivation in gastric tumors [70].

On the other hand, decreased E-cadherin (*CDH1*) expression leads to tumor cell invasion and it is common in poorly differentiated, advanced-stage carcinomas. E-cadherin is an indispensable protein for cell adhesion in normal epithelium and loss of its function leads to tumor invasion. Experimental data has demonstrated that DNA methylation plays an important role in the loss of E-cadherin expression in human breast and prostate carcinomas [71] and in digestive tumors such as gastric cancer [72].

DNA methylation is also one of the major mechanisms for gene silencing of other cell pathways. Some DNA repair genes are affected by aberrant promoter hypermethylation, such as the mismatch DNA repair gene hMLH1 that has shown an abnormal methylation pattern in endometrial and digestive tumors, which is associated with the microsatellite instability phenotype [66, 73, 74]. Another repair gene, O(6)-methylguanine DNA methyltransferase (*MGMT*), also undergoes methylation in different cancer types and its silencing is associated with the appearance of K-ras and p53 mutations [75, 76]. In addition, as MGMT plays a prominent role in DNA adduct repair that limits the mutagenic and cytotoxic effect of alkylating agents, gene downregulation by epigenetic silencing may predict the response to alkylating agent therapy in selected malignancies [77, 78].

We could mention many more examples of known or putative suppressor genes that are methylated in several cancer types, remarking on the importance of aberrant methylation status in the carcinogenesis process, its participation in the development (early stages) or progression of cancer as well as the cellular pathways that are affected by this silencing mechanism. However, so as not to overextend ourselves on a subject that deserves a review all its own, we present in Table 1 a selected list of genes that undergo CpG island hypermethylation in human cancer, mentioning the tumor profiles where this event is more frequent and the cellular consequences of their silencing.

Clinical Applications

One of the most attractive promises of DNA methylation is the evident potential of this event to serve as a biomarker in clinical approaches. The great amount of knowledge gained over the past few years about altered methylation patterns in human cancers can be applied to cancer diagnostics, prognosis and therapeutics.

The extraordinary advances in the techniques used to detect methylated sites across the genome have made it possible to evaluate the abnormal methylation status of the cells with high precision. These detection methods including powerful tools such as sodium bisulfite conversion, cDNA microarray, restriction landmark genomic scanning and CpG island microarrays [79]. Of all of them, the sodium bisulfite method is ideal for mapping the normal and aberrant patterns of methylation and, in addition, is an accessible technique for many laboratories that work in the molecular biology field. After bisulfite modification, there are a number of methods available to study CpG island methylation. These include sequencing, methylation-specific polymerase chain reaction, combined bisulfite restriction analyses, methylation-sensitive single nucleotide primer extension and methylation-sensitive single-strand conformational polymorphism [68, 79].

The methylation patterns can be evaluated from two general approaches: (1) by studying the 5-methylcytosine distribution along of DNA, representing a single individual DNA strand by means of bisulphite genomic sequencing of cloned polymerase chain reaction (PCR) products; (2) by detecting the presence of specific DNA patterns with very high sensitivity and specificity using methylation-specific PCR (MSP)-based techniques [62, 63]. From both approaches, alteration in methylation (be it hypermethylation or hypomethylation) has significant potential as a diagnostic or prognostic indicator. There are essentially four types of clinical application studies that can use the aberrant DNA methylation pattern as a biomarker for cancer: (1) detection of tumor cells from biological samples or free-circulating DNA in plasma; (2) prediction of tumor behavior associated with a particular methylator phenotype; (3) prediction of response to chemotherapies and/or hormone therapies; and (4) prediction of cancer risk by evaluating the presence of preneoplastic lesions markers [80, 81].

Methylation as a Diagnostic Indicator

Use of methylation as a clinical tool in the field of diagnosis is not yet extensively applied, but many findings support its potential in this area. For most types of cancer, DNA methylation changes occur early in carcinogenesis and therefore are potentially good indicators of early disease as well as of advanced cancers.

The efficacy of any biomarker is determined by its sensitivity and specificity, with clinical sensitivity being understood as the proportion of individuals with confirmed disease who are positive for the biomarker assay, whereas specificity is the proportion of individuals without the disease whose result is negative for the biomarker assay [79]. The sensitivity and specificity of DNA methylation markers in cancer diagnosis depends on several factors, including the type of cancer and the gene to be studied, the type of body fluid to be used and the technique involved [82].

In the case of genome-wide hypomethylation, this event does not appear to be particularly useful for initial detection of cancers mainly for two reasons. First, in some cancers of somatic tissues only a fraction of the cases display global hypomethylation and in the absence of disease the overall level of DNA methylation varies significantly between individuals and tissues. Second, the technical analysis of hypomethylation status is more difficult since the changes are quantitative rather than qualitative [83]. Some improvements in detection methods, based on using bisulfite treatment of DNA and simultaneous PCR, have allowed for an evaluation of the decreases in DNA methylation of multiple repetitive elements, such as Alu elements and long interspersed nucleotide elements (LINE) with greater efficiency and less labor [84]. This method can be used as a surrogate marker of genome-wide methylation changes and are clinically useful in evaluations of hypomethylation events that have been shown in some carcinomas, such as the LINE demethylation in both precancerous lesions and adenocarcinoma stages in colorectal cancer [85] and the hypomethylation of SATR-1 satellite sequence in breast cancer [86].

In contrast to hypomethylation, detection of cancers by hypermethylation analysis is currently being developed for routine use, particularly in cancers with consistent and highly prevalent hypermethylation of specific genes [79, 80]. Since most CpG islands are completely unmethylated in normal cells, hypermethylation may be a highly sensitive and specific marker, although for its application in the clinic it is necessary for one or more genes to be consistently hypermethylated in the cancer to be detected [80]. For example, in prostate cancer, detecting methylation in only one gene (*GSTP1*) is informative in 80-90% of the cases [87]; and the determination of aberrant methylation in gene groups can be informative in 100 % of cases [66].

Numerous studies have shown the aberrant hypermethylation patterns associated with different cancer types and thus more knowledge has been gained about its role in carcinogenesis. In order to translate laboratory findings into clinical application, there is a need for aberrant methylation to be easily and specifically detected in cancer cells from body fluids, such as urine, sputum, bronchoalveolar lavage (BAL), mammary aspiration fluids, saliva and stools [82]. Many reports describe detection of methylation in cancer-derived DNA using plasma/serum in lung, head and neck, esophageal, breast, liver, colorectal, stomach, prostate and bladder cancers and melanomas [88]. For instance, methylation of

MGMT, *GSTP1*, *RARb2* and *DAPK1* genes have been analyzed from different body fluids for detecting lung cancers [89], prostate cancer [87, 89-91], breast cancer [92, 93] and bladder cancer [94, 95]. Most of these studies report concordance between methylation analysis and conventional cytology.

Methylation as a Prognostic Indicator

The importance of DNA methylation as a prognostic biomarker in cancer comes from the close relationship between this event and the development and progression of carcinogenesis. It has been reported that the methylation status of individual genes or patterns of multiple genes can be associated with disease prognoses and responses to therapeutics. This would help in modifying initial treatment options, monitoring patient response to therapy and predicting survival [82].

In certain cancers, it has been observed that the hypomethylation levels of repeat sequences correlated with cancer progression. This provides clinically useful information for predicting disease outcome. In several cancers, strong hypomethylation is significantly more prevalent in advanced stage tumors, increasing from non-neoplastic tissue toward the carcinoma stage and showing correlation with a poor prognosis. The relation between certain hypomethylation events and metastasis could be particularly applicable to evaluating the presence of micrometastases in cancer patients and therefore help to choose more adequate therapies [39, 83]. In ovarian carcinogenesis, strong satellite DNA hypomethylation is an independent marker of poor prognosis with regard to overall survival as well as relapse-free survival [96]. In breast cancer, methylation of the uPA gene promoter (a protease only expressed in highly invasive human breast cancer cells) avoids the invasiveness of metastatic breast cancer cells. Demethylation levels of the uPA gene can be metastatic markers and indicate the need for alternative therapies for those patients who present a high expression of this gene [97].

Hypomethylation analysis could also be particularly suited to the detection of molecular therapy targets. For example, detection of hypomethylated MAGE genes could be useful prior to immunotherapy of disseminated melanomas [98]. Hypomethylation of some genes, such as the multidrug resistance 1 gene (MDR1), which encodes the cell surface P-glycoprotein that transports cytotoxic drugs out of the cell, could be assayed to determine the response to conventional chemotherapy in acute myeloid leukemia patients [99]. In addition, PCR-based assays have been used to monitor levels of hypomethylation in patients undergoing treatment with 5-Aza-2′-deoxycytidine and appear to provide a useful predictor of clinical response [100].

Again, the CpG island hypermethylation event has had greater impact as a biomarker with significant prognostic and predictive value. Hypermethylation events that lead to loss of function of genes involved in either cell cycle and apoptosis control or intercellular interactions are a reflection of both the growth advantage and malignant potential of cancer cells. The downregulation of certain genes is a recognized indicator of tumor behavior. For instance, hypermethylation of the E-cadherin gene (CDH1) has been significantly associated with a worse prognosis in several human cancers. In node-positive diffuse gastric cancer,

CDH1 hypermethylation also retains its prognostic role for disease-free survival [101]. Similar findings were reported for oral tongue carcinoma [102]. The combined use of more than one gene improves the prognosis value of hypermethylation event in some cases. Detection of aberrant methylation of CDH1/CDH13 (H-Cadherin) is a useful marker for selecting patients with cervical cancer at high risk for relapse [103]. Also, APC and E-cadherin methylation analysis in the serum of patients with gastric cancer allows for the identification of a group of patients with poor prognosis [104]. In some well documented cases, methylation-dependent inactivation of many other tumor suppressor genes maintains an important and strong relationship with the development of the disease and its clinical outcome. Methylation of DAPK, a positive mediator of apoptosis, is associated with early recurrence in bladder cancers [105] and with the presence of nodal metastasis, advance stages of the disease and a poorer event-free survival in gastric cancer [106]. Methylation of RASSF1A and/or APC in serum DNA is associated with poor prognosis in breast cancer cases [107].

Hypermethylation events occur in different classes of genes, including genes affecting the response to therapy. Two markers of clinical response are widely discussed in the literature: hypermethylation of both the O-6-methylguanine-DNA methyltransferase (MGMT) gene and the mismatch repair hMLH1 gene [108].

Loss of MGMT expression in the early stages of neoplastic transformation impairs cellular defense against carcinogenic DNA alkylating agents and, on the other hand, its methylation in cancer determines response to alkylating drugs [109]. Its predictive value has been confirmed in the clinical response of gliomas to alkylating agents [110]. In a prospective clinical trial, the predictive value of the methylation status of MGMT promoter for outcome was also determined in glioblastoma patients treated with the alkylating agent temozolomide, demonstrating that the epigenetic inactivation of the DNA repair gene MGMT is associated with a better outcome [111]. These findings may have important implications for the design of future trials and support efforts to deplete MGMT by O-6-benzylguanine, a non-cytotoxic substrate of this enzyme. However, further analyses are necessary to confirm the relationship between MGMT promoter hypermethylation and the response to alkylating agents because in another study no such correlation was observed [112].

The loss of function of mismatch repair (MMR) plays a role in the resistance to a number of clinically important anti-cancer drugs. Inactivation of this gene often occurs due to hMLH1 promoter hypermethylation and it is therefore to be expected that demethylation of hMLH1 can revert the resistance to chemotherapeutical drugs. In an experimental trial, inactivation of the hMLH1 gene by a DNA methyltransferase inhibitor reversed the resistance to cisplatin, carboplatin, temozolomide and epirubicin in ovarian and colon xenografts in bearing mice [113]. Similarly, it was demonstrated that *in vitro* resistance to the chemotherapeutic reagent 5-fluorouracil (5-FU) can be overcome by re-expression of the hMLH1 protein through demethylation induced by demethylating agent 5-Aza-2'-deoxycytidine (5 aza-dC), which could have a significant impact on the development of future chemotherapy strategies [114].

DNA Methylation and Cancer Therapeutics

A third clinical application of DNA methylation, in addition to its use as a diagnostic and prognostic biomarker, is the possible therapeutic benefit of reverting the methylator phenotype of the epigenetically silenced cancer genes and restoring their cell function.

Ten years ago it was shown that a deoxy version of 5-azacytidine (5-aza-2-deoxycytidine) is able to trap the DNA methyltransferase enzyme in a covalent complex with DNA, which results in a loss of DNA methylation with each round of cell division [115]. The exposure of human tumor cell lines to this demethylation agent showed that the activities of silencing genes (*p16*) can be restored, decreasing the rate of tumor growth in a heritable fashion [116].

Several inhibitors of DNMTs are known to modify epigenetic information, such as azacitidine, decitadine, MGI and zebularine [117]. Pre-clinical studies have shown that decitabine reverses methylation in a number of cell lines and in cells from human leukemia patients [118, 119]. However, these drugs have had the most significant impact and have been most successful on the treatment of hematologic malignancies, such as the myelodysplastic syndromes (MDSs) [120-122].

One relevant disadvantage of the DNMT inhibitors is their lack of specificity which does not allow them to be used to target particular genes. Furthermore, these agents seem to have toxic effects on normal cells, limiting their use for the activation of cancer genes silenced by methylation, although these effects could be overcome with low-dose treatment [123]. However, the novel findings on the functions of DNMT1 would to be considered at the time of to evaluate the use these drugs even if they are less toxic than 5-aza-deoxycytidine. The suggested relationship between hypomethylation and metastasis can affect the hoped-for clinical outcome after demethylation therapy. The use of 5-aza-2'-deoxycytidine may stimulate the invasive behavior of pancreatic cancer cells by reactivating invasion-promoting genes, in accordance with an in vitro assay using pancreatic cancer cell lines [124]. Another concern is that drug-induced hypomethylation could disturb the function of the immune system and induce autoimmunity because several immune cell specificities are maintained by DNA methylation [83].

Since methylation-based gene silencing occurs through a combined effect of DNMT, methyl-binding proteins, histone-modifying enzymes (deacetylases and methylases) and other chromatin remodeling factors, histone deacetylase (HDAC) inhibitors are also being tested as potential chemotherapeutic agents [125]. Although HDAC inhibitors fail to activate expression from densely methylated genes when used alone, they can synergize with a demethylating agent into inducing expression from methylated genes [126]. The combined treatment of DNA methylation inhibitors with histone deacetylase inhibitors and differentiating factors has achieved moderate success in several patients with acute promyelocytic leukemia [127].

Further laboratory studies and clinical trials will establish the therapeutic efficacy of those drugs in reverting the epigenetic changes present in cancer cells. Recent reviews have given detailed information about the advances in this field of research and the therapeutic impact of both DNA methylation and HDAC inhibitors [81, 88, 117, 128].

Conclusion

DNA methylation is one of the epigenetic mechanisms involved in gene expression regulation in mammals; it is necessary for embryonic development and plays a critical role in specific-gene silencing during cellular differentiation. This mechanism also participates in genome defense by repressing mobile elements, such as endogenous viral sequences and retrotransposons. Furthermore, DNA methylation is important for stability maintenance of our genome, X-chromosome inactivation in females, genomic imprinting, among other functions.

Alterations in DNA methylation patterns are associated with various pathologies, mainly those that implicate changes in cellular transformation processes. The experimental evidence has shown the importance of epigenetic mechanisms on transcriptional regulation. In the last 10 years, our knowledge about how DNA methylation modulates transcriptional repression, chromatin structure, genome stability and neoplastic process has advanced greatly. However, many aspects remain to be clarified in order to understand exactly the specific functions and targets of methylation machinery that create and maintain the specific-gene methylation patterns including DNA methyltransferases, DNA binding proteins, histone-modifying enzymes and other chromatin-remodeling factors. Recent findings seem to confirm that 5-methylcytosines are indispensable in maintaining transcriptional repression rather than in initiating it. Furthermore, since epigenetic mechanisms may be influenced by environmental factors, such as diet or potential carcinogens, it has been suggested that epigenetic changes are used by organisms as a response to external factors.

In human cancer cells, an imbalance in DNA methylation occurs in comparison with normal cells, which results in genomic hypomethylation and CpG island hypermethylation. Both events are relevant and they have been observed in several cancer types, facilitating the development and progression of tumors. These features have provided novel and attractive uses of DNA methylation in the field of cancer research.

Aberrant DNA methylation may serve as a potential marker for neoplastic disease due to the ubiquity of DNA methylation changes in human cancers. Early detection, prediction of tumor behavior, prediction of chemotherapy/hormonotherapy response, identification of therapeutic targets, risk prediction, selection of combined therapies, among others, are all promising areas of epigenetics that are already being considered in clinical trials. However, further optimization of methylation biomarkers needs the development of high-throughput, low-cost analytical procedures. If DNA methylation assays become a clinical reality, they will consist of a panel or array including both hypermethylation and hypomethylation markers.

Novel anti-cancer therapies based on epigenetic information are being intensively evaluated and they promise to contribute to cancer control. A high efficacy of demethylating agents has been reported mainly in hematological malignancies, but their use on solid tumors is now being tested. Since the epigenetic pathways are interconnected, the anti-cancer therapy may be more effective using combined strategies that include the use of epigenetic drugs (demethylating agents and histone deacetylase inhibitors) and conventional therapies (chemotherapy). This combined approach could be more effective than the use of individual drugs and, in addition, the side effects of each agent may be reduced if low doses are applied.

References

[1] Russo VE, Martienssen RA, Riggs A. Epigenetic mechanisms of gene regulation. Cold Spring Harbor, NY: Cold Spring Harbor Laboratory Press 1996.

[2] Feinberg AP, Tycko B. The history of cancer epigenetics. *Nature reviews.* 2004 Feb;4(2):143-53.

[3] Bird AP. CpG-rich islands and the function of DNA methylation. *Nature.* 1986 May 15-21;321(6067):209-13.

[4] Esteller M, Fraga MF, Guo M, Garcia-Foncillas J, Hedenfalk I, Godwin AK, et al. DNA methylation patterns in hereditary human cancers mimic sporadic tumorigenesis. *Hum. Mol. Genet.* 2001 Dec 15;10(26):3001-7.

[5] Antequera F, Bird A. Number of CpG islands and genes in human and mouse. Proceedings of the National Academy of Sciences of the United States of America. 1993 Dec 15;90(24):11995-9.

[6] Serman A, Vlahovic M, Serman L, Bulic-Jakus F. DNA methylation as a regulatory mechanism for gene expression in mammals. *Collegium antropologicum.* 2006 Sep;30(3):665-71.

[7] Siedlecki P, Zielenkiewicz P. Mammalian DNA methyltransferases. *Acta biochimica Polonica.* 2006;53(2):245-56.

[8] Bestor T, Laudano A, Mattaliano R, Ingram V. Cloning and sequencing of a cDNA encoding DNA methyltransferase of mouse cells. The carboxyl-terminal domain of the mammalian enzymes is related to bacterial restriction methyltransferases. *J. Mol. Biol.* 1988 Oct 20;203(4):971-83.

[9] Pradhan S, Bacolla A, Wells RD, Roberts RJ. Recombinant human DNA (cytosine-5) methyltransferase. I. Expression, purification, and comparison of de novo and maintenance methylation. *J. Biol. Chem.* 1999 Nov 12;274(46):33002-10.

[10] Li E, Beard C, Jaenisch R. Role for DNA methylation in genomic imprinting. *Nature.* 1993 Nov 25;366(6453):362-5.

[11] Okano M, Xie S, Li E. Cloning and characterization of a family of novel mammalian DNA (cytosine-5) methyltransferases. *Nature genetics.* 1998 Jul;19(3):219-20.

[12] Okano M, Bell DW, Haber DA, Li E. DNA methyltransferases Dnmt3a and Dnmt3b are essential for de novo methylation and mammalian development. *Cell.* 1999 Oct 29;99(3):247-57.

[13] Chen T, Ueda Y, Dodge JE, Wang Z, Li E. Establishment and maintenance of genomic methylation patterns in mouse embryonic stem cells by Dnmt3a and Dnmt3b. *Mol. Cell Biol.* 2003 Aug;23(16):5594-605.

[14] Kim GD, Ni J, Kelesoglu N, Roberts RJ, Pradhan S. Co-operation and communication between the human maintenance and de novo DNA (cytosine-5) methyltransferases. *The EMBO journal.* 2002 Aug 1;21(15):4183-95.

[15] Rhee I, Bachman KE, Park BH, Jair KW, Yen RW, Schuebel KE, et al. DNMT1 and DNMT3b cooperate to silence genes in human cancer cells. *Nature.* 2002 Apr 4;416(6880):552-6.

[16] Rhee I, Jair KW, Yen RW, Lengauer C, Herman JG, Kinzler KW, et al. CpG methylation is maintained in human cancer cells lacking DNMT1. *Nature*. 2000 Apr 27;404(6781):1003-7.

[17] Razin A, Riggs AD. DNA methylation and gene function. *Science*. 1980 Nov 7;210(4470):604-10.

[18] Bird AP, Wolffe AP. Methylation-induced repression--belts, braces, and chromatin. *Cell*. 1999 Nov 24;99(5):451-4.

[19] Tyler JK, Kadonaga JT. The "dark side" of chromatin remodeling: repressive effects on transcription. *Cell*. 1999 Nov 24;99(5):443-6.

[20] Tate PH, Bird AP. Effects of DNA methylation on DNA-binding proteins and gene expression. *Curr. Opin. Genet. Dev.* 1993 Apr;3(2):226-31.

[21] Lewis JD, Meehan RR, Henzel WJ, Maurer-Fogy I, Jeppesen P, Klein F, et al. Purification, sequence, and cellular localization of a novel chromosomal protein that binds to methylated DNA. *Cell*. 1992 Jun 12;69(6):905-14.

[22] Meehan RR, Lewis JD, Bird AP. Characterization of MeCP2, a vertebrate DNA binding protein with affinity for methylated DNA. Nucleic acids research. 1992 Oct 11;20(19):5085-92.

[23] Hendrich B, Bird A. Identification and characterization of a family of mammalian methyl-CpG binding proteins. *Mol. Cell Biol.* 1998 Nov;18(11):6538-47.

[24] Ballestar E, Paz MF, Valle L, Wei S, Fraga MF, Espada J, et al. Methyl-CpG binding proteins identify novel sites of epigenetic inactivation in human cancer. *The EMBO journal*. 2003 Dec 1;22(23):6335-45.

[25] Jenuwein T, Allis CD. Translating the histone code. *Science*. 2001 Aug 10;293(5532):1074-80.

[26] Peterson CL, Laniel MA. Histones and histone modifications. *Curr. Biol.* 2004 Jul 27;14(14):R546-51.

[27] Turner BM. Reading signals on the nucleosome with a new nomenclature for modified histones. *Nat. Struct. Mol. Biol.* 2005 Feb;12(2):110-2.

[28] Fuks F, Burgers WA, Brehm A, Hughes-Davies L, Kouzarides T. DNA methyltransferase Dnmt1 associates with histone deacetylase activity. *Nature genetics*. 2000 Jan;24(1):88-91.

[29] Rountree MR, Bachman KE, Baylin SB. DNMT1 binds HDAC2 and a new co-repressor, DMAP1, to form a complex at replication foci. *Nature genetics*. 2000 Jul;25(3):269-77.

[30] Fuks F. DNA methylation and histone modifications: teaming up to silence genes. *Curr. Opin. Genet. Dev.* 2005 Oct;15(5):490-5.

[31] Robertson KD, Wolffe AP. DNA methylation in health and disease. *Nat. Rev. Genet.* 2000 Oct;1(1):11-9.

[32] Feinberg AP, Vogelstein B. Hypomethylation distinguishes genes of some human cancers from their normal counterparts. *Nature*. 1983 Jan 6;301(5895):89-92.

[33] Jang H, Mason JB, Choi SW. Genetic and epigenetic interactions between folate and aging in carcinogenesis. *The Journal of nutrition*. 2005 Dec;135(12 Suppl):2967S-71S.

[34] Robertson KD. DNA methylation and human disease. *Nat. Rev. Genet.* 2005 Aug;6(8):597-610.

[35] Esteller M, Herman JG. Cancer as an epigenetic disease: DNA methylation and chromatin alterations in human tumors. *The Journal of pathology.* 2002 Jan;196(1):1-7.
[36] Gama-Sosa MA, Slagel VA, Trewyn RW, Oxenhandler R, Kuo KC, Gehrke CW, et al. The 5-methylcytosine content of DNA from human tumors. *Nucleic acids research.* 1983 Oct 11;11(19):6883-94.
[37] Jones PA, Laird PW. Cancer epigenetics comes of age. *Nature genetics.* 1999 Feb;21(2):163-7.
[38] Plass C. Cancer epigenomics. *Hum. Mol. Genet.* 2002 Oct 1;11(20):2479-88.
[39] Hoffmann MJ, Schulz WA. Causes and consequences of DNA hypomethylation in human cancer. *Biochemistry and cell biology = Biochimie et biologie cellulaire.* 2005 Jun;83(3):296-321.
[40] Narayan P, Starling J. Minimally invasive therapies for the treatment of symptomatic benign prostatic hyperplasia: the University of Florida experience. *Journal of clinical laser medicine and surgery.* 1998 Feb;16(1):29-32.
[41] Hansen RS, Wijmenga C, Luo P, Stanek AM, Canfield TK, Weemaes CM, et al. The DNMT3B DNA methyltransferase gene is mutated in the ICF immunodeficiency syndrome. Proceedings of the National Academy of Sciences of the United States of America. 1999 Dec 7;96(25):14412-7.
[42] Xu GL, Bestor TH, Bourc'his D, Hsieh CL, Tommerup N, Bugge M, et al. Chromosome instability and immunodeficiency syndrome caused by mutations in a DNA methyltransferase gene. *Nature.* 1999 Nov 11;402(6758):187-91.
[43] Lander ES, Linton LM, Birren B, Nusbaum C, Zody MC, Baldwin J, et al. Initial sequencing and analysis of the human genome. *Nature.* 2001 Feb 15;409(6822):860-921.
[44] Belgnaoui SM, Gosden RG, Semmes OJ, Haoudi A. Human LINE-1 retrotransposon induces DNA damage and apoptosis in cancer cells. *Cancer Cell Int.* 2006;6:13.
[45] Chalitchagorn K, Shuangshoti S, Hourpai N, Kongruttanachok N, Tangkijvanich P, Thong-ngam D, et al. Distinctive pattern of LINE-1 methylation level in normal tissues and the association with carcinogenesis. *Oncogene.* 2004 Nov 18;23(54):8841-6.
[46] Roman-Gomez J, Jimenez-Velasco A, Agirre X, Cervantes F, Sanchez J, Garate L, et al. Promoter hypomethylation of the LINE-1 retrotransposable elements activates sense/antisense transcription and marks the progression of chronic myeloid leukemia. *Oncogene.* 2005 Nov 3;24(48):7213-23.
[47] Schulz WA, Steinhoff C, Florl AR. Methylation of endogenous human retroelements in health and disease. *Curr. Top Microbiol. Immunol.* 2006;310:211-50.
[48] Costello JF, Plass C. Methylation matters. *Journal of medical genetics.* 2001 May;38(5):285-303.
[49] Ehrlich M. DNA methylation in cancer: too much, but also too little. *Oncogene.* 2002 Aug 12;21(35):5400-13.
[50] Schulz WA. DNA methylation in urological malignancies (review). *International journal of oncology.* 1998 Jul;13(1):151-67.
[51] Kaneda A, Tsukamoto T, Takamura-Enya T, Watanabe N, Kaminishi M, Sugimura T, et al. Frequent hypomethylation in multiple promoter CpG islands is associated with

global hypomethylation, but not with frequent promoter hypermethylation. *Cancer Sci.* 2004 Jan;95(1):58-64.

[52] De Smet C, Lurquin C, Lethe B, Martelange V, Boon T. DNA methylation is the primary silencing mechanism for a set of germ line- and tumor-specific genes with a CpG-rich promoter. *Mol. Cell Biol.* 1999 Nov;19(11):7327-35.

[53] Honda T, Tamura G, Waki T, Kawata S, Terashima M, Nishizuka S, et al. Demethylation of MAGE promoters during gastric cancer progression. *British journal of cancer.* 2004 Feb 23;90(4):838-43.

[54] Jang SJ, Soria JC, Wang L, Hassan KA, Morice RC, Walsh GL, et al. Activation of melanoma antigen tumor antigens occurs early in lung carcinogenesis. *Cancer research.* 2001 Nov 1;61(21):7959-63.

[55] Sood AK, Fletcher MS, Gruman LM, Coffin JE, Jabbari S, Khalkhali-Ellis Z, et al. The paradoxical expression of maspin in ovarian carcinoma. *Clin. Cancer Res.* 2002 Sep;8(9):2924-32.

[56] Akiyama Y, Maesawa C, Ogasawara S, Terashima M, Masuda T. Cell-type-specific repression of the maspin gene is disrupted frequently by demethylation at the promoter region in gastric intestinal metaplasia and cancer cells. *Am. J. Pathol.* 2003 Nov;163(5):1911-9.

[57] Sato N, Fukushima N, Matsubayashi H, Goggins M. Identification of maspin and S100P as novel hypomethylation targets in pancreatic cancer using global gene expression profiling. *Oncogene.* 2004 Feb 26;23(8):1531-8.

[58] Ogasawara S, Maesawa C, Yamamoto M, Akiyama Y, Wada K, Fujisawa K, et al. Disruption of cell-type-specific methylation at the Maspin gene promoter is frequently involved in undifferentiated thyroid cancers. *Oncogene.* 2004 Feb 5;23(5):1117-24.

[59] Gupta A, Godwin AK, Vanderveer L, Lu A, Liu J. Hypomethylation of the synuclein gamma gene CpG island promotes its aberrant expression in breast carcinoma and ovarian carcinoma. *Cancer research.* 2003 Feb 1;63(3):664-73.

[60] Nakamura N, Takenaga K. Hypomethylation of the metastasis-associated S100A4 gene correlates with gene activation in human colon adenocarcinoma cell lines. *Clinical and experimental metastasis.* 1998 Jul;16(5):471-9.

[61] Greger V, Passarge E, Hopping W, Messmer E, Horsthemke B. Epigenetic changes may contribute to the formation and spontaneous regression of retinoblastoma. *Human genetics.* 1989 Sep;83(2):155-8.

[62] Frommer M, McDonald LE, Millar DS, Collis CM, Watt F, Grigg GW, et al. A genomic sequencing protocol that yields a positive display of 5-methylcytosine residues in individual DNA strands. Proceedings of the National Academy of Sciences of the United States of America. 1992 Mar 1;89(5):1827-31.

[63] Herman JG, Graff JR, Myohanen S, Nelkin BD, Baylin SB. Methylation-specific PCR: a novel PCR assay for methylation status of CpG islands. Proceedings of the National Academy of Sciences of the United States of America. 1996 Sep 3;93(18):9821-6.

[64] Issa JP. CpG island methylator phenotype in cancer. *Nature reviews.* 2004 Dec;4(12):988-93.

[65] Paz MF, Fraga MF, Avila S, Guo M, Pollan M, Herman JG, et al. A systematic profile of DNA methylation in human cancer cell lines. *Cancer research.* 2003 Mar 1;63(5):1114-21.

[66] Esteller M, Corn PG, Baylin SB, Herman JG. A gene hypermethylation profile of human cancer. *Cancer research.* 2001 Apr 15;61(8):3225-9.

[67] Di Croce L, Raker VA, Corsaro M, Fazi F, Fanelli M, Faretta M, et al. Methyltransferase recruitment and DNA hypermethylation of target promoters by an oncogenic transcription factor. *Science.* 2002 Feb 8;295(5557):1079-82.

[68] Esteller M. Aberrant DNA methylation as a cancer-inducing mechanism. *Annual review of pharmacology and toxicology.* 2005;45:629-56.

[69] Esteller M, Cordon-Cardo C, Corn PG, Meltzer SJ, Pohar KS, Watkins DN, et al. p14ARF silencing by promoter hypermethylation mediates abnormal intracellular localization of MDM2. *Cancer research.* 2001 Apr 1;61(7):2816-21.

[70] Esteller M, Sparks A, Toyota M, Sanchez-Cespedes M, Capella G, Peinado MA, et al. Analysis of adenomatous polyposis coli promoter hypermethylation in human cancer. *Cancer research.* 2000 Aug 15;60(16):4366-71.

[71] Graff JR, Herman JG, Lapidus RG, Chopra H, Xu R, Jarrard DF, et al. E-cadherin expression is silenced by DNA hypermethylation in human breast and prostate carcinomas. *Cancer research.* 1995 Nov 15;55(22):5195-9.

[72] Wang L, Zhang F, Wu PP, Jiang XC, Zheng L, Yu YY. Disordered beta-catenin expression and E-cadherin/CDH1 promoter methylation in gastric carcinoma. *World J. Gastroenterol.* 2006 Jul 14;12(26):4228-31.

[73] Vaish M, Mittal B. DNA mismatch repair, microsatellite instability and cancer. *Indian J. Exp. Biol.* 2002 Sep;40(9):989-94.

[74] Wheeler JM. Epigenetics, mismatch repair genes and colorectal cancer. *Ann. R. Coll. Surg. Engl.* 2005 Jan;87(1):15-20.

[75] Esteller M, Risques RA, Toyota M, Capella G, Moreno V, Peinado MA, et al. Promoter hypermethylation of the DNA repair gene O(6)-methylguanine-DNA methyltransferase is associated with the presence of G:C to A:T transition mutations in p53 in human colorectal tumorigenesis. *Cancer research.* 2001 Jun 15;61(12):4689-92.

[76] Esteller M, Toyota M, Sanchez-Cespedes M, Capella G, Peinado MA, Watkins DN, et al. Inactivation of the DNA repair gene O6-methylguanine-DNA methyltransferase by promoter hypermethylation is associated with G to A mutations in K-ras in colorectal tumorigenesis. *Cancer research.* 2000 May 1;60(9):2368-71.

[77] Liu L, Gerson SL. Targeted modulation of MGMT: clinical implications. *Clin. Cancer Res.* 2006 Jan 15;12(2):328-31.

[78] Sabharwal A, Middleton MR. Exploiting the role of O6-methylguanine-DNA-methyltransferase (MGMT) in cancer therapy. *Curr. Opin. Pharmacol.* 2006 Aug;6(4):355-63.

[79] Laird PW. The power and the promise of DNA methylation markers. *Nature reviews.* 2003 Apr;3(4):253-66.

[80] Esteller M. Relevance of DNA methylation in the management of cancer. *The lancet oncology.* 2003 Jun;4(6):351-8.

[81] Esteller M. DNA methylation and cancer therapy: new developments and expectations. *Current opinion in oncology.* 2005 Jan;17(1):55-60.

[82] Das PM, Singal R. DNA methylation and cancer. *J. Clin. Oncol.* 2004 Nov 15;22(22):4632-42.

[83] Wilson AS, Power BE, Molloy PL. DNA hypomethylation and human diseases. *Biochimica et biophysica acta.* 2007 Jan;1775(1):138-62.

[84] Yang AS, Estecio MR, Doshi K, Kondo Y, Tajara EH, Issa JP. A simple method for estimating global DNA methylation using bisulfite PCR of repetitive DNA elements. *Nucleic acids research.* 2004;32(3):e38.

[85] Feinberg AP, Gehrke CW, Kuo KC, Ehrlich M. Reduced genomic 5-methylcytosine content in human colonic neoplasia. *Cancer research.* 1988 Mar 1;48(5):1159-61.

[86] Costa FF, Paixao VA, Cavalher FP, Ribeiro KB, Cunha IW, Rinck JA, Jr., et al. SATR-1 hypomethylation is a common and early event in breast cancer. *Cancer genetics and cytogenetics.* 2006 Mar;165(2):135-43.

[87] Jeronimo C, Usadel H, Henrique R, Oliveira J, Lopes C, Nelson WG, et al. Quantitation of GSTP1 methylation in non-neoplastic prostatic tissue and organ-confined prostate adenocarcinoma. *Journal of the National Cancer Institute.* 2001 Nov 21;93(22):1747-52.

[88] Miyamoto K, Ushijima T. Diagnostic and therapeutic applications of epigenetics. *Japanese journal of clinical oncology.* 2005 Jun;35(6):293-301.

[89] Topaloglu O, Hoque MO, Tokumaru Y, Lee J, Ratovitski E, Sidransky D, et al. Detection of promoter hypermethylation of multiple genes in the tumor and bronchoalveolar lavage of patients with lung cancer. *Clin. Cancer Res.* 2004 Apr 1;10(7):2284-8.

[90] Cairns P, Esteller M, Herman JG, Schoenberg M, Jeronimo C, Sanchez-Cespedes M, et al. Molecular detection of prostate cancer in urine by GSTP1 hypermethylation. *Clin. Cancer Res.* 2001 Sep;7(9):2727-30.

[91] Goessl C, Krause H, Muller M, Heicappell R, Schrader M, Sachsinger J, et al. Fluorescent methylation-specific polymerase chain reaction for DNA-based detection of prostate cancer in bodily fluids. *Cancer research.* 2000 Nov 1;60(21):5941-5.

[92] Evron E, Dooley WC, Umbricht CB, Rosenthal D, Sacchi N, Gabrielson E, et al. Detection of breast cancer cells in ductal lavage fluid by methylation-specific PCR. *Lancet.* 2001 Apr 28;357(9265):1335-6.

[93] Krassenstein R, Sauter E, Dulaimi E, Battagli C, Ehya H, Klein-Szanto A, et al. Detection of breast cancer in nipple aspirate fluid by CpG island hypermethylation. *Clin. Cancer Res.* 2004 Jan 1;10(1 Pt 1):28-32.

[94] Chan MW, Chan LW, Tang NL, Tong JH, Lo KW, Lee TL, et al. Hypermethylation of multiple genes in tumor tissues and voided urine in urinary bladder cancer patients. *Clin. Cancer Res.* 2002 Feb;8(2):464-70.

[95] Dulaimi E, Uzzo RG, Greenberg RE, Al-Saleem T, Cairns P. Detection of bladder cancer in urine by a tumor suppressor gene hypermethylation panel. *Clin. Cancer Res.* 2004 Mar 15;10(6):1887-93.

[96] Widschwendter M, Jiang G, Woods C, Muller HM, Fiegl H, Goebel G, et al. DNA hypomethylation and ovarian cancer biology. *Cancer research.* 2004 Jul 1;64(13):4472-80.

[97] Pakneshan P, Tetu B, Rabbani SA. Demethylation of urokinase promoter as a prognostic marker in patients with breast carcinoma. *Clin. Cancer Res.* 2004 May 1;10(9):3035-41.

[98] Sigalotti L, Fratta E, Coral S, Tanzarella S, Danielli R, Colizzi F, et al. Intratumor heterogeneity of cancer/testis antigens expression in human cutaneous melanoma is methylation-regulated and functionally reverted by 5-aza-2'-deoxycytidine. *Cancer research.* 2004 Dec 15;64(24):9167-71.

[99] Nakayama M, Wada M, Harada T, Nagayama J, Kusaba H, Ohshima K, et al. Hypomethylation status of CpG sites at the promoter region and overexpression of the human MDR1 gene in acute myeloid leukemias. *Blood.* 1998 Dec 1;92(11):4296-307.

[100] Yang AS, Doshi KD, Choi SW, Mason JB, Mannari RK, Gharybian V, et al. DNA methylation changes after 5-aza-2'-deoxycytidine therapy in patients with leukemia. *Cancer research.* 2006 May 15;66(10):5495-503.

[101] Graziano F, Arduini F, Ruzzo A, Bearzi I, Humar B, More H, et al. Prognostic analysis of E-cadherin gene promoter hypermethylation in patients with surgically resected, node-positive, diffuse gastric cancer. *Clin. Cancer Res.* 2004 Apr 15;10(8):2784-9.

[102] Chang HW, Chow V, Lam KY, Wei WI, Yuen A. Loss of E-cadherin expression resulting from promoter hypermethylation in oral tongue carcinoma and its prognostic significance. *Cancer.* 2002 Jan 15;94(2):386-92.

[103] Widschwendter A, Ivarsson L, Blassnig A, Muller HM, Fiegl H, Wiedemair A, et al. CDH1 and CDH13 methylation in serum is an independent prognostic marker in cervical cancer patients. *International journal of cancer.* 2004 Mar 20;109(2):163-6.

[104] Leung WK, To KF, Chu ES, Chan MW, Bai AH, Ng EK, et al. Potential diagnostic and prognostic values of detecting promoter hypermethylation in the serum of patients with gastric cancer. *British journal of cancer.* 2005 Jun 20;92(12):2190-4.

[105] Tada Y, Wada M, Taguchi K, Mochida Y, Kinugawa N, Tsuneyoshi M, et al. The association of death-associated protein kinase hypermethylation with early recurrence in superficial bladder cancers. *Cancer research.* 2002 Jul 15;62(14):4048-53.

[106] Chan AW, Chan MW, Lee TL, Ng EK, Leung WK, Lau JY, et al. Promoter hypermethylation of Death-associated protein-kinase gene associated with advance stage gastric cancer. *Oncology reports.* 2005 May;13(5):937-41.

[107] Muller HM, Widschwendter A, Fiegl H, Ivarsson L, Goebel G, Perkmann E, et al. DNA methylation in serum of breast cancer patients: an independent prognostic marker. *Cancer research.* 2003 Nov 15;63(22):7641-5.

[108] Paluszczak J, Baer-Dubowska W. Epigenetic diagnostics of cancer--the application of DNA methylation markers. *Journal of applied genetics.* 2006;47(4):365-75.

[109] Esteller M, Herman JG. Generating mutations but providing chemosensitivity: the role of O6-methylguanine DNA methyltransferase in human cancer. *Oncogene.* 2004 Jan 8;23(1):1-8.

[110] Esteller M, Garcia-Foncillas J, Andion E, Goodman SN, Hidalgo OF, Vanaclocha V, et al. Inactivation of the DNA-repair gene MGMT and the clinical response of gliomas to alkylating agents. *The New England journal of medicine.* 2000 Nov 9;343(19):1350-4.

[111] Hegi ME, Diserens AC, Godard S, Dietrich PY, Regli L, Ostermann S, et al. Clinical trial substantiates the predictive value of O-6-methylguanine-DNA methyltransferase promoter methylation in glioblastoma patients treated with temozolomide. *Clin. Cancer Res.* 2004 Mar 15;10(6):1871-4.

[112] Blanc JL, Wager M, Guilhot J, Kusy S, Bataille B, Chantereau T, et al. Correlation of clinical features and methylation status of MGMT gene promoter in glioblastomas. *Journal of neuro-oncology.* 2004 Jul;68(3):275-83.

[113] Plumb JA, Strathdee G, Sludden J, Kaye SB, Brown R. Reversal of drug resistance in human tumor xenografts by 2'-deoxy-5-azacytidine-induced demethylation of the hMLH1 gene promoter. *Cancer research.* 2000 Nov 1;60(21):6039-44.

[114] Arnold CN, Goel A, Boland CR. Role of hMLH1 promoter hypermethylation in drug resistance to 5-fluorouracil in colorectal cancer cell lines. *International journal of cancer.* 2003 Aug 10;106(1):66-73.

[115] Juttermann R, Li E, Jaenisch R. Toxicity of 5-aza-2'-deoxycytidine to mammalian cells is mediated primarily by covalent trapping of DNA methyltransferase rather than DNA demethylation. Proceedings of the National Academy of Sciences of the United States of America. 1994 Dec 6;91(25):11797-801.

[116] Bender CM, Pao MM, Jones PA. Inhibition of DNA methylation by 5-aza-2'-deoxycytidine suppresses the growth of human tumor cell lines. *Cancer research.* 1998 Jan 1;58(1):95-101.

[117] Stebbing J, Bower M, Syed N, Smith P, Yu V, Crook T. Epigenetics: an emerging technology in the diagnosis and treatment of cancer. *Pharmacogenomics.* 2006 Jul;7(5):747-57.

[118] Momparler RL, Bouchard J, Onetto N, Rivard GE. 5-aza-2'-deoxycytidine therapy in patients with acute leukemia inhibits DNA methylation. *Leukemia research.* 1984;8(2):181-5.

[119] Wilson VL, Jones PA, Momparler RL. Inhibition of DNA methylation in L1210 leukemic cells by 5-aza-2'-deoxycytidine as a possible mechanism of chemotherapeutic action. *Cancer research.* 1983 Aug;43(8):3493-6.

[120] Fenaux P. Inhibitors of DNA methylation: beyond myelodysplastic syndromes. *Nature clinical practice.* 2005 Dec;2 Suppl 1:S36-44.

[121] Gore SD. Combination therapy with DNA methyltransferase inhibitors in hematologic malignancies. *Nature clinical practice.* 2005 Dec;2 Suppl 1:S30-5.

[122] Wijermans P, Lubbert M, Verhoef G, Bosly A, Ravoet C, Andre M, et al. Low-dose 5-aza-2'-deoxycytidine, a DNA hypomethylating agent, for the treatment of high-risk myelodysplastic syndrome: a multicenter phase II study in elderly patients. *J. Clin. Oncol.* 2000 Mar;18(5):956-62.

[123] Lubbert M, Wijermans P, Kunzmann R, Verhoef G, Bosly A, Ravoet C, et al. Cytogenetic responses in high-risk myelodysplastic syndrome following low-dose treatment with the DNA methylation inhibitor 5-aza-2'-deoxycytidine. *British journal of haematology.* 2001 Aug;114(2):349-57.

[124] Sato N, Maehara N, Su GH, Goggins M. Effects of 5-aza-2'-deoxycytidine on matrix metalloproteinase expression and pancreatic cancer cell invasiveness. *Journal of the National Cancer Institute.* 2003 Feb 19;95(4):327-30.

[125] Thiagalingam S, Cheng KH, Lee HJ, Mineva N, Thiagalingam A, Ponte JF. Histone deacetylases: unique players in shaping the epigenetic histone code. *Annals of the New York Academy of Sciences.* 2003 Mar;983:84-100.

[126] Cameron EE, Bachman KE, Myohanen S, Herman JG, Baylin SB. Synergy of demethylation and histone deacetylase inhibition in the re-expression of genes silenced in cancer. *Nature genetics.* 1999 Jan;21(1):103-7.

[127] Lo Coco F, Zelent A, Kimchi A, Carducci M, Gore SD, Waxman S. Progress in differentiation induction as a treatment for acute promyelocytic leukemia and beyond. *Cancer research.* 2002 Oct 1;62(19):5618-21.

[128] Brown R, Strathdee G. Epigenomics and epigenetic therapy of cancer. *Trends Mol. Med.* 2002;8(4 Suppl):S43-8.

Index

D

E

F

G

H

I

N

Q

R

S

T